Lecture Notes in Electrical Engineering

Volume 295

For further volumes:
http://www.springer.com/series/7818

About this Series

"Lecture Notes in Electrical Engineering (LNEE)" is a book series which reports the latest research and developments in Electrical Engineering, namely:

- Communication, Networks, and Information Theory
- Computer Engineering
- Signal, Image, Speech and Information Processing
- Circuits and Systems
- Bioengineering

LNEE publishes authored monographs and contributed volumes which present cutting edge research information as well as new perspectives on classical fields, while maintaining Springer's high standards of academic excellence. Also considered for publication are lecture materials, proceedings, and other related materials of exceptionally high quality and interest. The subject matter should be original and timely, reporting the latest research and developments in all areas of electrical engineering.

The audience for the books in LNEE consists of advanced level students, researchers, and industry professionals working at the forefront of their fields. Much like Springer's other Lecture Notes series, LNEE will be distributed through Springer's print and electronic publishing channels.

Xue Wang · Li Cui · Zhongwen Guo
Editors

Advanced Technologies in Ad Hoc and Sensor Networks

Proceedings of the 7th China Conference on Wireless Sensor Networks

Springer

Editors
Xue Wang
Tsinghua University
Beijing
People's Republic of China

Zhongwen Guo
Ocean University of China
Qingdao
People's Republic of China

Li Cui
Institute of Computing Technology, CAS
Beijing
People's Republic of China

ISSN 1876-1100 ISSN 1876-1119 (electronic)
ISBN 978-3-642-54173-5 ISBN 978-3-642-54174-2 (eBook)
DOI 10.1007/978-3-642-54174-2
Springer Heidelberg New York Dordrecht London

Library of Congress Control Number: 2014931577

Printed on acid-free paper

Springer is part of Springer Science+Business Media (www.springer.com)

Preface

This volume of Lecture Notes in Electrical Engineering contains the proceedings of the 7th China Conference of Wireless Sensor Networks (CWSN 2013), which was held in Qingdao, China, during October 17–19, 2013. CWSN represents the highest research level of sensor networks in China. CWSN 2013 served as a forum for researchers, developers and users to compare their experiences of sensor network research and applications, and to discuss the key challenges and research directions facing the sensor network community.

CWSN 2013, with its focus on sensor network design and implementation, aimed to promote the exchange of the theories and applications surrounding sensor networks. In addition, the conference provides the opportunity to consider research on CPS and the Internet of Things. Six prominent experts were invited to attend and present keynote speeches, and a lot of industries had demonstrated their state-of-the-art products and technologies on sensor networks and Internet of Things. In total, 324 papers were submitted to CWSN 2013, among which 191 were completed in English. Thirty-six high-quality papers, focusing, among other things, on node systems, infrastructures, communication protocols, data management, etc., were recommended by the Program Committee to be presented at the conference and included in this volume of LNEE.

On behalf of the Organizing Committee, we would like to thank Springer for publishing the proceedings of CWSN 2013. We would also like to express our gratitude to the reviewers for providing extra help in the review process, and to the authors for contributing their research results to the conference.

We look forward to seeing all of you next year at CWSN 2014. With your support and participation, CWSN will continue its success for a longtime.

October 2013

Xue Wang
Li Cui
Zhongwen Guo

Organization

The 7th China Conference of Wireless Sensor Networks (CWSN2013) was held in Qingdao, China, and was organized by Ocean University of China.

General Chair

Dexing Wu	President of Ocean University of China
Jianzhong Li	Director of Sensor Network Special Committee

Honorary Chair

Hao Dai	Academician of Chinese Academy of Engineering

Co-Chair

Limin Sun	Institute of Information Engineering Chinese Academy of Sciences
Huadong Ma	Beijing University of Posts and Telecommunications

Organizing Chair

Zhongwen Guo	Ocean University of China

Organizing Vice-Chair

Junyu Dong	Ocean University of China
Bing Zheng	Ocean University of China

Local Chair

Feng Hong	Ocean University of China

Organizing Committee

Liyan Jiang	Ocean University of China
Yuan Feng	Ocean University of China
Yongguo Jiang	Ocean University of China
Haipeng Qv	Ocean University of China
Zuojuan Liang	Ocean University of China
Yanxiu Sheng	Ocean University of China
Hua Li	Ocean University of China
Chao Liu	Ocean University of China

Contents

Chapter 1
Mobile Anchor-Assisted Localization in 3D Wireless Sensor Networks with Hilbert Curve

Chuanai Zhou, Huanqing Cui, Xiaojing Meng and Rong Hua

Abstract Localization is one of the key technologies of wireless sensor networks (WSNs). This paper proposes a mobile anchor-assisted localization method for three-dimensional (3D) networks. In this method, an anchor is equipped with global positioning system (GPS) device, and traverses the network along the Hilbert space–filling curve. The ordinary nodes estimate their positions using trigonometric method based on the characteristics of Hilbert curve. We compare the localization error, communication cost, and computational complexity of our proposed algorithm and other existing mobile anchor-assisted algorithms. The simulation results show that our method is better than other methods.

Keywords Wireless sensor network · Localization · Mobile anchor · Hilbert curve

---

This study is supported in part by the Shandong Provincial Natural Science Foundation, China (ZR2010FQ014); Key technology development projects of Qingdao Economic and Technical Development Zone(2012-2-50); the Open Research Fund from Shandong Provincial Key Laboratory of Computer(SDKLCN-2012-01).

C. Zhou
School of Information Engineering, Qingdao Binhai University, Qingdao 266555, China

H. Cui (✉)
Shandong Provincial Key Laboratory of Computer Networks, Shandong Computer Science Center, Jinan 250014, China
e-mail: cuihq79@163.com

H. Cui · X. Meng · R. Hua
College of Information Science and Engineering, Shandong University of Science and Technology, Qingdao 266590, China

X. Wang et al. (eds.), *Advanced Technologies in Ad Hoc and Sensor Networks*, Lecture Notes in Electrical Engineering 295, DOI: 10.1007/978-3-642-54174-2_1,

1.1 Introduction

Recently, many localization methods for wireless sensor networks (WSNs) have been developed, and most of them are anchor-based. An *anchor* is a special sensor node being always aware of its own position through global positioning system (GPS), so it is more expensive and consumes more power than ordinary nodes. To mitigate such issues with anchor-based localization, many mobile anchor-assisted localization methods have been proposed.

The localization strategies can generally be divided into *range-based* and *range-free*. The former utilizes the node to node distance or angle measurement in location computation. In [1], received signal strength (RSS) is used to measure the distance between mobile anchor and ordinary nodes. Virtual ruler approach [2] employs a vehicle equipped with multiple ultrasound beacons to travel around network to measure distances between pairwise sensors. Reference [3] gave an algorithm based on multidimensional scaling (MDS) and local map registration. Two generalized geometrical localization algorithms provided in [4] aim to achieve an accurate estimation based on time-of-arrival measurements. Range-free algorithms do not need any distance or angle measurement. Because the mobile anchor moves along a straight line during a short period, a sensor node can obtain some lines containing itself, and takes the intersection point of these borderlines as the estimated position [5]. If the mobile anchor is fitted with four-directional antennas, the sensor nodes can apply the statistical median to compute their coordinates based upon the advertised positions of the mobile anchor nodes [6]. Rectangle overlapping approach [7] uses a mobile anchor with a rotatable antenna to provide sensor nodes with rotation angle and position to localize.

The moving trajectory of mobile anchor has an important effect on localization performance. The existing path planning methods can be classified into *static* and *dynamic* methods. The former designs the trajectory before localization, and the anchor must follow the given path during localization. Since the random paths such as random waypoint (RWP) have drawback of incomplete coverage, Ref. [8] presented scan, double-scan, and Hilbert, where Hilbert is the best trajectory in 2D region [9]. Circles and S-Curves [10] require the anchor to move along curves instead of straight lines to avoid collinearity, but they cannot cover the entire network. The static methods cannot make full use of the real-time information during localization, so [11] proposed a dynamic path planning method based on the received-beacon numbers of ordinary nodes and deployed nodes in different regions. Reference [12] proposed a path defined as the track of depth-first traversal of the network graph.

The above methods are designed for two-dimensional (2D) networks, but WSNs are often deployed in 3D terrains in real applications. The mobile anchor-based 3D localization method using MDS [13] takes full advantage of MDS with connectivity and measurements, but MDS is complicated. This paper proposes 3D Hilbert mobile anchor-assisted localization algorithm (3D-HMAL), where the anchor's moving trajectory is a Hilbert space–filling curve.

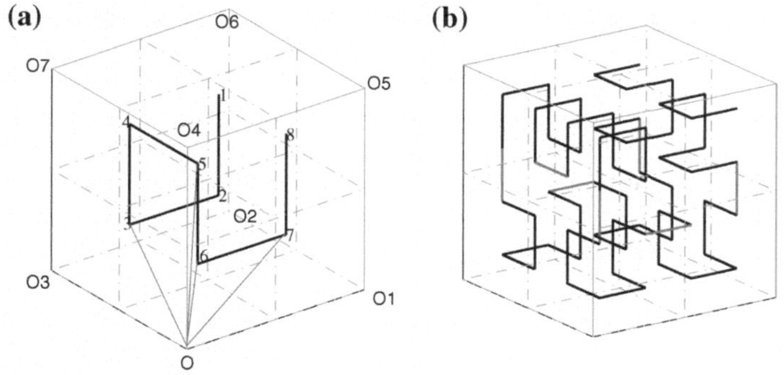

Fig. 1.1 Hilbert curves. **a** Order 1. **b** Order 2

1.2 3D-HMAL

1.2.1 Mobile Anchor Trajectory

A Hilbert curve maps any point in the space to a corresponding index, which denotes the position of the point on the curve from the start of the curve. The basic Hilbert curve for a $2 \times 2 \times 2$ cubic grid has an *order* of 1, and Hilbert curve of order m is constructed from eight copies of the Hilbert curve of order $m - 1$, properly rotated and/or reflected and connected by seven connector lines. The points order in the linear ordering going from 1 to 8^m are called *H-Keys*, denoted by h. The *scale of the grid*, denoted by s, is the distance between two successive H-keys. A *unit cube* is a sub-cube-section of a given 3D cubic region where it encloses the basic Hilbert curve with eight successive H-keys, where the last H-key must be divisible by 8. Assuming that every sensor node belongs to one and only one unit cube, and the radius of the communication range of the mobile anchor is R, we can draw the following conclusion:

Theorem 1. *Each sensor node in a unit cube is localizable iff* $s \leq \sqrt{\frac{4}{11}}R$.

Proof. As shown in Fig. 1.1a, a sensor node is localizable iff it receives four non-coplanar packets of mobile anchor. Because the sensor nodes at O, O_1, O_2, O_3, O_4, O_5, O_6 and O_7 are set at the unit cube extremities, all the sensor nodes in the given cube grid are localizable iff the sensor node at O is localizable. Let $dist(A, B)$ be the distance between two points A, B in a 3D region, then, $dist(O, 6) < dist(O, 3) = dist(O, 5) = dist(O, 7) < dist(O, 2) = dist(O, 4) = dist(O, 8) < dist(O, 1)$. Therefore, the sensor node at O can be localized iff $dist(O, 3) \leq R$. Because $dist(O, 3) = \frac{\sqrt{11}}{2}s$, $dist(O, 3) \leq R$ iff $s \leq \sqrt{\frac{4}{11}}R$.

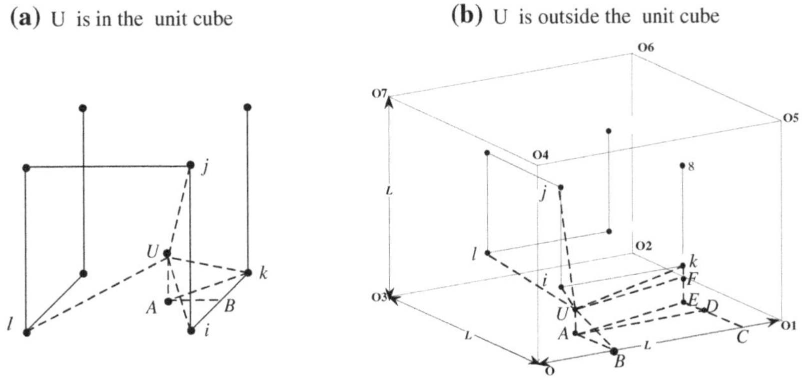

Fig. 1.2 Proof of theorem 2. **a** U is in the unit cube. **b** U is outside the unit cube

Based on theorem 1, $m \geq \left\lceil log_2\left(\frac{\sqrt{11}L}{2R}\right)\right\rceil$, where L is the cube side length. We assume the anchor broadcasts packets only at each Hilbert key, and the packets contain the H-key and its corresponding coordinates.

1.2.2 Localization Algorithm

The general procedure of our proposed localization algorithm consists of three steps. (1) The anchor propagates in the given cubic region while broadcasting packets. (2) The sensor node receives and records the anchor's packets in a list, then it estimates the distance to each packet of anchor. Further, it chooses the best four noncoplanar packets. (3) The sensor node calculates its position based on the chosen packets.

Theorem 2. *In a unit cube, if the nearest H-key to sensor node U is i, U can be localized by H-keys i, j, k, l, where* $dist(i, j) = dist(i, k) = dist(i, l) = s$.

Proof. As shown in Fig. 1.2, U may be inside or outside the cube Δ formed by eight H-keys. If U is inside Δ (Fig. 1.2a), let A be the projection of U in the plane determined by i, k, l, $AB \perp ik$. Because i is the nearest H-key to U, $dist(U, i) \leq \frac{\sqrt{3}}{2}s$. Then, $dist(A, B) \leq \frac{s}{2}$, $dist(U, A) \leq \frac{s}{2}$, $dist(B, k) \leq s$. Hence, $dist(U, k)^2 = dist(U, A)^2 + dist(A, B)^2 + dist(B, k)^2 \leq \left(\frac{s}{2}\right)^2 + \left(\frac{s}{2}\right)^2 + s^2 \leq \frac{3}{4}s^2$, So, $dist(U, k) \leq \frac{\sqrt{3}}{2}s \leq \frac{\sqrt{3}}{\sqrt{11}}R < R$. If U is outside Δ (Fig. 1.2b), let A and E be respectively the projection of U and 7 in the plane determined by $\{O, O_1, O_2, O_3\}$, $AB \perp OO_1$, $EC \perp OO_1$, $AD \perp EC$, $UF \perp 7E$. Because i is the nearest H-key to U, $dist(U, i) \leq \frac{\sqrt{3}}{2}s$. Then, $dist(A, B) \leq \frac{3}{2}s$, $dist(E, D) \leq \frac{s}{2}$, $dist(k, F) \leq \frac{s}{2}$. Hence, $dist(U, k)^2 = dist(A, B)^2 + dist(E, D)^2 + dist(F, k)^2 \leq \left(\frac{3s}{2}\right)^2 + \left(\frac{s}{2}\right)^2 + \left(\frac{s}{2}\right)^2 \leq \frac{11}{4}s^2$, So, $dist(U, k) \leq \frac{\sqrt{11}}{2}s \leq R$.

Similarly, $dist(U, l) \leq R$ and $dist(U, j) \leq R$. Therefore, U can be localized by H-keys i, j, k, l.

The following algorithm presents the choice procedure based on Theorem 2:

```
program H_Keys_choice ()
{Hk_list record the H-keys list of sensor node U};
begin
  i := the nearest H-key in Hk_list to U
  index := round((i-1)/8)+1;
  S := {i};
  for A := 8 * index - 7 to 8 * index do
     if dist(i,A) = s then
       S := S + {A};
     end if
  end for
  return S;
end
```

After determining the best tetrahedron to localize the sensor node U, the location of U can be estimated by some methods. Since the nodes are equipped with radios to perform communications, the distance estimation based on RSS has attracted enough attention, and we assume that the distance between U and H-keys can be measured by RSS. Let the tetrahedron Δ formed by $\{i, j, k, l\}$ be the best one to localize U, and the real and measured distances between U and H-key i be d_{ui} and $\hat{d}_{ui}$ respectively. Let $\hat{U}(\hat{x}_u, \hat{y}_u, \hat{z}_u)$ and $U(x_u, y_u, z_u)$ be, respectively, the estimated and real locations of U, and $H_i(x_i, y_i, z_i)$ be the location of H-key i.

(1) *Multilateration*. Multilateration is a basic method. Suppose U can be localized by tetrahedron formed by $\{i, j, k, l\}$, so

$$\begin{cases} \hat{d}_{ui}^2 = (x_i - \hat{x}_u)^2 + (y_i - \hat{y}_u)^2 + (z_i - \hat{z}_u)^2 \\ \hat{d}_{uj}^2 = (x_j - \hat{x}_u)^2 + (y_j - \hat{y}_u)^2 + (z_j - \hat{z}_u)^2 \\ \hat{d}_{uk}^2 = (x_k - \hat{x}_u)^2 + (y_k - \hat{y}_u)^2 + (z_k - \hat{z}_u)^2 \\ \hat{d}_{ul}^2 = (x_l - \hat{x}_u)^2 + (y_l - \hat{y}_u)^2 + (z_l - \hat{z}_u)^2 \end{cases} \tag{1.1}$$

Then, we can use least square method to solve it to obtain the estimated location.

(2) *Geometrical method*. Figure 1.3 is an example of geometrical method. Let A be the projection of sensor node U in plane formed by H-keys i, k, l, $AB \perp li$, and $\angle UlB = \alpha$. Moreover, il parallels to axis x, and ik parallels to axis y. Based on cosine theorem, $\cos(\alpha) = \frac{\hat{d}_{ul}^2 + s^2 - \hat{d}_{ui}^2}{2 \times s \times \hat{d}_{ul}}$. Hence, $\hat{x}_u = x_A = x_B = x_l + \hat{d}_{ul} \times \cos(\alpha)$. Similarly, we can find $\hat{y}_u$ and $\hat{z}_u$.

(3) *Weighted centroid algorithm*. Based on weighted centroid localization (WCL) [14], $\hat{U}\left(\hat{x}_u, \hat{y}_u, \hat{z}_u\right)^T = \frac{\sum_{b \in \Delta} w_{ub} \times H_b(x_b, y_b, z_b)}{\sum_{b \in \Delta} w_{ub}}$, where $\Delta = \{i, j, k, l\}$ and

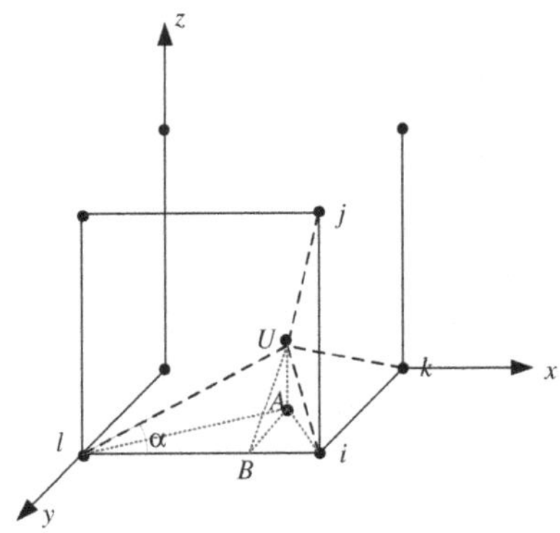

Fig. 1.3 Geometrical method of computing location of U

$w_{ub} = \frac{1}{(\hat{d}_{ub})^g}$. A very small transmission range and a degree $g = 1$ produces the best localization results, but a degree $g = 3$ yields the best results in other configurations.

1.3 Performance Evaluation

The performance of mobile anchor-assisted localization method can be evaluated according to the following criteria: (1) Localizable sensor nodes in one-hop. (2) Localization accuracy, defined as $e_i = \frac{\sqrt{(x_{u_i} - \hat{x}_{u_i})^2 + (y_{u_i} - \hat{y}_{u_i})^2 + (z_{u_i} - \hat{z}_{u_i})^2}}{R}$. For simplicity, we denote the average and standard deviation of localization errors of all sensor nodes are μ_e and σ_e respectively. (3) Trajectory length. Because the length of different trajectories is detailed compared in [15], we do not analyze this metric in our paper. (4) Scalability. Scalability means that the performance of localization approach is independent of the nodes density.

Our simulations are performed by Matlab, and the deployment region of WSN is a cube of side length $L = 100$ m. Three moving trajectories are compared in this section: RWP, Layered-Scan, and Hilbert, and three methods, including multilateration, geometric method, and WCL are considered. For simplicity, we use "MUL" to represent multilateration, "LS" to represent Layered-Scan, and "GEO" to represent geometric method. Combining different estimation method and trajectory, there are seven localization methods: RWP-WCL, RWP-MUL, LS-WCL, LS-MUL, HMAL-MUL, HMAL-WCL, and HMAL-GEO.

In our simulations, we varied the communication range R of mobile anchor from 10 to 50 m by a step of 10 m, and we varied the ranging error E_{ranging} between sensor node and anchor from 0 m to 0.3 m by a step of 0.05 m. Moreover, suppose $\hat{d}_{ui} \sim N(d_{ui}, E_{\text{ranging}})$. We set $v_{\min} = 1$ m/s and $v_{\max} = \frac{R}{2}$ of RWP. And we set both the distance between adjacent layers and velocity of anchor of Layered-Scan are $\frac{R}{2}$.

Table 1.1 Number of localized nodes in one-hop of anchor

R/m	10	20	30	40	50
RWP	50	72	79	88	96
Layered-scan	89	91	94	92	94
Hilbert	100	100	100	100	100

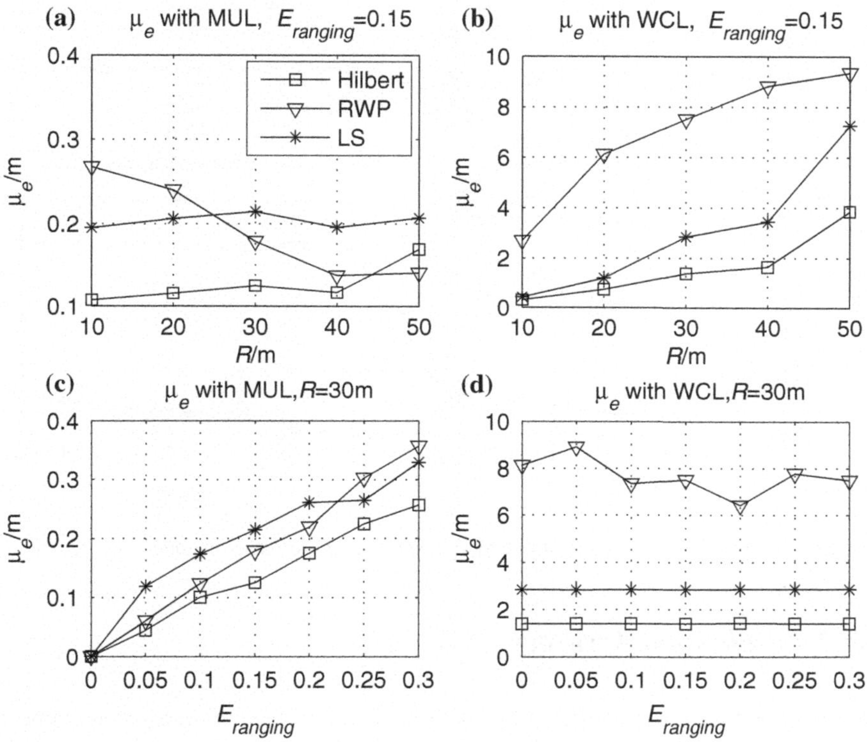

Fig. 1.4 Comparison of μ_e with different trajectories

1.3.1 Localizable Sensor Nodes in One-Hop

As shown in Table 1.1, RWP and Layered-Scan cannot localize all sensor nodes, whereas all sensor nodes can be localized using Hilbert. RWP cannot cover the whole network, which leads to some sensor nodes that cannot be localized. The reason that some sensor nodes cannot be localized by LS is that the packets received by a sensor node may be coplanar.

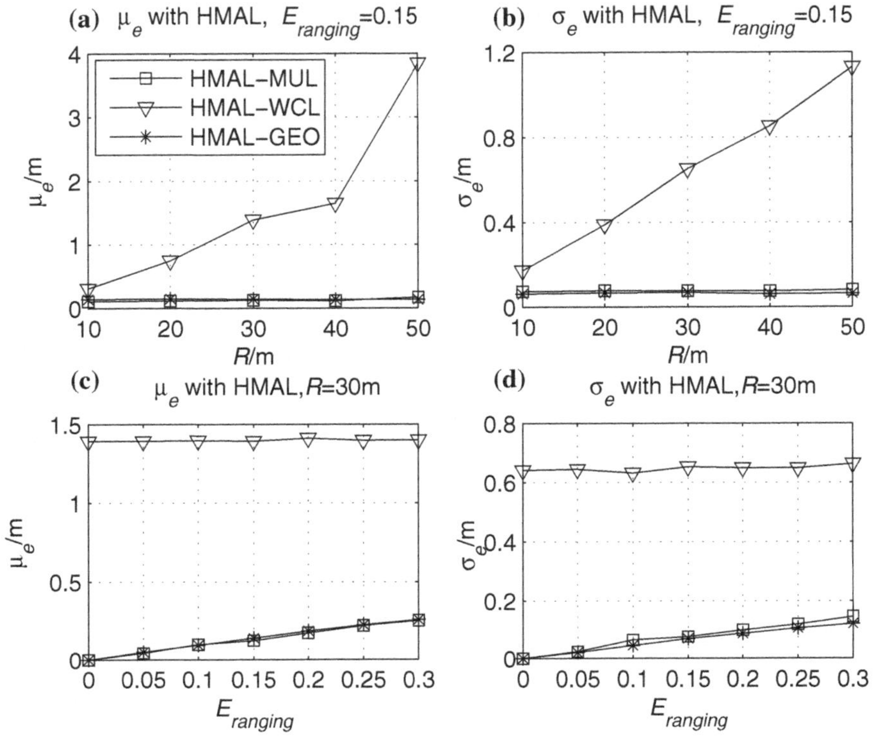

Fig. 1.5 Comparison of μ_e and σ_e with HMAL and different estimation methods

1.3.2 Localization Accuracy

As shown in Fig. 1.4, regardless of estimation methods, the localization error under Hilbert is lower than RWP and LS. In fact, μ_e of Hilbert is $44 \sim 93\,\%$ of that of RWP, and $35 \sim 81\,\%$ of that of LS, if the positions of sensor nodes are estimated by MUL. And if WCL is utilized, μ_e of Hilbert is $10 \sim 38\,\%$ of that of RWP, and $47 \sim 76\,\%$ of that of LS. According to Fig. 1.4a, if MUL is used, R has little effect on μ_e, but R has a great influence on μ_e in WCL according to Fig. 1.4b. On the contrary, E_{ranging} has a great impact on μ_e with MUL, while it has a little effect with WCL, based on Fig. 1.4c, d. Moreover, MUL provides a higher localization accuracy than WCL using the same trajectory. Therefore, Hilbert is better that RWP and LS.

Based on Fig. 1.5a, c, μ_e using WCL is far higher than MUL and GEO methods, but HMAL-MUL and HMAL-GEO have almost the same localization accuracy. In fact, the difference of μ_e between HMAL-MUL and HMAL-GEO is only $0.0027 \sim 0.0813$ m with different R and $E_{\text{ranging}} > 0$, but μ_e of HMAL-GEO is $0.6 \sim 3.8$ m lower than HMAL-WCL. Moreover, Fig. 1.5a shows that μ_e of HMAL-WCL becomes larger when R increases, but μ_e of HMAL-MUL and HMAL-GEO changes little with different R. By contrast, Fig. 1.5c shows that μ_e of HMAL-WCL

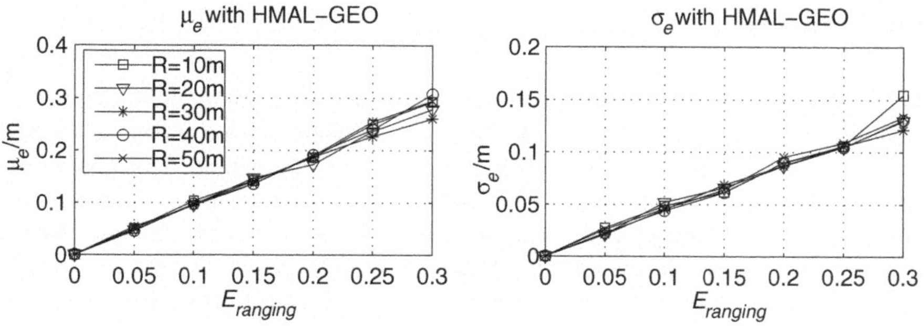

Fig. 1.6 Localization accuracy of HMAL-GEO

Table 1.2 Scalability of HMAL-GEO

N	20	40	60	80	100
μ_e/m	0.2861	0.2602	0.3188	0.2948	0.2776
σ_em	0.1362	0.1472	0.1423	0.1470	0.1294

changes small with different E_{ranging}, but μ_e of HMAL-MUL and HMAL-GEO becomes larger if E_{ranging} increases. Comparing σ_e, we can draw similar conclusions, which means that most of the localization errors using HMAL-MUL and HMAL-GEO concentrate in the near μ_e, while localization errors using HMAL-WCL is a more dispersed distribution than the other two methods.

Based on the above comparison, and taking into account that the complexity of HMAL-GEO is lower than HMAL-MUL, HMAL-GEO is the best method. Further, Fig. 1.6 illustrates that μ_e and σ_e of HMAL-GEO increase with E_{ranging}, but their changes are very small. For example, the difference between the maximum and minimum μ_e is only 0.3073 m.

1.3.3 Scalability

We deploy 20–100 sensor nodes by a step of 20 in the cubic region, and set $R = 30$ m, $E_{\text{ranging}} = 0.15$ m. As shown in Table 1.2, the localization error remains almost the same for the different network density, where the maximum μ_e is 0.3188 m larger than the minimum μ_e. We have also noted that all nodes were localized by HMAL while some sensor nodes cannot be localized by RWP and Layered-Scan. This result clearly validated the scalability of our approach.

1.4 Conclusion

Anchor-based localization methods provide more accurate results than anchor-free methods, but anchors are more expensive than ordinary sensor nodes, so mobile anchor-assisted localization methods become a promising method. This paper proposed 3D-HMAL, a mobile anchor-assisted localization method using Hilbert space–filling curve as the trajectory of mobile anchor. We presented the shortest Hilbert curve to cover the whole network deployment region. Moreover, we gave a geometric method to estimate the positions of sensor nodes, which is fully decentralized and has low complexity. We developed a detailed simulations to evaluate our approach. The experiment results showed that our algorithm is better than other methods for stability. The scalability of our method was also validated by simulations.

In the future, we plan to investigate the dynamic Hilbert space–filling curve to localize sensor nodes in the region including obstacles. Moreover, we plan to improve the localization accuracy by exploiting the rest of the received packets by each senor node.

References

1. Liu K, Xiong J (2010) A fine-grained localization scheme using a mobile beacon node for wireless sensor networks. J Inf Proces Syst 6:147–162
2. Ding Y, Wang C, Xiao L (2010) Using mobile beacons to locate sensors in obstructed environments. J Parall Distr Comput 70:644–656
3. Zhou Yi, Lamont L (2011) A mobile beacon based localization approach for wireless sensor network applications. In: 5th international conference on sensor technologies and applications, IEEE Press, New York, pp 243–247
4. Chen H, Liu B, Huang P, Liang J, Gu Y (2012) Mobility-assisted node localization based on TOA measurements without time synchronization in wireless sensor networks. Mobile Netw Appl 2012(17):90–99
5. Guo Z, Guo Y, Hong F, Jin Z, He Y, Feng Y, Liu Y (2010) Perpendicular intersection: locating wireless sensors with mobile beacon. IEEE T Veh Technol 59:3501–3509
6. Ou C-H (2011) A localization scheme for wireless sensor networks using mobile anchors with directional antennas. IEEE Sensors J 11:1607–1616
7. Wu Y, Chen W (2010) Localization using a mobile beacon with directional antenna for wireless sensor networks. IEICE T Inf Syst 94:2370–2377
8. Koutsonikolas D, Das SM, Hu YC (2007) Path planning of mobile landmarks for localization in wireless sensor networks. Comput Commun 30:2577–2592
9. Bahi JM, Makhoul A, Mostefaoui A (2007) A mobile beacon based approach for sensor network localization. In: Proceedings of 3rd IEEE international conference on wireless and mobile computing, networking and communications, IEEE Press, New York, pp 44–51
10. Huang R, Záuba GV (2007) Static path planning for mobile beacons to localize sensor networks. In: 5th annual IEEE international conference on pervasive computing and communications, IEEE Press, New York, pp 323–330
11. Li J, Yang K (2011) Improvement of path planning in mobile beacon assisted positioning. In: 7th international conference on intelligent computing, Springer, Heidelberg, pp 309–316
12. Li X, Mitton N, Simplot-Ryl I, Simplot-Ryl D (2012) Dynamic beacon mobility scheduling for sensor localization. IEEE T Parall Distr Syst 23:1439–1452

13. Kim E, Lee S, Kim C, Kim K (2010) Mobile beacon-based 3D-localization with multidimensional scaling in large sensor networks. IEEE Commun Lett 14:647–649
14. Blumenthal J, Grossmann R, Golatowski F, Timmermann D (2007) Weighted centroid localization in zigbee-based sensor networks. In: IEEE international symposium on intelligent signal processing, IEEE Press, New York, pp 1–6
15. Cui H, Wang Y, Lv J (2012) Path planning of mobile anchor in three-dimensional wireless sensor networks for localization. J Inf Comput Sci 9:2203–2210

Chapter 2
Human Tracking System Based on PIR Sensor Network and Video

Ji Xiong, Fang-Min Li and Jing-Yuan Zhang

Abstract To detect and locate the human target motion precisely, this paper intends to present a tracking algorithm based on pyroelectric sensor network and video analysis technologies. According to the advantages of pyroelectric sensor network system and video system, this paper uses weighted least squares to fuse the data which is collected by multiple heterogeneous sensor nodes to realize human target real-time tracking. Moreover, the data can also be collected by pyroelectric sensor network system and video cameras. Through simulation, the error of tracking results is analyzed. The results show that the method using homogeneous and heterogeneous sensors to fuse the measured vector obtain better human target real-time tracking effect.

Keywords Human tracking · Pyroelectric infrared sensors network · Video · The weighted least squares method

2.1 Introduction

Pyroelectric infrared sensor, a noncontact form used to detect the infrared radiation changes in the environment, has a relative sensitivity to the human movement. In addition, it has other advantages such as wide application range, strong concealment performance and little influence of the ambient light. Therefore, the technology using PIR (Pyroelectric Infrared) to locate and track targets has received special attention. In relevant literature [1–4], Qi Hao and other researchers have made an in-depth study in this field and have utilized PIR to realize human targets real-time tracking. They designed the PIR nodes to detect human movement by using PIR sensors.

J. Xiong · F-M. Li (✉) · J-Y. Zhang
Key Laboratoryof Fiber Optical Sensing Technology and Information Processing, Ministry of Education, School of Information Engineering, Wuhan University of Technology, Wuhan 430070, China
e-mail: lifangmin@whut.edu.cn
http://i.whut.edu.cn/

X. Wang et al. (eds.), *Advanced Technologies in Ad Hoc and Sensor Networks*, Lecture Notes in Electrical Engineering 295, DOI: 10.1007/978-3-642-54174-2_2,

According to the relative angle changes of human body movement and based on the space-time conversion model, the PIR nodes completed the human target tracking and recognition. They also used the fusion algorithm based on kalman and particle filter algorithm to fuse the human data which was collected by the four fixed PIR nodes and have conducted an experiment in a real environment on human tracking. Because of the limited accuracy of pyroelectric sensor, the tracking effect is relatively poor in multibody tracking.

In literature [5–7], these papers explained that the tracking and recognition based on video were relatively mature and in multibody tracking, and the effect was quite good. However, the tracking based on video information requires a higher ability for data processing and at the same time, the amount of collected tracking data is relatively big. In the process of using wireless network to transfer the tracking information, because the wireless network bandwidth is relatively limited, the wireless network cannot play an effective real-time video tracking function in transferring the tracking information. Therefore, using pyroelectric sensors to mix with the video system for surveillance become the focus of our research. The paper discusses the sensors' advantages and disadvantages, respectively, in the system of heterogeneous sensor network, which has both the active sensors (camera with a PTZ system) and the passive sensors (pyroelectric sensor node). The pyroelectric sensor nodes can measure the azimuthal angle of the target, but cannot measure the target distance, while the video camera can. Two kinds of sensors to get complementary measurements at the same time, and merging forms a system which is beneficial for human tracking [8, 9].

The literature [10] used HMM algorithm to estimate the target location and single human target tracking. HMM tracking method, an advanced solution method, solves the target trajectory through iteration. However, this approach must plan the tracking area first and get the prior distribution for the human body target trajectory tracking. In order to obtain more accurate position estimation, the method needs more iterations and a large amount of calculation. According to the respective advantages of heterogeneous sensors, our study makes use of the redundant information from the sensors to measure data and complement the incomplete measurement to a complete one. Then the target position is calculated by using extended weighted least squares method. Meanwhile, in order to facilitate the discussion, we assume that the observation noise is independent of Gauss noise, and in the position measurement, there would be at least two sources of measurement from the pyroelectric sensors. What is more, using the video camera as one kind of sensor in the heterogeneous sensor networks will play an important role in the target identification and multihuman target tracking.

2.2 Description of the Problem and the Overall Framework

In the process of real-time target tracking, high-end video cameras are used in real-time target tracking. While in this case, the data backup and network transmission requires better bandwidth, especially in the wireless environment. Therefore, it is

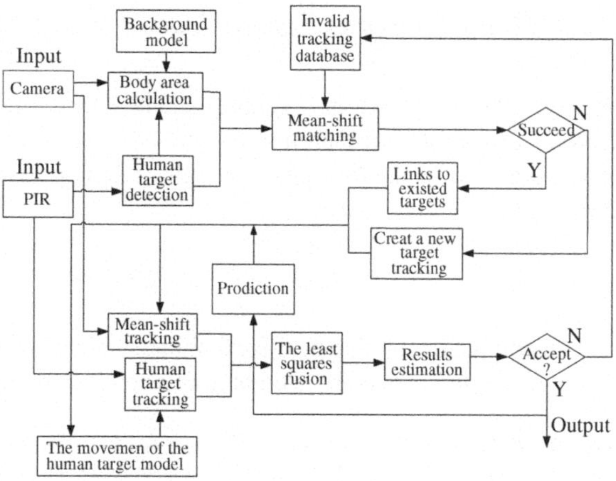

Fig. 2.1 Overall functional framework

difficult to achieve good results. At the same time, the high-end video cameras will not play their proper function when they are in the relatively hostile environments. Some of the researchers have proposed that according to the characteristics of the released red infrared from human body, the technology of pyroelectric sensor can be used for human target tracking and recognition. But the tracking accuracy is not very satisfactory, and it needs to optimize the tracking environment. It cannot achieve the desired effect in the actual environment. In multihuman target tracking, the error rate is higher. Based on the previous studies, this paper has explored a new method to add a camera to the pyroelectric sensors network which can not only make the pyroelectric network systems and camera systems work individually when tracking human body, but also can fuse the measurement obtained by the two different systems to track the human target. This method can overcome the problems which exist in the previous studies and has the potential for wide application.

In order to overcome the uncertainty of target state and limitations of the individuals or single type of sensor measurement, the paper proposes a way which can improve the effect of the sensors network system and accurately describe the measured target by fusing the data that are obtained by a variety of sensors. The paper [11, 12] used the least squares method to fuse and estimate the data which are measured by the same kind of sensors. In the data acquisition system, different kinds of sensors information fusion are important. By using the data measured by multiple heterogeneous sensors, the aim of monitoring various parameters and tracking multiple targets has been achieved.

In this paper, the fusion method based on the principle of least squares is simple in calculation. It can objectively reflect the reliability of each sensor. In data processing, it does not need to know a priori information about the data, and the fusion result has high accuracy. As Fig. 2.1 shows, the functional framework which is composed by the pyroelectric network and video system and it expounds the general concept.

2.2.1 Heterogeneous Sensor Data Fusion Method

We used the weighted least squares method to get the estimate vector $\widehat{x}$ which is computed from the measurement vector y_i. The criterion of weighted least squares estimation is to make the sum of weighted squared error as shown in the following formula to reach its minimum.

$$J_w(\widehat{x}) = (y_i - H_i\widehat{x})^T W(y_i - H_i\widehat{x}) \tag{2.1}$$

In formula (2.1), W is a positive diagonal weighting matrix, and $W = \mathrm{diag}(w_1, w_2, \ldots, w_i, \ldots, w_n)$. W is the weight coefficient matrix, and $w_i \in R^{n_i \times n_i}$.

Theorem 1. *The ith sensor and the jth sensor consist of rangefinders matrix, which are $H_i \in R^{n_i \times m}$ and $H_j \in R_{n_j \times m}$ ($H_i \neq H_j$) the random vector $\widehat{x}$ is the least squares estimation.*

$$\widehat{x} = \left[\sum_{i=1}^{n} H_i^T V_i^{-1} H_i\right]^{-1} \sum_{i=1}^{n} H_i^T V_i^{-1} y_i \qquad i = 1, 2, \ldots, n \tag{2.2}$$

In formula (2.2), V_i is error variance matrix of the ith sensor, to prove the theorem 1 as follows:

Proof: In order to reduce the weighted error sum of squares to minimum, we assume that $\frac{\partial J_w}{\partial x} = 0$, so we can get formula (2.3)

$$-H_i^T(W + W^T)^T W(y_i - H_i\widehat{x}) = -H_i^T(W + W^T)y_i + H_i^T(W + W^T)H_i\widehat{x} = 0 \tag{2.3}$$

The nature of the diagonal matrix shows

$$W = W^T, \quad \widehat{x} = \left[\sum_{i=1}^{n} H_i^T W_i H_i\right]^{-1} \sum_{i=1}^{n} H_i^T W_i y_i \tag{2.4}$$

Using the principle of least squares weighted fusion algorithm is also an unbiased estimation algorithm in the heterogeneous sensors, so we can get the estimated variance matrix R of $\widehat{x}$.

$$R = E[e(\widehat{x})e^T(\widehat{x})] = E[(x - \widehat{x}^2)] = E\left\{\left[x - \left(\sum_{i=1}^{n} H_i^T w_i H_i\right)^{-1} \sum_{i=1}^{n} H_i^T w_i y_i\right]^2\right\}$$

$$= E\left\{\sum_{i=1}^{n}\left[\left(\frac{w^i}{\sum\limits_{i=1}^{n}} w_i\right)^2 H_i^T (H_i x - y_i)(H_i x - y_i)^T H_i\right]\right.$$

$$\left. + \sum_{i=1}^{n}\sum_{j=1, j\neq i}^{n}\left[\frac{w_i w_j}{\left(\sum\limits_{i=1}^{n} w_i\right)^2} H_i^T (H_i x - y_i)(H_j x - y_j)^T H_i\right]\right\} \tag{2.5}$$

Because V_i and V_j are independent of each other, therefore

$$E[(H_i x - y_i)(H_j x - y_j)^T] = 0 \quad i \neq j \tag{2.6}$$

$$R = E[e(\widehat{x})e^T(\widehat{x})] = E\sum_{i=1}^{n}\left[\left(\frac{w_i}{\sum\limits_{i=1}^{n}}\right)^2 H_i^T (H_i x - y_i)(H_i x - y_i)^T H_i\right] \tag{2.7}$$

$$R = \sum_{i=1}^{n}\left[\left(\frac{w_i}{\sum_{i=1}^{n} w_i}\right)^2 H_i^T V_i H_i\right] \tag{2.8}$$

Through finding the minimum error, taking the partial derivative of W_i, and obtaining $\partial R/\partial w_i = 0$, we can get W_i as follow:

$$w_i = H_i V_i^{-1} H_i^T, \quad i = 1, 2, \ldots, n \tag{2.9}$$

And we can get R and $\widehat{x}$

$$R = \left[\sum_{i=1}^{n} H_i^T V_i^{-1} H_i\right]^{-1} \tag{2.10}$$

$$\widehat{x} = \left[\sum_{i=1}^{n} H_i^T V_i^{-1} H_i\right]^{-1} \cdot \sum_{i=1}^{n} H_i^T V_i^{-1} y_i = R \cdot \sum_{i=1}^{n} H_i^T V_i^{-1} y_i \tag{2.11}$$

2.3 Heterogeneous Sensors Hybrid Tracking Model Design

This paper has proposed a mixed tracking model of heterogeneous sensors, which comprises a camera system and a pyroelectric infrared sensor system. According to the physical properties of pyroelectric infrared sensors, the direction angle of

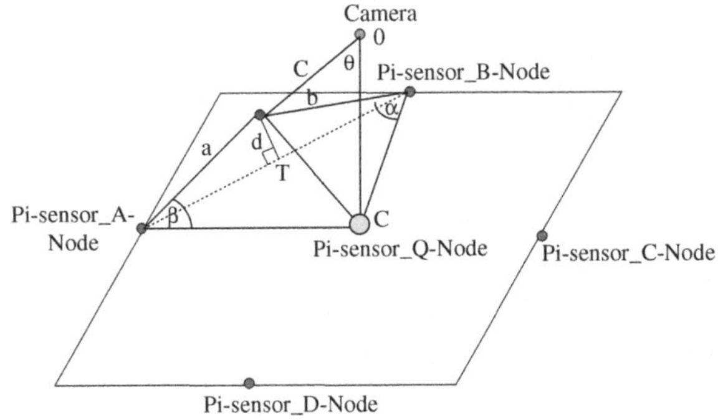

Fig. 2.2 Design for mixed tracking model

human body in the area of target tracking can be obtained. When the camera gets the human information, it can conduct the simple image processing. Thus we can get the relative distance between the target and camera. As the Fig. 2.2 shows, the tracking area is composed of five pyroelectric infrared sensor nodes and a PTZ camera. The heterogeneous pyroelectric sensor nodes system is comprised of nodes A, B, and Q. As Fig. 2.2 shows, it can get the primary positioning and tracking of human target. Meanwhile, we can transfer the location data of human to camera [13], so that we can make use of the heterogeneous tracking mode which consists of a camera and nodes A and B for tracking and recognition.

2.3.1 Homogeneous Sensor Positioning Method

Pyroelectric sensor location system belongs to the homogeneity sensor location system. The AOA location model that the system uses is simple and has strong real-time performance. We assume that the position of pyroelectric sensor node is (x_i, y_i) and that of the human target position is (x, y). Each value of θ_i measured by using AOA mode is opposite. Therefore, the expression is as follows:

$$\tan\theta_i = \frac{y - y_i}{x - x_i}, \quad i = 1, 2, \ldots, m \tag{2.12}$$

And formula (2.12) can be converted to: $\tan(\theta_i) \times x - y = \tan(\theta_i) \times x_i - y_i$, $i = 1, 2, \ldots, m$.

Error equation:

$$\psi = H - Y \times x \tag{2.13}$$

$$H = \begin{bmatrix} \tan\theta_1 & -1 \\ \tan\theta_2 & -1 \\ \vdots & \vdots \\ \tan\theta_m & -1 \end{bmatrix} \quad x = \begin{bmatrix} x \\ y \end{bmatrix} \quad Y = \begin{bmatrix} \tan(\theta_1)x_1 & -y_1 \\ \tan(\theta_2)x_2 & -y_2 \\ \vdots & \vdots \\ \tan(\theta_m)x_m & -y_m \end{bmatrix} \tag{2.14}$$

If we introduce formula (2.14) into the homogeneous sensor fusion model, then we can compute the position data of a human target in the pyroelectric sensors system.

2.3.2 Supplementary Method for Heterogeneous Sensor

As camera or pyroelectric sensors cannot measure the complete data of the target location separately, we use the redundant information to supply the incomplete position measurement until position measurements are integrated. Considering that we cannot get the satisfactory data of real-time human target tracking by pyroelectric sensors, measurement information can be supplemented by the camera which satisfies the real-time location and tracking of human target.

The pyroelectric sensor positions are $A(X_A, Y_A)$ and $B(X_B, Y_B)$, measurements of azimuth angle are β_A and α_B, and the human target coordinate is $Pt(X_{Pt}, Y_{Pt})$. At the same time, the $X-Y$ coordinate of the camera projection position is $C(X_C, Y_C)$. The distance between the human target and camera is L [14, 15]; so we can get that the distance between human target and three passive sensors are a, b, and L.

First, according to two-dimensional space, formula of distance between two points is:

$$d_i = \sqrt{(x - x_i)^2 + (y - y_i)^2} \tag{2.15}$$

Three distance equations are established:

$$\begin{cases} (X_{Pt} - X_C)^2 + (Y_{Pt} - Y_C)^2 = L^2 \\ (X_{Pt} - X_A)^2 + (Y_{Pt} - Y_A)^2 = a^2 \\ (X_{Pt} - X_B)^2 + (Y_{Pt} - Y_B)^2 = b^2 \end{cases} \tag{2.16}$$

So the coordinates of the human target can be obtained as follows:

$$\begin{bmatrix} X_{Pt} \\ Y_{Pt} \end{bmatrix} = \begin{bmatrix} 2(X_A - X_C) & 2(Y_A - Y_C) \\ 2(X_B - X_C) & 2(Y_B - Y_C) \end{bmatrix}^{-1} \cdot \begin{bmatrix} X_A^2 - X_C^2 + Y_A^2 - Y_C^2 + L^2 - a^2 \\ X_B^2 - X_C^2 + Y_B^2 - Y_C^2 + L^2 - b^2 \end{bmatrix} \tag{2.17}$$

According to the pyroelectric node measured angles α and β_A, the following formula can be derived and the following results are deduced:

$$\begin{cases} \tan(\beta_A - 45) = \frac{d}{AT} \\ \tan(\alpha_B - 45) = \frac{d}{BT} \\ AT + BT = AB \\ AB = \sqrt{2} \cdot CA \\ \begin{cases} (OPt)^2 = L^2 + OC^2 \\ AT = \frac{\tan(\alpha_B - 45) \cdot \sqrt{2} \cdot AC}{\tan(\beta_A - 45) + \tan(\alpha_B - 45)} \end{cases} \\ BT = \sqrt{2} \cdot AC - AT \\ \begin{cases} a = \frac{AT}{\cos(\beta_A - 45)} \\ b = \frac{BT}{\cos(\alpha_B - 45)} \end{cases} \end{cases} \tag{2.18}$$

Substituting the above equation into the heterogeneous sensor fusion model, we can get the human target location data of pyroelectric and video mixing systems.

2.4 Simulation Examples

Assuming that the camera and PIR sensors track the same object, a camera, four PIR sensor nodes consist of dual-column and the sensor nodes with a field view of 360° are at the same platform, the synchronous sampling, data latency is zero. Mostly, the camera and pyroelectric node perform well. A PIR node has an azimuth measurement accuracy of 1 rad, accuracy of the camera is 0.5 m. Actually, PTR sensors work as pure azimuth observation. The camera in the process of tracking can be independent and can also cooperate with pyroelectric track together. The velocities deviation of PIR sensor nodes and the distance measured deviation of camera are related to the actual distance which is between the human body target and these fixed sensor nodes. Calculation and location methods would bring errors to the positioning result. However, these errors reflect the credibility of the node location. Therefore, we could implement the weighting of location node by using the errors, and the values of weighting would demonstrate the credibility so that we could improve the accuracy such as formula (2.9), W_i is the relative credibility, and the value is decided by the deviation and location precision. Before the simulation starts, it is necessary to set the weighting system. The values are determined by prior knowledge.

The PIR sensor tracking model is composed of four PIR sensor nodes with the structure of dual-column and the sensor nodes have a field view of 360°. The field of monitoring is 30 × 30 m and the measurement cycle is 1 s. First, to track by using the least square method with Gaussian noise, then, to implement the efficient tracking by using the expanded Kalman filter. The nodes have higher precision distribution, relatively higher weight, because the tracking precision of surrounding nodes is higher than the central nodes. The simulation results are Fig. 2.3 as follows:

1. The azimuth angle of observation noise variance is $\pi/180°$
2. The azimuth angle of observation noise variance is $2 \times \pi/180°$

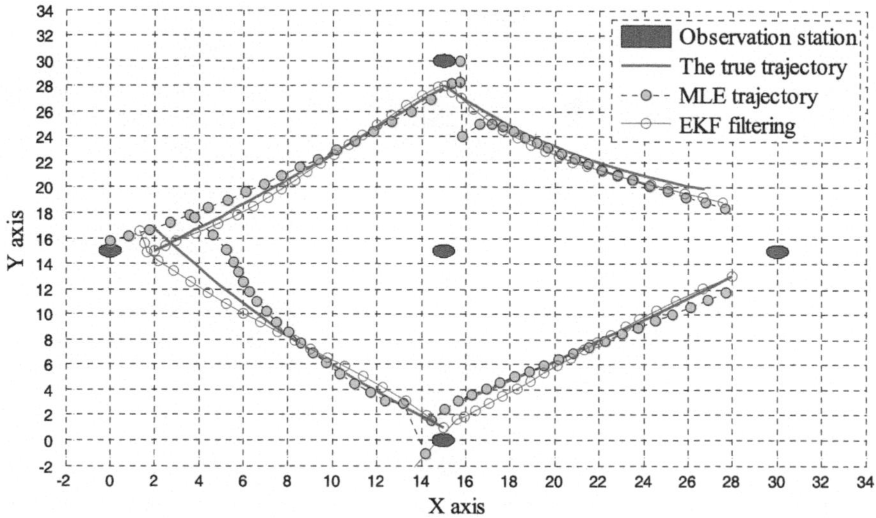

Fig. 2.3 Human tracking model

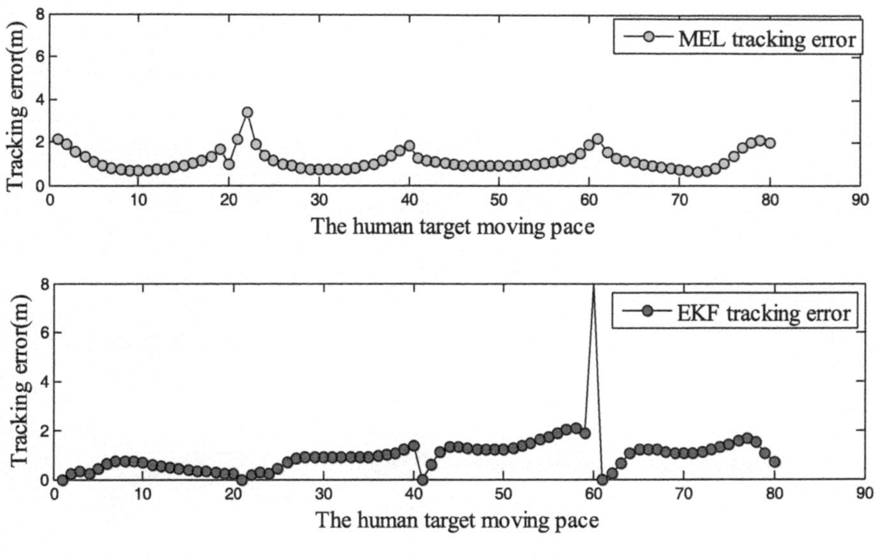

Fig. 2.4 Tracking error

With the increase of azimuth measurement noise variance, tracking error is increasing; the simulation results are shown in Fig. 2.4. After several tests, when azimuth measurement noise variance is 1°. As follows: *X* axis error mean of Least squares is 0.31 m and *Y* axis error mean is 0.42 m. After EKF filtering, *X* axis error mean is 0.22 m and Y axis error mean is 0.34 m. When azimuth measurement noise variance is 2°, *X* axis error mean of Least squares is 1.15 m and *Y* axis error mean

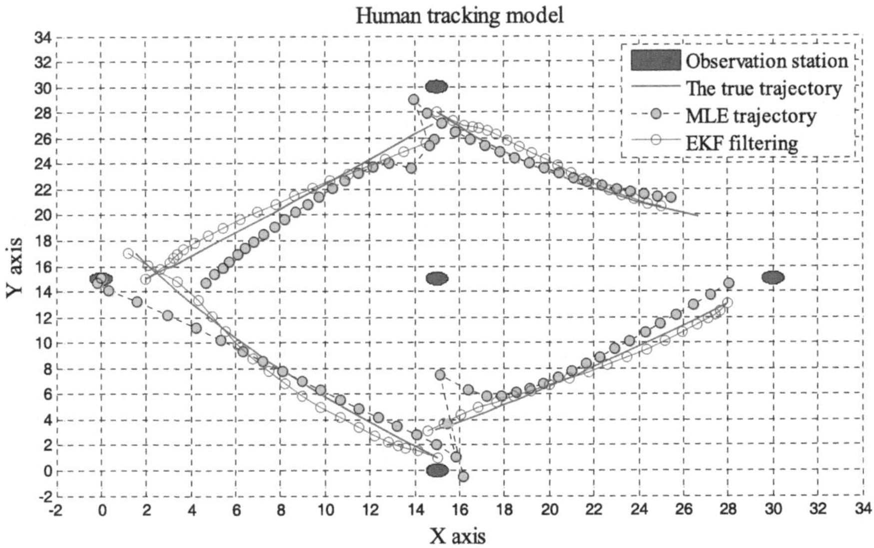

Fig. 2.5 Human tracking model

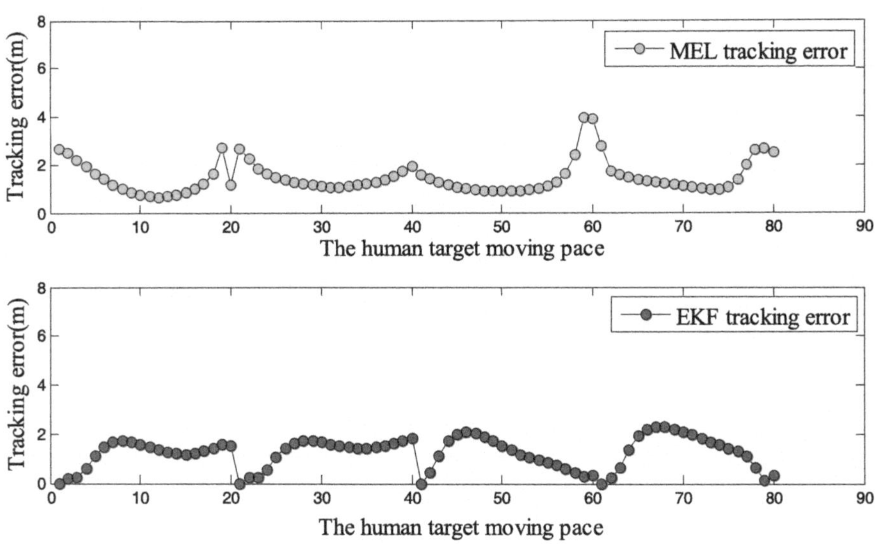

Fig. 2.6 Tracking error

is 1.43 m. After EKF filtering, X axis error mean is 1.01 m and Y axis error mean is 1.13 m (Figs. 2.5 and 2.6).

When pyroelectric nodes and cameras are used as a reference point to establish a fixed coordinate system, human target is doing the uniform motion in the mixing monitored area. Monitoring area is 30×30 m, measurement cycle is 1 s, the azimuth

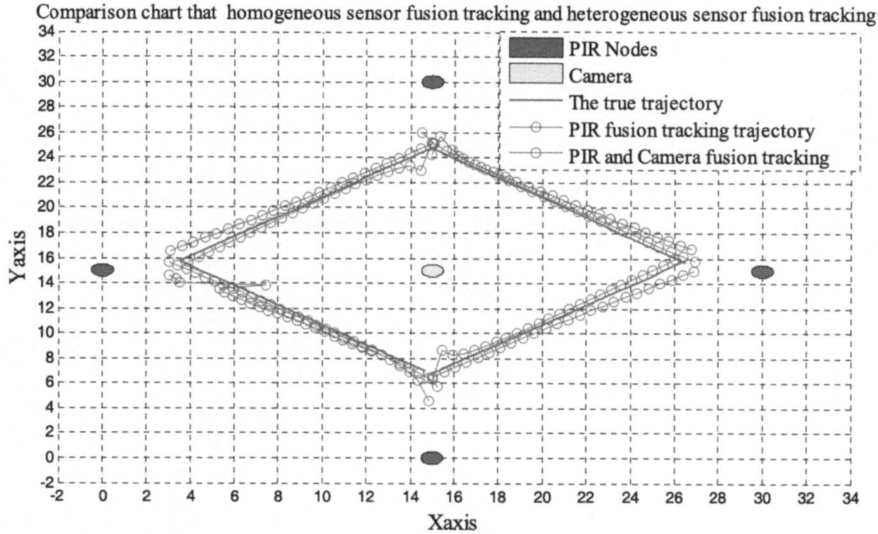

Fig. 2.7 Human tracking model

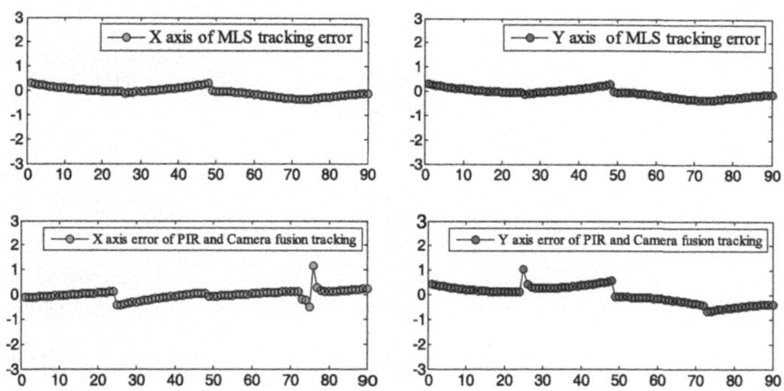

Fig. 2.8 Error-comparison chart of X-axis and Y-axis

angle of observation noise variance is $0.5 \times \pi/180°$, and distance of measurement noise variance is 0.5. After several simulation experiments, pyroelectric and camera fusion tracking error mean is 0.26 m. Tracking error mean of pyroelectric fusion with the camera is 0.37 m. The simulation results are Figs. 2.7 and 2.8. Comparing the hybrid tracking algorithm with $Qi.Hao.$ tracking algorithm, we can get the simulation results. It is shown in Figs. 2.9 and 2.10 simulation diagram.

Simulation area is 10×10 m, measurement cycle is 1 s, the azimuth angle of observation noise variance is $0.5 \times \pi/180°$, and the distance of measurement noise variance is 0.5.

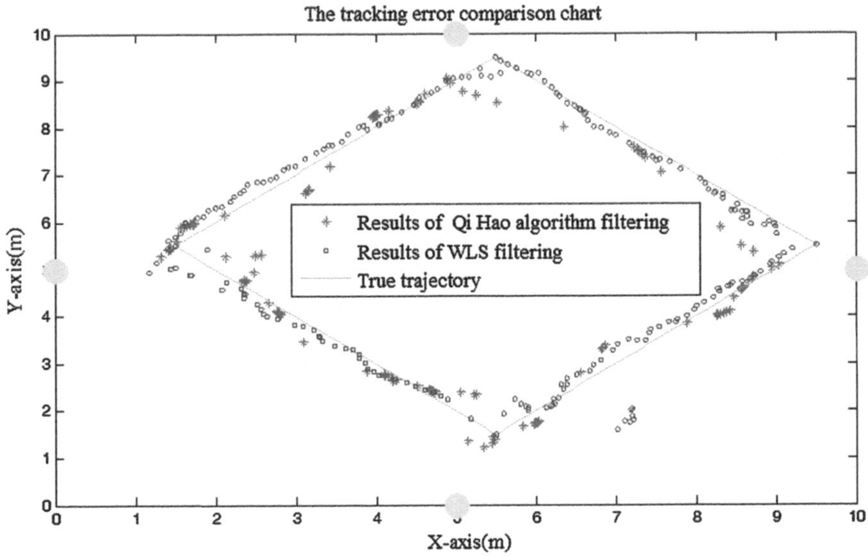

Fig. 2.9 Comparison chart of different tracking model

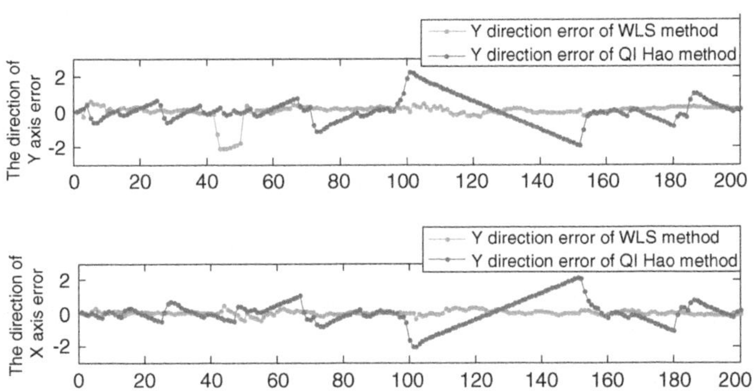

Fig. 2.10 Error-comparison chart of X-axis and Y-axis

2.5 Summary

This paper presents a system based on wireless pyroelectric infrared sensors network and video system. We have established the platform that realizes real-time human tracking and aggregates the information of distributed homogeneous sensors and heterogeneous sensors by using the weighted least square method, implementing the various patterns of human tracking. The hybrid monitoring system perfects the predecessor's research and solidifies the foundation of multihuman tracking. Meanwhile, the scheme presented in our paper would be more unprecedented because it combines the PIR sensor networks with the video systems.

Acknowledgments This work was supported by the National Natural Science Foundation of China under Grant No. 61170090

References

1. Hao Q, Brady DJ, Guenther BD, Burchett J, Shankar M, Feller S (2006) Human tracking with wireless distributed pyroelectric sensors. IEEE Sens J 6:1683–1696
2. Hao Q, Hu F, Xiao Y (2009) Multiple human tracking and identification with wireless distributed pyroelectric sensors. IEEE Syst J 3:428–439
3. Hao Q, Hu F, Lu J (2010) Distributed multiple human tracking with wireless distributed pyroelectric sensors. In: Proceedings of IEEE Conference on Sensors 946–950
4. Lu J, Gong J, Hao Q, Hu F (2012) Space encoding based compressive multiple human tracking with distributed binary pyroelectric infrared sensor networks. In: Proceedings of IEEE Conference on MFI 181–185
5. Hoffmann M, Wittke M, Bernard Y, Soleyman R (2008) Dmctrac: distributed multi camera tracking. In: Proceedings of IEEE conference on ICDSC 600–607
6. Wang X, Wang S (2007) Collaborative signal processing for target tracking in distributed wireless sensor networks. J Parallel Distrib Comput 67:501–515
7. Aguilar-Ponce R, Kumar A, Tecpanecatl-Xihuitl JL, Bayoumi M (2007) A network of sensor-based framework for automated visual surveillance. J Netw Comput Appl 30:1244–1271
8. Cheng W, Shaohong L, Huai H (2002) Multitarget measurement data association for passive location systems based on direction of arrival. Syst Eng Electron 24:104–106
9. Chongquan Z, Liyong Z, Suying Y, Zhuohan L (2003) A weighted fusion algorithm of multi-sensor based on the principle of least squares. Chin J Sci Instrum 24:427–430
10. Hao Q (2006) Mulitple Human Tracking and Identification With Pyroelectric Sensors. Duke University, Durham
11. Shufeng W, Yibin H, Zhangqin H, Yong Z, Rui C (2009) Error analysis of least squares method and optimization for wsn. J Syst Simul 21:6211–6216
12. Jianshu L, Renhou L, Hong C (2006) Multi-sensor data fusion based on correlation function and least square. Control Decis 21:714–717
13. Desai P, Rattan KS (2008) System level approach for surveillance using wireless sensor networks and ptz camera. In: Proceedings of IEEE conference on National aerospace and electronics 353–357
14. Comaniciu D, Ramesh V, PM (2003) Kernel based object tracking. Pattern analysis and machine intelligence, IEEE transactions on 25 564–577
15. Tang-wen Y, Jian-da H, Hong-bo W, Qiu-qi R (2009) Object measurement based on spatial geometric constrains with monocular camera. J Nanjing Univ Sci Technol (Natural Science) 33:210–214

Chapter 3
A Novel Skeleton Extraction Algorithm in Sensor Networks

DongHui Zhu, Yubao Wang, Jing Xing, Wenping Liu, Hongbo Jiang and Gang Wu

Abstract This paper analyzes the problem of location-free skeleton extraction in sensor networks. Different from most of the previous solutions, the proposed algorithm has no dependency on boundary information. Our work is based on the proposed index of a node, named *centrality*, which can reflect the centeredness of the node. We first identify the node having the maximal centrality as the root skeleton node, based on whether a skeleton arc is obtained. Secondly, each node then computes its hop count distance to the skeleton arc, and the network is decomposed into a set of level sets. The node with the maximal centrality is identified as a skeleton node. Thirdly, these skeleton nodes are connected properly to form a coarse skeleton possibly with spurious branches. Finally, a pruning operation is conducted on the coarse skeleton, and the final skeleton is generated. Extensive simulations show the efficiency of the proposed algorithm.

Keywords Sensor networks · Skeleton · Location-free · Boundary-free · Centrality

D. Zhu · Y. Wang · J. Xing · W. Liu (✉)
Hubei University of Economics, Wuhan, China
e-mail: wenpingliu2009@gmail.com

Y. Wang
e-mail: wyb7907@gmail.com

J. Xing
e-mail: jingxing0618@gmail.com

D. Zhu
e-mail: donghuizhu2013@gmail.com

W. Liu · H. Jiang · G. Wu
Huazhong University of Science and Technology, Wuhan, China

H. Jiang
e-mail: hongbojiang2004@gmail.com

G. Wu
e-mail: 598106@qq.com

X. Wang et al. (eds.), *Advanced Technologies in Ad Hoc and Sensor Networks*,
Lecture Notes in Electrical Engineering 295, DOI: 10.1007/978-3-642-54174-2_3,

3.1 Introduction

More recently, there has been growing interests in topology recognition of sensor networks, such as segmentation [1–3] and skeleton extraction [4–9], etc. On top of these topological features, many algorithms, e.g., routing [2, 4], navigation [10, 11], and localization [3, 12], have witnessed an improved performance.

In this research, we are primarily interested in boundary-free skeleton extraction in sensor networks based on connectivity information only. In continuous domain, skeleton (aka medial axis), is defined as the set of the locus of the centers of maximal inscribed disks, or alternatively, the collection of points with locally maximal distance transform. There are a handful of studies on skeleton extraction in sensor networks. Assuming that the boundary nodes are recognized by [13], Gao et al. [4] proposed MAP, which first identifies the medial axis nodes with two or more nearest boundary nodes which are not nearby, and then connects these nodes to form a curve. Finally, the medial axis is used for designing a load-balanced and delivery-guaranteed routing protocol. Jiang et al. [5, 6] proposed to segment the boundaries into branches by identifying so-called *corner nodes*, and a node is a skeleton node such that it has two nearest boundary nodes on different boundary branches. Note that these two algorithms both assume that the boundaries are given. This potentially limits their applicabilities owing to the fact that in sparse networks, existing boundary recognition schemes such as [13] often do not work well. As such, Liu et al. [7, 8] proposed DIST, a distance transform based skeleton extraction algorithm which uses the method in [14] to identify incomplete boundaries. However, in complex networks, DIST delivers a skeleton bent toward the concave boundary nodes which cannot be identified as boundary nodes by [14]. A state-of-the-art is the work by Liu et al. [9], which identifies skeleton nodes based on the k-hop neighborhood size of the l-hop neighbors. The beauty is that it can extract the skeleton without any boundary information. However, its performance heavily relies on the properly chosen parameters k and l, otherwise it might incur many flooding operations.

In this paper, we first examine the properties of the skeleton point in continuous domain, and design an index, named *centrality*, to reflect the centeredness of a point. We then adapt it to the discrete analog, i.e., the sensor networks, and propose a novel, distributed, and light-weight skeleton extraction algorithm based on pure connectivity information, without reliance on any boundary information.

The remainder of the paper is organized as follows. In Sect. 3.2, we study the properties of the skeleton in continuous domain, and then design a distributed skeleton extraction algorithm in Sect. 3.3. We validate the proposed algorithm by extensive simulations in Sect. 3.4, and conclude the paper in Sect. 3.5.

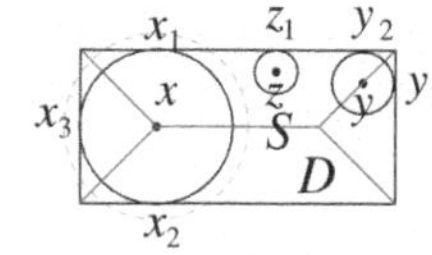

Fig. 3.1 Illustration of the skeleton

3.2 Theoretical Foundation

As mentioned earlier, the skeleton of a 2D object is the set of the locus of the centers of the maximal inscribed disks (**MIDs** for short), or equivalently, the set of the points with locally maximal distance transform. In this section, we first take a closer look at the property of a skeleton point, and then derive a sufficient (but not necessary) condition for skeleton point identification.

We denote by D a 2D object with boundary ∂D and skeleton S. Please see Fig. 3.1 for some intuition. The red curve is the skeleton, and points x, y are both skeleton points while z is not. Denote by $D(p, r(p))$ the MID centered at skeleton point p with radius $r(p)$. Clearly, the disk $D(x, r(x))$(or $D(y, r(y))$), indicated by the solid black circle, is tangent to the boundary at three (or two) tangent points x_1, x_2, x_3 (or y_1, y_2). If radius $r > r(x)$, then disk $D(x, r)$, shown by the dashed green circle, will not lie totally inside D. As such, we have the following observations.

Observation 1. A MID has two or more tangent points.

Observation 2. Let $r(p)$ denote the radius of the MID centered at skeleton point p. For a positive $r \leq r(p)$, $D(p, r)$ will be included completely in D, and the area of the intersection of $D(p, r)$ and D, denoted by $\lambda(D(p, r) \bigcap D)$, equals πr^2; and if $r > r(p)$, the disk $D(p, r)$ will not lie completely inside D, and $\lambda(D(p, r) \bigcap D) < \pi r^2$.

With these observations, we then present a series of lemmas and theorems as follows.

Lemma 1. *For a skeleton point p and a sufficiently small real $\epsilon > 0$, if there is a real number $r > 0$ such that*

$$\frac{\lambda(D(p, r + \epsilon) \bigcap D)}{(r + \epsilon)^2} < \pi \tag{3.1}$$

then $D(p, r)$ is a MID.

Lemma 2. *For an arbitrary point $q \in D$, if there exists a positive real r, such that for any real value $\delta > 0$,*

$$\frac{\lambda(D(q, r) \bigcap D)}{r^2} = \pi, \tag{3.2}$$

$$\frac{\lambda(D(q, r + \delta) \bigcap D)}{(r + \delta)^2} < \pi \tag{3.3}$$

then $D(q, r)$ is an inscribed disk with radius r.

Lemma 2 provides a necessary (but not sufficient) condition for skeleton point. That is, for a skeleton point p, there is a positive real r satisfying Eqs. (3.2) and (3.3).

Lemma 3. *If $D(p, r)$ is an inscribed disk with one tangent point $q \in \partial D$, then the inscribed disk centered at any point on the chord pq must be tangent with the object at point q.*

Proof. We prove this claim by contradiction. Let v be a point on the chord pq. Suppose that the inscribed disk centered at v is not tangent to the boundary at q. Without loss of generality, we assume that q' is the tangent point. Obviously, we have $|vq'| < |vq|$ where $|\cdot|$ denotes the length of a line segment. Then $|pq| = |pv| + |vq| > |pv| + |vq'| = |pq'|$ holds. That is to say, q is not a tangent point, which is a contradiction. □

Corollary 1. *If two inscribed disks $D(p_1, r_1)$ and $D(p_2, r_2)$ share one common tangent point, and $r_1 < r_2$, then $D(p_1, r_1) \subset D(p_2, r_2)$.*

Theorem 1. *Let $D(q_i, r_i)(i = 1, 2, \cdots)$ be a series of inscribed disks with one common tangent point, and $r_j = max_i\{r_i\}$. Then, $D(q_j, r_j)$ is a maximally inscribed disk, and q_j is a skeleton point.*

Proof. We prove this by contradiction. Suppose that $D(q_j, r_j)$ is not a maximally inscribed disk, and thus q_j is not a skeleton point. Then there exists at least one inscribed disk $D(q_m, r_m)(m \neq j, m = 1, 2, \cdots)$ such that $D(q_j, r_j) \subset D(q_m, r_m)$. That is, $r_m > r_j$, contradicting with $r_j = max_i\{r_i\}$. □

Theorem 1 says that the inscribed disk with the largest radius, among all inscribed disks sharing one common tangent point, is a maximally inscribed disk, and accordingly, the center of the largest inscribed disk is a skeleton point.

Corollary 2. *If the radius of an inscribed disk $D(p, r)$ is locally maximal, then p is a skeleton point.*

Theorem 1 and Corollary 2 provide a sufficient, yet not necessary, condition for skeleton point identification, which will be adapted to sensor networks detailed in Sect. 3.3.

3.3 Algorithm Description

In this section, we will adapt the aforementioned principle of a skeleton point to sensor networks, and propose a connectivity-based and distributed skeleton extraction algorithm. The proposed algorithm has three main steps, i.e., skeleton node identification, network segmentation, and skeleton establishment.

3.3.1 Skeleton Node Identification

As in sensor networks, the locations of sensors are usually unknown and the distance between nodes can only be estimated by hops, it still faces nontrivial challenges to identify skeleton nodes in sensor networks. To address this, for ease of statement, we assume that the sensors are uniformly distributed in the sensing field, and each node has a unique ID. The communication radio range of each node is assumed to be the same, denoted by r.

For a sensor node p, let $N_k(p)$ be its k-hop neighbors, namely, the set of nodes at most k hops from p, and $|N_k(p)|$ denote the k-hop neighborhood size. Clearly, $N_k(p)$ is a discrete analog of the circle $D(p, kr)$ in continuous domain. The larger the value of k, the larger the area of $D(p, kr)$ is, and thus the larger the k-hop neighborhood size $|N_k(p)|$. Note that in continuous domain, if $D(p, kr)$ is a maximally inscribed disk or included by an inscribed disk, then clearly we have $\frac{\lambda(D(p,kr)\cap D)}{(kr)^2} = \pi$, otherwise, $\frac{\lambda(D(p,kr)\cap D)}{(kr)^2} < \pi$. We observe that $\frac{|N_k(p)|}{k^2}$ also exhibits such a behavior. That is, the value of $\frac{|N_k(p)|}{k^2}$ will be stable when k is smaller than a threshold corresponding to the radius of maximally inscribed disk, and then decreases as k increases. As such, we have

Definition 1. *For a node p and a given $\delta > 0$, if there is an integer k such that*

$$\frac{|N_k(p)|}{k^2} \Big/ \frac{|N_{k+1}(p)|}{(k+1)^2} > 1 + \delta, \tag{3.4}$$

We call k as the centrality of p, denoted by $k(p)$.

In fact, here $k(p)$ corresponds to the radius of the inscribed disk centered at p, which can reflect how central a point is to the object. The larger the value of $k(p)$, the more likely the node p is a skeleton node.

Definition 2. *If $k(p)$ is locally maximal, then p claims itself as a skeleton node.*

Clearly, if $k(p)$ is the largest centrality among all nodes, then p is a skeleton node. Such a skeleton node is unique, and we call it as a root skeleton node, denoted by $\mathcal{RSN}$.

Now we propose to identify skeleton nodes in a distributed way. More specifically, each node initiates a limited flooding within its h-hop neighbors to collect the information of its neighborhood size, where h is a predefined integer and is assigned a relatively large value, e.g., 10. Then each node p computes its centrality $k(p)$ according to Eq. (3.4), and by flooding within the network, the node with maximal centrality identify itself as the unique $\mathcal{RSN}$. As $\mathcal{RSN}$ is a skeleton node, there must be at least two tangent nodes, which can be identified based on their neighborhood sizes as described in [14]; two farthest tangent nodes can be connected with $\mathcal{RSN}$ to derive a curve C, referred to as *skeleton arc*, as shown in Fig. 3.2a.

Fig. 3.2 An illustration of the proposed algorithm. **a** The connecting of the root skeleton node and its two tangent nodes forms a skeleton arc. **b** Level sets. The nodes within different level sets are shown in different colors. **c** Skeleton nodes. **d** Coarse skeleton. **e** The final skeleton

3.3.2 Network Segmentation

With the established skeleton arc, we now propose to decompose the network into components, and then identity a skeleton node within each component. To be specific, the nodes on the skeleton arc $\mathcal{C}$ flood simultaneously in the network, and each node p records its hop count distance to the skeleton arc. Let $F : V \rightarrow \mathcal{N}$ be the hop count distance function to $\mathcal{C}$, where V is the set of the sensors. Clearly, F serves as a feature function assigning a nonnegative value to each sensor. The sensors with the same hop count distance to $\mathcal{C}$ form a set of connected components (referred to as *level set*). Please see Fig. 3.2b for some intuition. Within each level set, the node with the largest centrality identifies itself as a skeleton node, as shown in Fig. 3.2c.

3.3.3 Skeleton Establishment

Due to the discrete nature of sensor networks, the identified skeleton nodes will not be self-connected. As such, in this subsection, we propose to connect these skeleton nodes to form a meaningful representation of the network. To that end, we connect the identified skeleton nodes according to their feature function values. More specifically, two skeleton nodes p_1 and p_2 will be connected if their feature function values differ by one, and the two level sets of p_1 and p_2, denoted by $ls(p_1)$ and $ls(p_2)$, are adjacent, i.e., there exists at least two nodes $q_1 \in ls(p_1)$ and $q_2 \in ls(p_2)$ such that q_1 and q_2 are neighbors. As a result, all skeleton nodes, including the $\mathcal{RSN}$, can be correctly connected to form a coarse skeleton, possibly with redundant skeleton branches, i.e., with small length, and/or skeleton loops, as shown in Fig. 3.2d. We finally refine the coarse skeleton by the method described in [5, 6] to deliver the final skeleton, as shown in Fig. 3.2e.

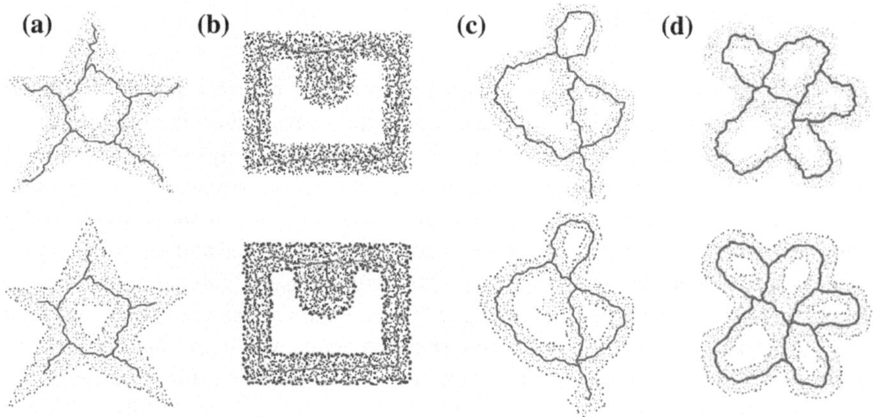

Fig. 3.3 Skeleton extraction under different scenarios. The *red curves* represent the derived skeletons. *Row 1* The proposed algorithm; *Row 2* MAP (the boundary nodes are marked in *blue*). **a** *Star-shaped* network, 1394 nodes, average degrees 6.59. **b** *OneHole-shaped* network, 2374 nodes, average degrees 6.54. **c** *Music-shaped* network, 1301 nodes, average degrees 6.5. **d** *Flower-shaped* network, 2422 nodes, average degrees 5.95

3.4 Performance Evaluation

Previously, we have proposed a distributed algorithm for skeleton extraction in sensor networks, without reliance on boundary and location information. To evaluate the performance of the proposed algorithm, we conduct extensive simulations under scenarios with different shapes (see Fig. 3.3), and compare our algorithm with MAP [4], which serves as a ground truth.

3.4.1 Simulation Settings

In our simulations, we randomly deployed the sensors in the sensing region. The communication radio range r of each sensor is assumed to be the same, and the communication radio model is Unit Disk Graph (UDG) model by default. That is, two nodes are connected if their separation is no greater than r. The parameters are $h = 6$ and $\delta = 0.5$ by default. The average degrees are all below 7 where existing boundary recognition solutions do not work well. Note that MAP requires boundary information. For fair comparison, we manually identify the boundary nodes for MAP; in order to control boundary noise, in our simulations MAP regards the unstable skeleton nodes, whose nearest boundary nodes have a separation less than 6, as nonskeleton nodes.

3.4.2 Simulation Results

Figure 3.3a shows the results by the proposed algorithm (the first row) and MAP (the second row) under Star-shaped network, where the convex inner hole is caused by nodes failure or an obstacle. Note that the average degree is only 6.59. Under such a low node density, existing boundary recognition schemes can hardly accurately identify the complete boundary nodes. By manually marking the boundary nodes, MAP delivers a skeleton which can reflect the underlying geometrical and topological features. We also see that the proposed algorithm, without using boundary information, still obtains a comparable result. Our algorithm yields longer skeleton branches than MAP since MAP disregards the unstable skeleton nodes which are actually genuine skeleton nodes. That is, our skeleton captures more topological information of the network. Different from Fig. 3.3a, the OneHole-shaped network in Fig. 3.3b has one concave hole, with average node degree 6.54. But our algorithm still accurately identifies the skeleton nodes, and delivers a comparable result with MAP.

The Music-shaped network in Fig. 3.3c and Flower-shaped network in Fig. 3.3d have more complicated inner holes with different sizes and shapes, potentially posing a challenge for skeleton extraction without boundary information. From Fig. 3.3c, d we see that both MAP and our algorithm obtain a good skeleton, which is well-centered and reflects the major topological features of the underlying network. In summary, even without reliance on any boundary information, the proposed algorithm still yields a good skeleton comparable to that by MAP.

3.5 Conclusion

We study the problem of boundary-free skeleton extraction in sensor networks based on connectivity information only. We first study the properties of the skeleton to construct an index, named centrality, which can reflect the centeredness of a point, and then propose a distributed and location-free skeleton extraction algorithm in sensor networks, without reliance on any boundary information. The simulation results show the efficiency of the proposed algorithm.

Acknowledgments This work was supported in part by the National Natural Science Foundation of China under Grant 61202460.

References

1. Zhu X, Sarkar R, Gao J (2007) Shape segmentation and applications in sensor networks. In: Proceedings of the IEEE INFOCOM
2. Tan G, Bertier M, Kermarrec AM (2009) Convex partition of sensor networks and its use in virtual coordinate geographic routing. In: Proceedings of the IEEE INFOCOM

3. Liu W, Wang D, Jiang H, Liu W, Wang C (2012) Approximate convex decomposition based localization in wireless sensor networks. In: Proceedings of the IEEE INFOCOM
4. Bruck J, Gao J, Jiang AA (2005) Map: medial axis based geometric routing in sensor networks. In: Proceedings of the ACM MOBICOM
5. Jiang H, Liu W, Wang D, Tian C, Bai X, Liu X, Liu W (2009) Case: connectivity-based skeleton extraction in wireless sensor networks. In: Proceedings of the IEEE INFOCOM
6. Jiang H, Liu W, Wang D, Tian C, Bai X, Liu X, Liu W (2010) Connectivity-based skeleton extraction in wireless sensor networks. IEEE Trans Parallel Distrib Syst 21(5):710–721
7. Liu W, Jiang H, Bai X, Tan G, Wang C, Liu W, Cai K (2012) Skeleton extraction from incomplete boundaries in sensor networks based on distance transform. In: Proceedings of the IEEE ICDCS
8. Liu W, Jiang H, Bai X, Tan G, Wang C, Liu W, Cai K (2013) Distance transform-based skeleton extraction and its applications in sensor networks. IEEE Trans Parallel Distrib Syst 24(9):1763–1772
9. Liu W, Jiang H, Wang C, Liu C, Yang Y, Liu W, Li B (2012) Connectivity-based and boundary-free skeleton extraction in sensor networks. In: Proceedings of the IEEE ICDCS
10. Buragohain C, Agrawal D, Suri S (2006) Distributed navigation algorithms for sensor networks. In: Proceedings of the IEEE INFOCOM
11. Li M, Liu Y, Wang J, Yang Z (2009) Sensor network navigation without locations. In: Proceedings of the IEEE INFOCOM
12. Lederer S, Wang Y, Gao J (2008) Connectivity-based localization of large scale sensor networks with complex shape. In: Proceedings of the IEEE INFOCOM
13. Wang Y, Gao J, Mitchell JSB (2006) Boundary recognition in sensor networks by topological methods. In: Proceedings of the ACM MOBICOM
14. Fekete SP, Kroller A, Pfisterer D, Fischer S, Buschmann C (2004) Neighborhood-based topology recognition in sensor networks. In: Proceedings of the 1st International workshop on algorithmic aspects of wireless sensor, networks

Chapter 4
A Real-Time Information Perceived Vehicle Opportunity Route: RIPVO

Kelei Li, Xinhua Wang, Mianmian Lv and Zhen Wang

Abstract Considering that VANET may reflect the characters of opportunity network, a routing scenario with real-time information perceiving is proposed in this paper. First, a node executes limited and directed data flooding and then carries out feedback with modified greedy algorithm using hot area. In this strategy, nodes dynamically adjust parameters by self-learning properly. The experiment result under ONE simulator shows its promotion comparing to classic Epidemic, etc.

Keywords VANET (vehicular ad hoc network) · DTN (delay-tolerant network) · MANET (mobile ad hoc network)

4.1 Introduction

VANET (Vehicle ad hoc network) communicates with sensors or other signal sending and receiving equipments on moving vehicles, and its nodes make up a self-organizing, heteroid and multihop network. In some situations (e.g. early morning or midnight, and disasters happening), VANET displays communication features like intermittent connectivity, long delay, high error rate, etc., which usually occur in DTN. So we call VANET with DTN features Vehicle Opportunity Network (VONET).

The main problem VONET faces in data routing is frequent changes in the network structure caused by high-speed mobility of nodes, namely unstable network topology. Proximity Graphs (Network topology structures) plays a decisive influence on network performance [1]. Considering the mobility of VONET, the way to build real-time and efficient Proximity Graphs will greatly affect the route efficiency and

K. Li (✉) · X. Wang · M. Lv · Z. Wang
Information Science and Engineering, Shandong Normal University, Jinan 250014, China
e-mail: 52c88@sohu.com

Shandong Provincial Key Laboratory for Novel Distributed Computer Software Technology, Jinan 250014, China

X. Wang et al. (eds.), *Advanced Technologies in Ad Hoc and Sensor Networks*, Lecture Notes in Electrical Engineering 295, DOI: 10.1007/978-3-642-54174-2_4,

message delivery reliability. To solve this question, RIPVO (A Real-time Information Perceived Vehicle Opportunity Route) is proposed in this paper. The remainder of this paper is organized as follows: In Sect. 4.1, we introduce the recent research status of VANET and DTN. RIPVO is described in Sect. 4.2. In Sect. 4.3, we evaluate the algorithm performance with ONE simulator. We conclude this paper in Sect. 4.4.

4.2 Relative Works

Opportunity network mainly involves node-centric route, location-centric route, data-centric route, mobility-centric route, etc. Node-centric route tries to sufficiently make use of nodes broadcasting and calculating ability to delivery message and build network topology. In addition, these routes think over classifying different nodes for better message transmission. Authors of paper [2] analyze the routes based on delay nodes, and an optimized delay node distribution algorithm for maximizing data delivery is proposed in [3].

Location-centric route can balance the unstable network topology caused by nodes' mobility. It tries its best to avoid using fixed message transmission path, and consider about forwarding according to certain weights like physical distance. For example, GPRS [4] always forwards messages to the node closer to the destination; besides nodes location, [5] takes nodes' connectivity into account to choose delivery path and anchor for message forwarding. GeoDTN+Nav [6] study and design route by thinking of the effect of vehicle moving direction. Location-centric route usually gets nodes' current location with GPS for decision-making. Its advantage lies in calculating with real-time information without maintaining a global routing table. Nowadays, more and more designs start to get road topology with GIS. With it, VADD [7] designs three different transfer patterns for specific road situations, which are L-VADD, H-VADD, and D-VADD.

Data-centric route focuses on the usage of highly infectious data and the content of messages. Recalling the message content, Rumor [8] balances the building of routing path and delivery reliability for quicker message delivery. Mobility-centric route takes nodes mobility into account and tries to balance problems caused by it. For example, MDDV [9] set Head Node to accelerate message transmission.

Recently, more and more designers combine VANET route scheme with vehicle nodes' sociality. Sueur et al. [10] show some characters in social networks through their studies. In RMR [11], designers consider the reliability of area and some researchers come up with Hot Area in thinking about nodes' mobility.

4.3 Algorithm Description

System assumption In this paper, the protocol RIPVO tries to dynamically get a transmission path by the vehicles' absolute self-organization with GPS for location obtaining only, when no information around is known in advance. Specific scheme goes to a limited quick flooding query for location confirmation first and then a directed convergence for data delivery. On the whole, messages get close to the destination area dynamically; while in details, a node executes local optimal routing algorithm with real-time information collected by itself and calculation result to delivery messages. As base station on wayside cannot always be useful and the node's moving track is irregular, the station and ferryman will be dynamically obtained by periodical query, in order to accelerate data transmission and improve delivery rate. Nodes calculation and storage ability are hoped to be used sufficiently so that nodes can know the information around them and save them dynamically for further usage.

Nowadays, most vehicles are equipped with GPS. GPS can provide free service for location with little geographical influence. So we assume that nodes use GPS to get geographical location information for route path building and message directed transmission. Especially, we do not use urban electronic maps mentioned in many studies to get urban road topology and do not forward messages through information station on the roadside. Because heavy natural disasters may destroy city roads and stations they cannot be useful. On the other hand, info-station can be obtained by dynamical query.

Limited quick flooding query Random nodes do not know each other (connection relation in detail) in variable network topology surroundings. When message transmission occurs, we may know who the message should be transferred to while we do not know where it is, so we call it nodes' "double-blinded."

While nodes' mobility is always in control, and there exists no teleportation, we can calculate an area according to nodes' moving speed, direction,, and habit. However a node moves in a limited accessible time slot, but the node will not move out of the area with accessible radio signal. So as long as a message is delivered into the area it can be sent to its receiver directly or indirectly. Therefore, we get a certain direction for message delivery so that we can transfer data "directly" to a certain area to realize intended goals.

Two nodes with long distance ensure each other's location by flooding query. The simplest way is use Epidemic to query. But this will waste nodes' calculation and storage ability, and may lead to congestion. In this paper, Location obtaining is realized by periodical query and wait. If the message is very small, the message can be delivered to its destination during the location obtaining process, and the feedback becomes transmission acknowledgment.

For convenience of message delivery, every node has an area in this paper, and hot area is also defined.

Node Area: Every node maintains a cyclo-area with center location O and radius R. Location O is the node's current position coordinate (x_current, y_current) got by GPS, and $R = n*c$. c is the node's radio communication ability. n is hop count. Given initial value, the node will control n by its own knowledge.

The hop count is initialized as $n = 5$ and will be adjusted automatically according to the comparison result of the time of successful message delivery (the time of failed delivery is defined as infinite). During the real operation, after comparison of query time request-time and time control $T\_\lim$, if request_time > $n*T\_\lim$, then $n = n + 1$; else $n = n - 1$.

The time control is obtained by formula $T\_\lim = (R+c)/\bar{v}$. The $\bar{v}$ is the average of nodes' speed within the area, and its initial value is the node's own speed; R is the radius of vehicle's area and c is the node's radio communication ability.

Assume that there exists a node A, a source node or an intermediate node, and it is broadcasting a message. If the coordinate of the last intermediate node is involved in the message, neighbors of node A will receive the message after several broadcasting. Guessing that node B receives the message, node B will check if it is the first destination. If it is, it will receive the message and feedback to the source node, otherwise it continues to calculate.

Node B checks if the message exists in its buffer, if it does, then B executes directed broadcast for the message. We set a tolerant existence time called $T_$tolerance for every node. Within the time, if needed, the message will be broadcast. If the time is used up, message will be broadcast with decreased hop counts. If the distance between node B and intermediate node is less than the area radius of B, B will receive the message and then broadcast it; otherwise, B will receive the message and then update it; at the same time B saves the information about the message in its buffer and then broadcasts the updated message.

Directed broadcast means a node broadcasts a message only to nodes in the same direction with itself before it moves out of the area marked in the message. This is to reduce the copies of messages.

Collection and storage of area information The calculating ability of vehicles is great and it is impossible that nodes transmit all the time, so we can use this spare time to dynamically collect area information and save them. Here, we try to let moving nodes persistently collect neighbor information around and exchange them with each other. With this information, a node can ensure the number of nodes within its area, and mark its area with its current location coordinate and radius. Every node will do this periodically; the information exchanged by vehicles include position, moving speed and direction, and neighbor links. As a node knows the distribution of nodes around, it can transmit message with a local optimal routing algorithm.

Through periodical information exchange nodes can know the vehicle density and the distribution of vehicles speed within its area. The information around can be easily obtained by periodical long-distance message exchange. During the local message transmission, GPRS can be taken, or a certain local optimal routing method. With this limited local real-time information, we can both build local link graph and define hot area. In this way, hot area which is closer to the destination area can be taken as intermediate area to execute local transmission.

As a node receives neighbor's broadcast message, it will update its link information in order to know the real-time information around. During the updating, time value will be taken to decide wether to update the old message or not and the new neighbor will be added into the neighbor link buffer if it is in the node's area. Every node maintains neighbors within the area with radius $R + c$, namely a hop bigger than the node's area radius. In addition, if the record time of neighbor in the neighbor buffer is bigger than time control $T\_\lim$, the neighbor will be deleted, and the same to neighbors connected through it.

Hot Area Every node maintains its own area and includes the real-time nodes link in that area. When a node knows the destination node's position through quick limited flooding query, it dynamically queries intermediate hot area to transmit messages. Hot area is simply defined as all connected area whose node number n follows formula 4.1.

$$n \geq N\_\text{lim},\ N\_\text{lim} = \frac{(2R')^2}{\pi c^2} = \frac{4(n+1)^2}{\pi},\ R' = R + c. \tag{4.1}$$

For every hot area, we need to maintain relative information as follows: (1) coord: the coordinate of area center; (2) R: radius of hot area; (3) ρ: density of nodes in the area; (4) $\bar{v}$:average speed of nodes in the area; (5) time: occurrence time of hot area; (6) life: existence time of hot area.

These properties of hot area are used to select the intermediate area and the merging, updating, and abandon of area. The property life and the average speed are obtained by local calculation. The others are obtained in real time. As formula 4.2 shows

$$\text{life} = \frac{R+c}{\bar{v}},\ \bar{v} = \frac{1}{\text{num\_node}} \sum v\_\text{node}\_i. \tag{4.2}$$

Every node periodically checks wether the area it maintains meets the definition of hot area. If it does, the area will be added or updated in the node's buffer. The node exchanges its hot area buffer with its neighbor periodically. For the new detected hot area the node will check its buffer first and then compare the value of Dis. If Dis $< R\_\text{new} + R\_\text{temp}$, then Dis $= |\text{coord\_new} - \text{coord\_temp}|$.

The R_new is the radius of the new hot area. The R_temp is the radius of cached hot area. The coord presents their center coordinate. The D is the distance between two hot areas. If fitted, update will be carried out; on the contrary, a new data will be added. There may exist several updatable areas and a situation with three areas is shown in Fig. 4.1 (left). The update of hot area is executed with formulas 4.3 and 4.4.

$$\text{Attr\_new} = \sum \left(\frac{\text{time\_}left\_i}{\sum \text{time\_}left\_i} \cdot \text{Attr\_}i \right),\ \text{Attr\_}i \in \{\text{coord}, \rho, \bar{v}\}, \tag{4.3}$$

$$R\_\text{new} = \frac{1}{\text{num}} \sum R\_i,\ \text{time\_new} = \text{time\_current},\ \text{life\_new} = \frac{R\_\text{new} + \bar{c}}{\bar{v}\_\text{new}}. \tag{4.4}$$

Message feedback and re-transmission The destination node will execute local calculation after receiving a query. The same is done to source node as it receives a feedback, in order to find out a better hot area as intermediate area for directed

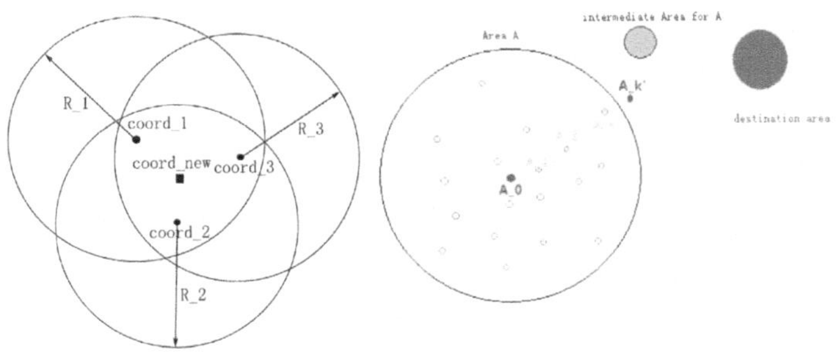

Fig. 4.1 Hot area update (*left*); Transmission with feedback (*right*)

message transmission. Here, we choose the hot area located in the same direction of destination as intermediate area with formula lim = ρ+ Ds-d/Ds-temp.

ρ is the number of nodes with the area, and Ds-d is the distance between the source node area (or intermediate area) and destination area. Ds-temp is the distance between the source node area (or intermediate area) and the next accessible intermediate area. We use this because the closer the intermediate area is, the more likely the transmission will be successful, and the greater the density is, the more reliable the area is.

S denotes the node executing, and *D* denotes the center of destination area, and *T* denotes the center of hot area stored in the current node. With calculation, if $\theta \leq 45^\circ \left(\theta = \arccos\frac{\overrightarrow{ST}\cdot\overrightarrow{SD}}{\|\overrightarrow{ST}\|\cdot\|\overrightarrow{SD}\|}\right)$, the hot area can be chosen as candidate area for the next calculation of value lim.

A node will confirm the transmission direction as it gets the value lim. Node S transmits data to node N_next which is closest to the next area stored in its buffer. Node S adds a time control in the message as formula: time_ctrl $=$ Dis(S $\rightarrow$ N_next)$/\bar{v_s}$.

As node S receives the message, S will recalculate the transmission direction if it moves out of S area within time_ctrl; Even if S does not move out, it will still execute the operation. If S fails to receive the feedback from N_next it will execute the operation, too.

As Fig. 4.1 (right) shows, the node A_0 calculates and gets the green area as its intermediate area, it forwards its local stable neighbor map to the nearest node in that area. If the node A_k moves out of A_0 area during the control time, the message will be updated and the intermediate area will be recalculated. As the control time is used up, if node A_k is still in the A_0 area, recalculation will also be executed and feedback to A_0, and node A_0 re-queries as receiving the feedback. If node A_0 cannot receive the feedback within the control, re-query will also be executed.

Fig. 4.2 Local route implement

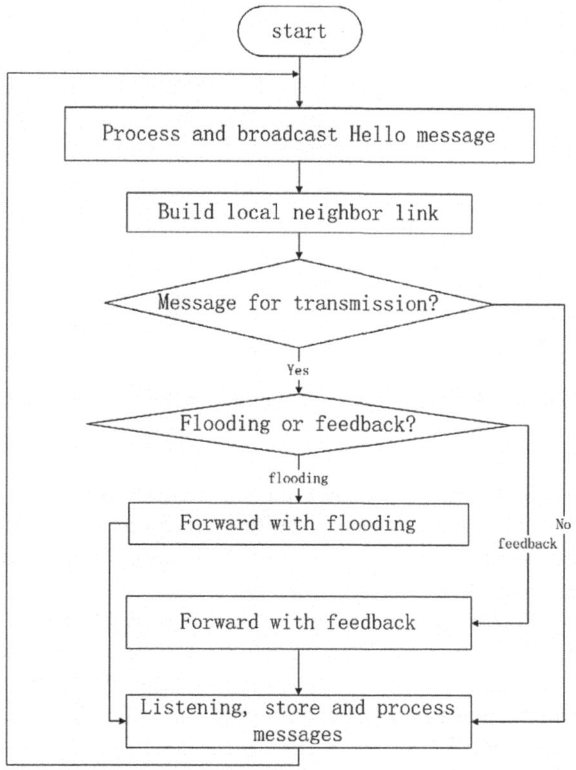

4.4 Experiment and Analysis

Experimental design and flowchart Theoretically, we divide every period into several parts for message broadcast, receiving, etc. The operation procedure of routing is shown in Fig. 4.2.

During the local routing, a node needs to maintain several data buffers, involves neighbor link buffer, hot area buffer, broadcast message buffer, and forward message buffer.

Every node maintains two kinds of messages with message type in the message structure, and both have their own spaces. For every stored message, the node records its broadcast or forwarding times that the node will delete the message with the biggest number; the number will not be added to the message and used to local control.

We overwrite the message event generator, Message receiving and forwarding method. In addition, the measurement of message and the collection of nodes information are realized with the algorithm mentioned ahead.

Experimental results and analysis During our experiment, Map Of Jinan city is used, the scope is 10,000 · 8,000 m, and two kinds of moving nodes are set to present bus, taxi, or private vehicle. Their speed limits are 35 km/h, 50 km/s, 40 km/s and

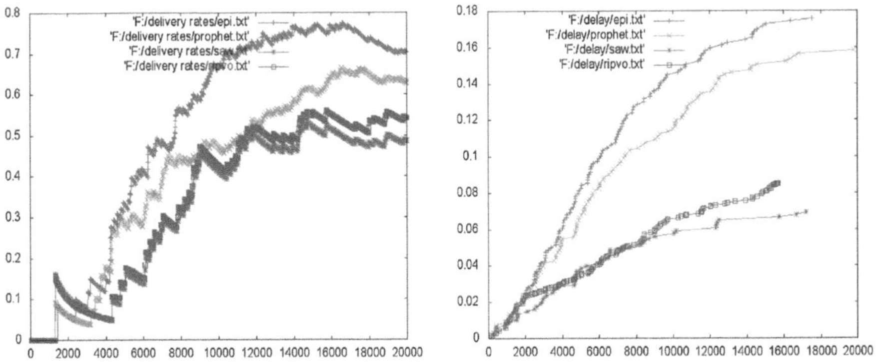

Fig. 4.3 Message delivery rate comparison (*left*), accumulative delay rate comparison (*right*)

60 km/s. The former runs with map-based movement and the latter runs with working day movement. Message buffers are 50 and 9 M. The nodes transmission speed is 256 Kbps and all nodes use Wi-Fi as communication interface and open GPS. Message TTL is set to 1,300 s.

With the same settings, we compare RIPVO with classic algorithm Epidemic, Prophet, Spay, and Wait on delivery rate and accumulative delay rate.

On delivery rate, as shown in Fig. 4.3 (left), RIPVO is better than Spay and Wait while not as Epidemic and Prophet; on accumulative delay rate, as shown in Fig. 4.3 (right), RIPVO is better than Epidemic and Prophet while not as Spay and Wait.

Overall, RIPVO has its advantage as a new route algorithm but still has room for improvement.

References

1. Lu G, Zhou MT, Niu XZ, She K, Tang Y, Qin K (2008) A servery of proximity graphs in wireless networks. J Softw 19(4):888–911
2. Farahmand F, Cerutti I, Patel AN, Jue JP, Rodrigues JJPC (2009) Performance of vehicular delay-tolerant networks with relay nodes. Commun Mob Comput 1–11
3. Sharma S, Shi Y, Hou YT, Kompella S (2011) An optimal algorithm for relay node assignment in cooperative ad hoc networks. IEEE/ACM Trans Netw (archive) 19(3):879–892
4. Karp B, Kung HT (2000) GPSR: greedy perimeter stateless routing for wireless networks. ACM MOBICOM, Boston, pp 243–254
5. Ramanathan R, Hansen R, Basu P, Rosales-Hain R, Krishnan R (2007) Prioritized epidemic routing for opportunistic networks. In: MobiOpp '07 proceedings of the 1st international MobiSys workshop on mobile opportunistic networking, pp 62–66
6. Cheng P-C, Lee KC, Gerla M (2010) GeoDTN+Nav geographic DTN routing with navigator predition for urban vehicle environment. Mob Netw Appl 15(1):61–82
7. Zhao J, Student Member, IEEE, Cao G (2008) VADD: vehicle-assisted data delivery in vehicular ad hoc networks. IEEE Trans Veh Technol 57(3):1910–1922

8. Braginsky D, Estrin D (2002) Rumor routing algorithm for sensor networks. In: WSNA '02 proceedings of the 1st ACM international workshop on wireless sensor networks and applications, pp 22–31
9. Wu H, Fujimoto R (2004) MDDV: a mobility-centric data dissemination algorithm vehicular networks. ACM SIGMOBILE, New York, pp 47–56
10. Sueur C, Jacobs A, Amblard F, Petit O, King AJ (2011) How can social network analysis improve the study of primate behavior? Am J Primatol 73:703–719
11. Gohari AA, Pakbaz R, Melliar-Smith PM, Moser LE, Rodoplu V (2011) RMR: reliability map routing for tactical ad hoc networks. Selected areas in communications. IEEE J 29(10):1935–1947

Chapter 5
Enabling Reliable and Efficient Network Reboot in Wireless Sensor Networks

Wenbin Wu, Yi Gao, Xuefeng Zhang, Chao Huang and Wei Dong

Abstract Wireless sensor networks often encounter unexpected failures due to a variety of reasons. Since detecting the failed nodes is normally a nontrivial work, rebooting all the nodes in the network is an effective mechanism to recover from failures. This paper presents NR, a network reboot mechanism based on TinyOS, which enables reliable and efficient network reboot in WSNs. NR ensures 100 % reliability by incorporating an eventual consistency dissemination protocol. It avoids flash I/Os by storing data items on persistent RAM. Evaluation results show that NR can ensure 100 % reliability and is efficient in I/O operations.

Keywords Wireless sensor networks · Dissemination protocols · Network reboot · Persistent storage

5.1 Introduction

Wireless sensor networks (WSNs) have been proposed for a wide range of applications such as ecological surveillance [1], habitat monitoring [2], infrastructure protection [3], etc. Applications commonly specify a network should operate unattended for months or years [4]. Software dependability and reliability are therefore critical concerns [5].

In practice, however, WSN systems often encounter unexpected failures due to a variety of reasons, e.g., unpredictable environment, link dynamics, and complex network interactions. Network reboot is an effective mechanism to recover from failures. For example, in a recent deployment on Reventador Volcano, a software error was recovered by a network reboot [6].

W. Wu · Y. Gao · X. Zhang · C. Huang · W. Dong (✉)
Zhejiang Key Laboratory of Service Robot, CS College, Zhejiang University, Hangzhou, Zhejiang, China
e-mail: dongw@zju.edu.cn

X. Wang et al. (eds.), *Advanced Technologies in Ad Hoc and Sensor Networks*, Lecture Notes in Electrical Engineering 295, DOI: 10.1007/978-3-642-54174-2_5,

Interestingly, reliable and efficient network reboot is not addressed sufficiently in the literature. Most real-world systems rely on network flooding to issue the reboot command. Network flooding can hardly guarantee 100 % reliability. A few malfunctioned nodes may cause severe performance degradations to the entire network. To ensure a high reliability in practice, flooding the reboot command requires a significant number of broadcasting messages. Moreover, nodes that received the command need to wait a sufficient time before rebooting for the purpose of forwarding. In other words, current network reboot mechanisms are not reliable or efficient enough.

An obvious solution to ensure 100 % reliability is to employ a dissemination protocol such as Drip [7] and DIP [8]. Such protocols employ ADV-REQ-DATA three-way handshake to ensure the eventual consistency property, but they cannot be directly applied in practical systems. This is because the status information will be lost after reboot and the command will never be propagated to distant nodes.

A workaround to data loss after reboot is to use persistent memory such as the program flash and external flash. However, data storage on the flash is more costly than RAM. In particular, writing to flash requires a much higher voltage than the minimum operational voltage for the commonly used TelosB nodes [9]. The MSP430F1611 microcontroller for the TelosB nodes can operate down to 1.8 V, while the minimum required voltage during a flash write or erase is 2.7 V. If the voltage falls below 2.7 V during a write or erase operation, the result of the write or erase will be unpredictable. As a result, based on study on the battery discharge curve [10], data writing to the flash requires being performed in the starting 23 % duration of the entire network lifetime.

In this paper, we propose NR, a novel network reboot mechanism to enable reliable and efficient network reboot for WSNs. First, we employ an eventual consistency dissemination protocol to reliably propagate the reboot command. All network nodes are ensured to receive the command after the sink issues. Second, we provide interfaces and components for persistent storage on RAM. This is achievable by the observation that the RAM data does not lose as long as the nodes are powered on [9]. Specifically, the reboot status is stored on persistent RAM to enable reliable and efficient network reboot.

We implement NR based on TinyOS 2.1.1. We evaluate our mechanism with TOSSIM 2.x. Results show that NR reliability reboots all network nodes. In addition, RAM I/O operation is faster and more energy-saving than I/O on program flash and external flash.

The rest of this paper is organized as follows: Sect. 5.2 describes the background and related work. Section 5.3 presents the design details. Section 5.4 shows the evaluation results. Finally, Sect. 5.5 concludes the paper.

5.2 Related Work

In this section we briefly discuss the necessary backgrounds and related works.

5.2.1 WSN Deployments

In recent years, many large WSNs are deployed in the real world. These deployments often encounter unexpected failures. Many deployments rely on network reboot to recover from failure.

Researchers at the University of Harvard deploy a 19-day sensor network at Reventador, an active volcano in Ecuador [6]. It is used to trigger on interesting volcanic activity. It initiates reliable data transfer to the base station. During the deployment, the network recorded 229 earthquakes, eruptions, and other seismographic acoustic events. For 20 days the system rebooted twice and once the deployers had to use radio command instead of protocol to reboot the network.

GreebOrbs [1] is a real-word WSN deployed in the forest in China. GreenOrbs aims at year-round ecological surveillance in the forest, collecting various sensory data such as temperature, humidity, illumination, and content of carbon dioxide. In the study of packet loss, reboot has become one of the reasons [11].

As mentioned, there is a need for reliable and efficient reboot mechanism.

5.2.2 Dissemination Protocols

Existing dissemination protocol can be classified into two categories: flooding protocols and dissemination protocols that ensure eventual consistency.

The literature in flooding protocol designs can be further classified into two categories: deterministic approaches and probabilistic approaches [12]. In the deterministic approaches, a fixed node within a connected dominating set is determined as a forwarding node. The connected dominating set is calculated by using global or local information. In probabilistic approaches, when a node receives a packet, it forwards the packetwith probability p. The value of p is determined by relevant information gathered at each node.

Double-Covered-Broadcast (DCB) [13] is a typical deterministic flooding protocol. It provides a simple, reliable broadcast algorithm. It takes advantage of broadcast redundancy to improve the delivery ratio in the environment that has rather high transmission error rate. Collective Flooding [12] is a recent flooding work, exploits the link correlation to achieve flooding reliability using the concept of collective ACKs, CF requires only 1-hop information from a sender, making the design highly distributed and scalable with low complexity.

In practical networks with unreliable wireless links, these protocols have reliability-cost tradeoffs: with a small number of broadcast messages, the protocols cannot guarantee a high reliability; to ensure a high reliability, they require much transmission cost. It is practically challenging to ensure a high reliability with too many transmissions because of huge chances of packet collisions.

Network reboot requires the dissemination process to be 100 % reliable. Even a small number of non-rebooting nodes can cause a malfunction in the entire network.

An obvious solution to ensure 100 % reliability is to employ a dissemination protocol that ensures eventual consistency.

Drip [7] is a dissemination protocol for small data items. Drip provides a transport layer interface to multiple channels of reliable message dissemination. Recent work DIP [8] is another dissemination protocol for wireless networks. DIP uses a hybrid approach of randomized scanning and tree-based directed searches to identify new items with O(log(T)) packets while maintaining a O(1) detection latency in a wide spectrum of network configurations.

The above protocols cannot directly be applied in practical systems. This is because these protocols store the values of data in a RAM cache. Rebooting a node will cause data loss. Therefore, the reboot command will never be propagated to distant nodes.

5.2.3 Reboot Mechanisms

Typically, nodes rely on the watchdog mechanism to force a hardware reboot. In setting the watchdog, users need to write a predefined password to a register. If the password is incorrect, the node will reboot immediately.

There are other mechanisms to reduce the rebooting cost. Microreboots [14] is a mechanism for rebooting fine-grained system components in Java Enterprise Edition Internet services. Microreboots depends on individual operations being idempotent and stateless, and requires a backing transactional store for completed operations. Furthermore, Microreboots focuses on high-level Internet services. In order to reduce the reboot cost in a sensor node, Neutron [5] provides a reboot mechanism that only reboots the recovery unit instead of rebooting an entire node. Neutron automatically enforces run-time memory safety and thus can efficiently recover from memory bugs. However, these mechanisms do not address the issue of reliably propagating the reboot command to all network nodes.

5.3 Design

In this section, we present the design of NR, a reliable and efficient network reboot mechanism for WSNs.

5.3.1 Overview

Figure 5.1 shows an overview of NR. Each node contains four components: the application component, the dissemination component that ensures eventual consistency, the persistent RAM component, and the reboot component. The black arrows show

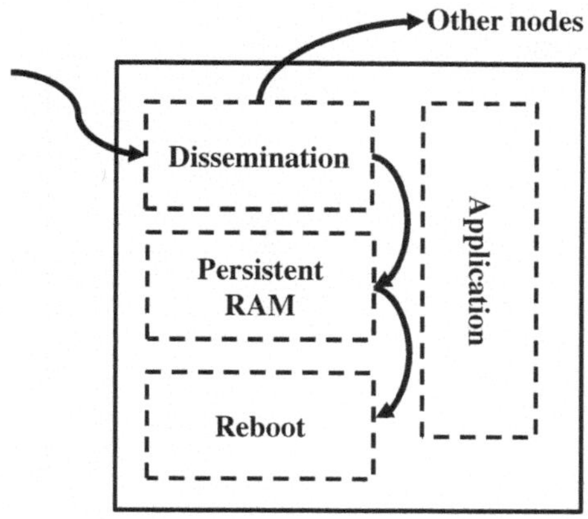

Fig. 5.1 NR overview

how the reboot command propagates in the network. First, when the node receives the reboot command in the network,it stores the reboot command in persistent RAM. Then the node calls the Reboot component to reboot itself. If the node receives a REQ message (asking for reboot command), it propagates the reboot command to other nodes.

We first introduce how to use NR in Sect. 5.3.2. We then present our design details of dissemination protocol, persistent RAM, and reboot mechanism in Sects. 5.3.3–5.3.5, respectively. We discuss several important issues of NR's design in Sect. 5.3.6.

5.3.2 Application Usage

In TinyOS 2.x, the DisseminationValue interface and DisseminationUpdate interface are used for dissemination. The DisseminationUpdate interface is used by producers while the DisseminationValue is for consumers.

The command DisseminationUpdate.change() should be called each time the producer wants to disseminate a new value, passing this new value as a parameter. The event DisseminationValue.changed() is signaled each time the disseminated value is changed (the producer has called change), and the command get() allows to obtain this new value.

We implement the above two interfaces in the component NRDisseminatorC, but we change the data storage scheme. In traditional dissemination protocols, the data and its version are stored on non-persistent RAM which will be lost after reboot. For example, rebooting a Drip node causes a node to lose all values of the data items, which need to be gathered again from other nodes. In our case, it causes the reboot command cannot propagate to distant nodes.

In order to address this issue, our dissemination protocol accesses the PersistentRAM interface (see Fig. 5.2) for storing the version of command on persistent RAM.

Fig. 5.2 Interface provided by NR

```
interface PersistentRAM <t> {
    command error_t store (t* data);
    command const t get ();
}
interface DisseminationUpdate <t> {
    command void change(t* newVal);
}
interface DisseminationValue <t> {
    command const t* get();
    event void changed();
}
interface Reboot {
    command error_t reboot( uint32_t time);
}
```

The PersistentRAM.store() command is used to store data. The PersistentRAM.get() command returns the current value of data. It is used by NR for storing the version of the reboot command. It is worth mentioning that the interface PersistentRAM is a general interface. Anyone can use it when he wants to store data on persistent RAM.

A programmer who wants to use NR only needs to wire two components: NRDisseminatorC and NRRebootC. The interfaces provided by the NR's components are summarized in Fig. 5.2.

5.3.3 Dissemination Protocol

NR uses an eventual consistency dissemination protocol to ensure that all nodes receive the reboot command of the newest version. The dissemination protocol employs Trickle mechanism [15] to guarantee 100 % reliability. While protocols like Drip offer such a property, it cannot be directly incorporated in NR. The reason is that the propagation of the reboot command could be interrupted by node reboots. We then modify the dissemination protocol: storing the data items on persistent RAM section instead of regular sections of RAM. By such modification, the reboot command is not lost after the reboot operation and the propagation will not be interrupted by reboot.

5.3.4 Persistent RAM

The storage in current sensor nodes can be classified into volatile memory and persistent memory. Volatile memory includes the RAM. Persistent memory includes the program flash and external flash.

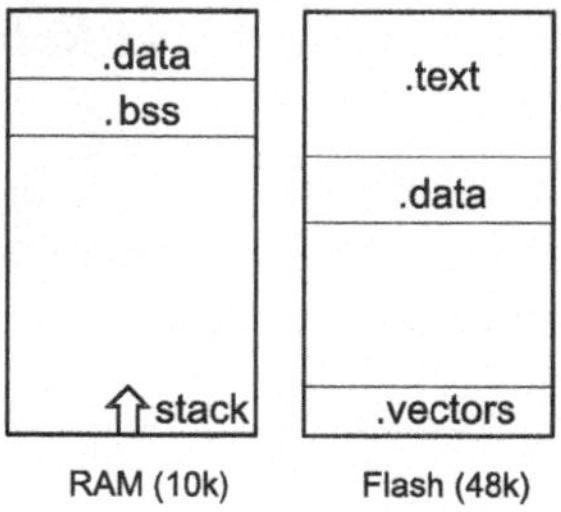

Fig. 5.3 Typical TelosB memroy layout

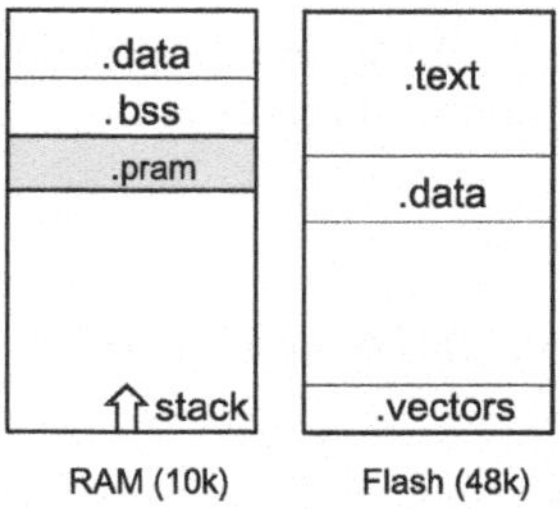

Fig. 5.4 NR's TelosB memroy layout

While we can use the program flash or external flash for persistent storage, the operations, however, is costly. First, I/O operations on flash are much more energy-hungry than RAM. For example, flash I/Os on TelosB nodes consume 5–12 mA, which is much larger than 0.5 mA energy consumption when the RAM I/Os. Second, flash usually enforces a limited number of I/O operations. Finally, the writing to flash usually requires a much higher voltage than the node's operational voltage.

We exploit a key observation: hardware reboot will not cause RAM data losses when the power is on [9]. It is the initializer provided by the GCC toolchain that reinitializes the data.

The TelosB node has 10 KB data RAM (from address 0x1100 to address 0x38FF) and 48 KB program flash (from address 0x4000 to address 0xFFFF). The highest 32 bytes of code space (from address 0xFFE0 to address 0xFFFF) are reserved for storing the interrupt vectors.

Figure 5.3 shows the memory layout of a typical program. There are two sections of the program in the RAM: .data section (for initialed data) and .bss section (for un-initialed data). Usually, three sections are placed on the program flash: .text section (for program code), .data section (for initializing the .data section in RAM), and .vectors section (for interrupt vectors). When the program starts execution, it first executes an initializer (usually starts at address 0x4000) which is responsible for initializing the .data section and the .bss section in RAM.

As Fig. 5.4 shows, we place a separate section called .pram following the .bss section so that the initializer will not perform initialization on this section and the data in this section keep persistent across reboots.

5.3.5 Node-Level Reboot

As mentioned before, in order to reboot the node, we just need to provide the watchdog an incorrect password. Once the password is incorrect, the node will reboot immediately. NR uses the following code snippet to reboot a node.

```
#define REBOOT_NOW() {
atomic{
WDTCTL = 0; while(1); }
}
```

We use randomized delay in order to avoid some critical issues caused by synchronized reboot.

5.3.6 Disscussion

Critical node failures. It is possible that critical failure causes some nodes unable to receiving the reboot command. In such a case, the original mechanism cannot force all nodes to reboot. We can improve the system design by employing a watchdog timer.

Multi-image systems. Multi-image systems store multiple program images on the external flash. The node can selectively execute one program image. In such systems, a program image with a large amount of data variable may overlap the .pram section, causing the data in .pram initialized. In order to avoid this situation, we can place the .pram at $\max\{\&\_\_bss\_end_i\}$ which i denotes the i-th program image.

Dynamic OSes. Dynamic OSes such as SOS [16], LiteOS [17], use dynamic memory allocation strategies. The dynamic memory allocator starts allocating memory after &__bss_end. NR can be ported to dynamic OSes by first allocating the .pram section and then modifying &__bss_end to be the end of .pram.

5.4 Evaluation

In this section, we evaluate NR's performance. Section 5.4.1 describes the evaluation methodology. Section 5.4.2 examines the reliability of NR's dissemination protocol compared with other flooding protocols. Section 5.4.3 shows the efficiency of PersistentRAM within NR.

5.4.1 Methodology

We evaluate NR in two main aspects. First, we examine NR's dissemination reliability. Second, we evaluate NR's overhead in I/O operations.

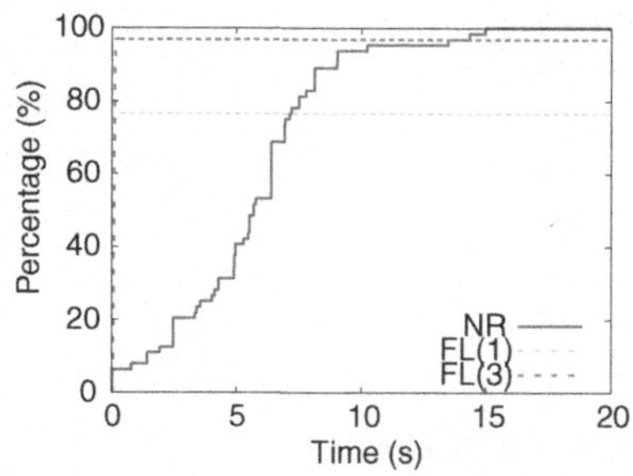

Fig. 5.5 Percentage of rebooted nodes with in the clean environment. NR reboots all nodes in the network

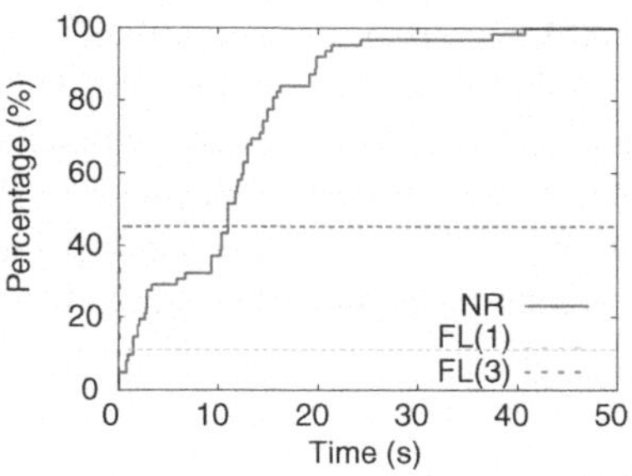

Fig. 5.6 Percentage of rebooted nodes with in the noisy environment. NR reboots all nodes in the network

We compare with traditional flooding protocols. We use FL(N) to denote a specific flooding protocol in which each node broadcasts the message N(N≥1) times. We conduct experiments in TOSSIM 2.x [18]. We use an 8*8 grid topology.

5.4.2 Reliability

Figure 5.5 shows the result with −105 dBm as the noise floor. This represents a "clean" environment. The x-axis is the time in seconds and the y-axis shows the percentage of rebooted nodes. We can see that (1) FL(1) has the lowest reliability, i.e., only 80 % nodes receive the reboot command; (2) while FL(3) achieves a higher reliability, it still cannot guarantee 100 % reliability; (3) NR's dissemination achieves 100 % reliability.

Figure 5.6 shows the results with meyer-heavy.txt (it is distributed with TinyOS 2.1.1) being the noise floor. This represents a noisy environment. We can see that (1) the reliability of both FL(1) and FL(3) decreases rapidly; (2) NR's dissemination still guarantees 100 % reliability, at the cost of longer completion time.

The above results show that current flooding protocols cannot be directly applied in real systems for reliable network reboot.

Table 5.1 Storage access speed in TelosB

Storage	Read speed (ms/byte)	Write speed (ms/byte)
RAM	0.0005	0.001
Program flash	0.020	0.027
External flash	0.056	0.059

5.4.3 *Efficiency*

The I/O cost of NR is low since it accesses data on RAM. Compared to flash I/Os, RAM operations are very efficient.

Table 5.1 shows the times needed for reading and writing on RAM, program flash, and external flash. We can see that the cost of RAM write operation is 27 times faster than write operations on program flash, and 59 times faster than write operations on external flash. RAM read operation is 40 times better than read operations on program flash, and 100 times better than read operations on external flash.

5.5 Conclusion

This paper presents NR, a network reboot mechanism based on TinyOS. NR enables reliable and efficient network reboot in WSNs. NR ensures 100 % reliability by incorporating an eventual consistency dissemination protocol. It avoids flash I/Os by storing data items on persistent RAM. Evaluation results show that NR can ensure 100 % reliability and is efficient in I/O operations.

Acknowledgments This work is supported by the National Science Foundation of China under Grant No. 61202402, the Fundamental Research Funds for the Central Universities (2012QNA5007), the Research Fund for the Doctoral Program of Higher Education of China (20120101120179).

References

1. Liu Y, He Y, Li M, Wang J, Liu K, Mo L, Dong W, Yang Z, Xi M, Zhao J, yang Li X (2011) Does wireless sensor network scale? a measurement study on GreenOrbs. In: Proceedings of INFOCOM, 2011
2. Szewczyk R, Polastre J, Mainwaring A, Culler D (2004) Lessons from a sensor network expedition. In: Proceedings of EWSN, 2004
3. Kim S, Pakzad S, Culler D, Demmel J, Fenves G, Glaser S, Turon M (2007) Health monitoring of civil infrastructures using wireless sensor networks. In: Proceedings of ACM/IEEE IPSN, 2007
4. Selavo L, Wood AD, Cao Q, Sookoor TI, Liu H, Srinivasan A, Wu Y, Kang W, Stankovic JA, Young D, Porter J (2007) Wireless sensor network for environmental research. In: Proceedings of ACM Sensys, 2007

5. Chen Y, GnawaliC O, Kazandjieva M, Levis P, Regehr J (2009) Surviving sensor network software faults. In: Proceedings of ACM SOSP, 2009
6. WernerAllen G, Lorincz K, Johnson J, Lees J, Welsh M (2006) Fidelity and yield in a volcano monitoring sensor network. In: Proceedings of USENIX OSDI, 2006
7. Tolle G, Culler D (2005) Design of an application-cooperative management system for wireless sensor networks. In: Proceedings of EWSN, 2005
8. Lin K, Levis P (2008) Data discovery and dissemination with DIP. In: Proceedings of ACM/IEEE IPSN, 2008
9. Dong W, Liu Y, Wu X, Gu L, Chen C (2010) Elon: enabling efficient and long-term reprogramming for wireless sensor network. In: Proceedings of ACM SIGMETRICS, 2010
10. Lachenmann A, Marrón PJ, Minder D, Rothermel K (2007) Meeting lifetime goals with energy levels. In: Proceedings of ACM SenSys, 2007
11. Dong W, Liu Y, He Y, Zhu T (2013) Measurement and analysis on the packet delivery performance in a large scale sensor network. In: Proceedings of INFOCOM, 2013
12. Zhu T, Zhong Z, He T, li Zhang Z (2010) Exploring link correlation for efficient flooding in wireless sensor networks. In: Proceedings of USENIX NSDI, 2010
13. Lou W, Wu J (2004) Double-covered broadcast (DCB): a simple reliable broadcast algorithm in MANETs. In: Proceedings of IEEE INFOCOM, 2004
14. Candea G, Kawamoto S, Fujiki Y, Friedman G, Fox A (2004) Microreboot: a technique for cheap recovery. In: Proceedings of USENIX OSDI, 2004
15. Levis P, Patel N, Culler D, Shenker S (2004) A self-regulating algorithm for code maintenance and propagation in wireless sensor networks. In: Proceedings of USENIX/ACM NSDI, 2004
16. Han C, Kumar R, Shea R, Kohler E, Srivastava M (2005) A dynamic operating system for sensor nodes. In: Proceedings of ACM MobiSys, 2005
17. Cao Q, Abdelzaher T (2008) The LiteOS operating system: towards Unix-like abstractions for wireless sensor networks. In: Proceedings of ACM/IEEE IPSN, 2008
18. Levis P, Lee N, Welsh M, Culler D (2003) TOSSIM: simulating large wireless sensor networks of tinyos motes. In: Proceedings of ACM SenSys, 2003

Chapter 6
An Automated Unit Testing Framework for Wireless Sensor Networks

Dongyu Yang, Fan Zhang and Jun Lin

Abstract Over 90 % of developers regard unit testing as most effective in reducing software bugs. However, current unit testing method which is writing test stub manually for Wireless Sensor Networks (WSN) is designed for different applications. Combining Embedded Remote Procedure Call (ERPC) with web service, this paper presents a new Automated Unit Testing Framework (AUTF) of unit testing for WSN. First, tester uploads source code through web service interfaces. At the time of compiling, the server stub of ERPC, web service, and test case are generated automatically for the corresponding function unit. The AUTF automatically runs test case through dynamic and asynchronous web service invocation, and then converts web service invocations into remote function invocations from ERPC Client to ERPC Server. Finally, automated unit testing is realized on sensor nodes under the AUTF.

Keywords Wireless sensor networks · Automated unit testing · Embedded remote procedure call · Web service · TinyOS

6.1 Introduction

Wireless Sensor Networks (WSN) have been widely applied in many fields, such as health monitoring, environmental monitoring, target tracking, security monitoring, smart home, industrial monitoring, precision agriculture, and intelligent grid [1, 2]. Comparing with traditional network system, the application environment of WSN is

D. Yang · F. Zhang · J. Lin (✉)
China Electronic Product Reliability and Environmental Testing Research Institute, No.110, Dongguanzhuang Street, Tianhe District, City of Guangzhou, China
e-mail: linjun@ceprei.com
http://www.ceprei.com

D. Yang
e-mail: dy.ceprei@gmail.com

F. Zhang
e-mail: zhangfan@ceprei.com

X. Wang et al. (eds.), *Advanced Technologies in Ad Hoc and Sensor Networks*, Lecture Notes in Electrical Engineering 295, DOI: 10.1007/978-3-642-54174-2_6,

more complex. Sensor nodes contain limited resources and energy and are deployed distributed [1, 2]. Their stability and reliability assessment is a difficult problem.

Currently, software testing is a key step to guarantee the quality of a software system. The testing can help identify software errors and help developers to find out and correct bugs. The Unit Testing is the most important and complex part of software testing [3]. A small problem in the unit could cause the collapse of the entire system.

There is no mature unit testing solution for WSN, which generally requires writing different testing scripts and stub codes for different applications. For WSN, testers manually download the testing program to actual hardware node one by one. This greatly increases the workload and affects work efficiency.

Currently, the Operating System for WSN mainly contains TinyOS [4], Contiki [5] and Z-Stack [6]. TinyOS has the features of small code, component-based, easy extended, high energy efficiency, and low resource consumption. It has been accepted worldwide and used to set up different applications. This paper describes the TinyOS-based Automated Unit Testing technology for WSN.

6.2 Related Works

Currently, researchers have raised some WSN test platforms [7, 10]. They were biased in favor of network protocols and algorithms testing, and focused on the network performance such as throughput, time delay, network setup time, packet loss rate, etc.

MoteLab [8] is a popular WSN testbed solution. It supports development and testing of sensor network programming environments, communication protocols, system design, and applications. The interaction between the users and the testbed is batch oriented and is controlled via a dynamic web interface supported by a backend database. Like MoteLab, the Kansei [9] testbed also uses Stargate boards, but it allows richer interaction with the SUE. Kansei allows evaluation of both flat and hierarchical WSN with different communication technologies. WTS [10] is a test suite for WSN. It has three parts: sensor nodes, server, and clients. At running time, sensor nodes report data to sink node in active or command-driven way. The Server receives data from sink node via serial port. The client access server through network collects data from Sink node, and displays the network topology, link status, the data of sensor nodes, etc.

These test platforms do not discuss how to test the function units. In TUnit [11], the authors proposed a Unit Testing Framework for TinyOS. Through predefined interfaces, programmers can define unit tests by creating new TinyOS programs that include and interface with an existing TinyOS module, and provides inputs to the module and checks its output. In Marionette [12], Whitehouse et al. proposed a debugging method for WSN, whose core is based on fat-client/thin-server Embedded Remote Procedure Call (ERPC) [14, 15]. It allows the program on PC side to directly access variable values on a sensor node, and even call the functions on remote sensor

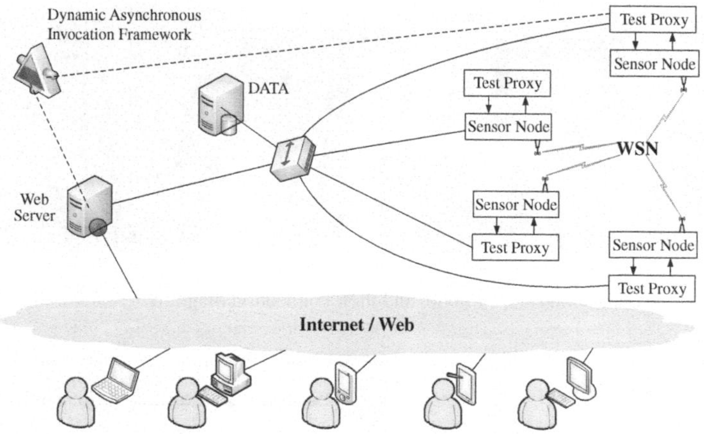

Fig. 6.1 The system environment of WSN automated unit testing framework

node. Michael Okoal et al. proposed Unit Testing Framework MUnit [13] for WSN based on Marionette. MUnit applies Marionette for WSN's Unit Testing for the first time, however, it does not discuss how to automate control and test.

6.3 System Environment

As shown in Fig. 6.1, the system environment of the proposed AUTF mainly contains:

(1) the Server which supports Web Service;
(2) the data server for data analysis and management;
(3) high efficient dynamic and asynchronous framework for web service invocation;
(4) the testing proxy for testing control and data forward;
(5) the sensor node which actually runs test cases.

The Apache Tomcat Application Server and Apache Axis2 Web Service support environment are deployed on the Server. The open-source database MySQL is used to control and manage data. As shown in Fig. 6.2, the BeagleBone, an embedded board is used as Test Proxy. Sensor node and test proxy are connected though USB interface, and transmit data through serial communication protocol. One test proxy can connect with multiple sensor nodes with USB-HUB. Test proxy, web server, and data server are connected and transmit data through the Ethernet.

6.4 System Design and Architecture

The AUTF for WSN is shown in Fig. 6.3. Through a web browser, tester can upload source code, set testing parameters, and select the function units. The server schedules the testing through the automation console. At the time of compiling, the server stub

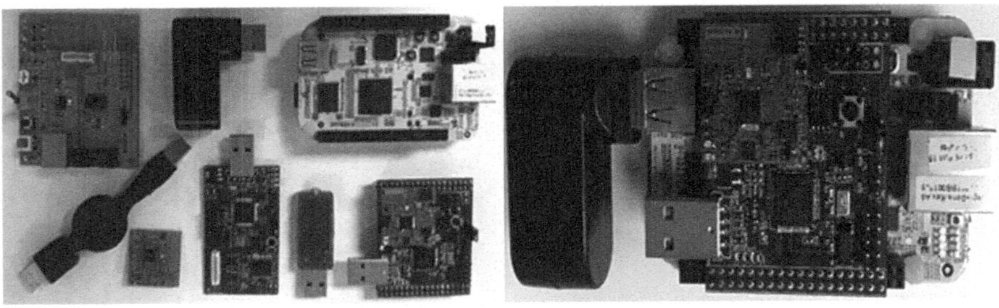

Fig. 6.2 The test proxy and sensor node, and their connection diagram

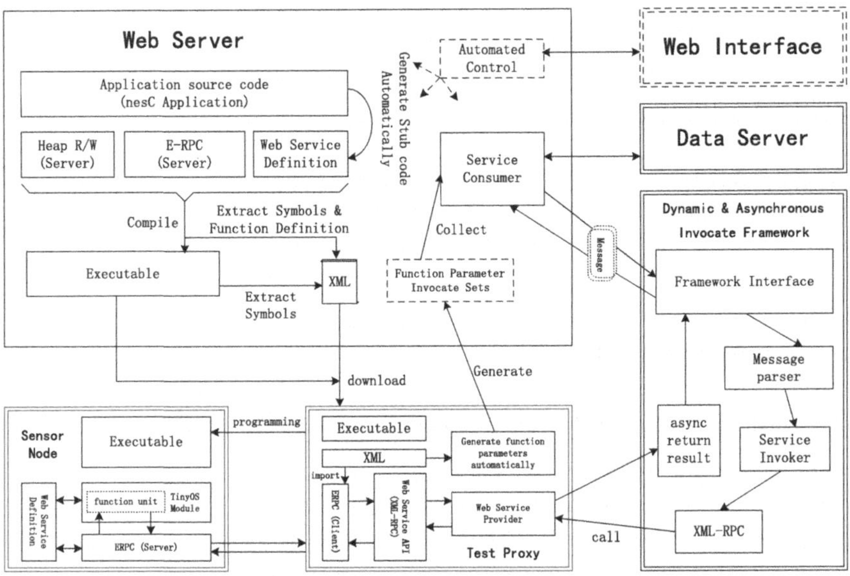

Fig. 6.3 The automated unit testing framework for WSN

of ERPC, web service, and test case are generated automatically for corresponding function unit. The automation console compiles the source code which contains server stub of ERPC and web service definitions. The test proxy receives executable image and the XML file, and automatically generates web service interfaces. Then, the test proxy downloads executable image to sensor node, and automatically generates the test cases and actual parameters for functions. The automation console runs test case step-by-step and transmits the functions' actual parameters to the Dynamic and Asynchronous Invocation Framework (DAIF). After the DAIF calls a web service asynchronously the test proxy converts web service invocation into remote function invocation from ERPC client to server. The test proxy returns results to the automation console through the DAIF asynchronously. The full process of unit testing can be monitored through web page in real-time.

6.4.1 Call the Function Units Through ERPC

Remote Procedure Call (RPC) is an abstraction of procedure call, which is cross-language, cross-platform, and cross-machine boundaries. It allows the user to call the function or procedure on remote platform [14]. ERPC is designed for resource-constrained embedded systems, which uses fat-client/thin-server architecture. Most of the work is placed on the client, and the server only handles simple service processes [15]. In this paper, we select the debugging method of Marionette to call the remote function units on sensor node.

6.4.2 Automatically Generate Web Services for Function Units

Web service is a technology of Service-Oriented Architecture (SOA). It is used to ensure interaction of applications among different platforms [16]. When the application's source code is obtained, the automation console realize function units invoke through web service by using the method in the literature [17].

6.4.3 Dynamic and Asynchronous Invocation Framework

In order to ensure the system's adaptability, a dynamic and asynchronous method is used to invoke the generated web services. Multithread is an effective method to achieve dynamic and asynchronous invocation. After web service interfaces, actual parameters are set and testing case set for functions are generated by test proxy. The automation console creates multithread program to execute testing case.

In each thread, the automation console transmits function's actual parameters to Dynamic and Asynchronous Invocation Framework (DAIF) for dynamic invocation request. After that, DAIF parses the input information and starts a standard web service request to invoke the web service interface, which is defined on the test proxy asynchronously through the actual XML-RPC protocol stack [18]. Finally, dynamical and asynchronous web services invocation is realized.

6.5 System Workflow Analysis

In this section, we give a simple example to analyze the AUTF's workflow.

Program 1. An example of TinyOS Application's source code.

```
module UnitTestModuleC{
    provides interface CommandInterface;
    uses     interface EventInterface;
```

```
}implementation{
    uint8_t test = 123;
    command error_t CommandInterface.testCommand(
            uint8_t val){
        test += val;  return SUCCESS;
    }
    event error_t EventInterface.testEvent(uint8_t val){
        test -= val;  return SUCCESS;
    }
}
```

6.5.1 Automatically Generate Test Case Program

When starting a unit test, user uploads the application's source code to server, as shown in Program 1. The automation console processes different testing requests. When a request is received, it automatically generates the corresponding test case program for source code. As shown in Program 2, the automation console automatically adds 'Boot' and 'SplitControl' interface, and sets '@ rpc ()' flag for them. Then, the top-level configuration file is created for this test case program and the connection is automatically completed among different module interfaces.

Program 2. Test Case Application Program and top-level Configuration File.

```
includes Rpc;
module AutoUnitTestC{
    provides interface CommandInterface @rpc();
    uses interface EventInterface @rpc();
    uses interface Boot;
    uses interface SplitControl as AutoUnitTestCtrl;
}implementation{
    uint8_t test = 123;
    command error_t CommandInterface.testCommand(
            uint8_t val){
        test += val;  return SUCCESS;
    }
    event error_t EventInterface.testEvent(uint8_t val){
        test -= val;  return SUCCESS;
    }
    event void Boot.booted(){
        call AutoUnitTestCtrl.start();
    }
    event void AutoUnitTestCtrl.startDone(error_t err){}
    event void AutoUnitTestCtrl.stopDone(error_t err){}
}

configuration AutoUnitTestAppC{
}implementation{
```

```
    components MainC, AutoUnitTestC, RpcC, RamSymbolsM;
    AutoUnitTestC.Boot -> MainC.Boot;
    AutoUnitTestC.AutoUnitTestCtrl -> RpcC;
}
```

6.5.2 Automatically Generate the Stub Code and XML File

After test case program is created the automation console compiles the TinyOS application. At the time of compiling, corresponding ERPC server stub code and platform-independent XML file are automatically generated by Marionette. The XML file contains detailed information for TinyOS application, such as variables, structure definitions, interfaces, and function definitions.

6.5.3 Automatically Create Function's Parameters

Marionette provides nescDecls package, which includes the dynamic typing system to mirror all types in TinyOS nesC applications. There are four types of applications:

- nescType: basic types like uint8_t, char, uint16_t, uint32_t, int8_t, etc;
- nescArray: arrays of any of four types;
- nescPointer: pointers to any of four types;
- nescStruct: structs of any of the four types.

Therefore, all basic types, typedefs, and structs defined in the nesC application can use one tree to describe, whose leaf nodes are nescType or the pointer of nescType. By setting the actual value for each leaf nodes, function's parameters can be created.

6.5.4 Dynamically and Asynchronously Invoke Web Services

When the web service interface is invoked, the automation console starts a new thread and immediately destroys this thread after invocation. It is shown in Program 3. Each thread is independent; current thread invocation does not affect others. Even if the invocation of current thread is blocked, it just only sleeps the current thread and does not affect the system to process other requests. Therefore, dynamical and asynchronous invocation can be realized and the system's concurrency is guaranteed in the meantime.

Program 3. Thread for Web Service Dynamically and Asynchronously Invoked.

```
class DaifThread extends Thread {
    String server;
    List<StructTree> structTreeList, valueTreeList;
    public DaiowsThread (String funcName, String serv,
        List<StructTree> sTree, List<StructTree> vTree){
        Server = serv;
        structTreeList = sTree;
        valueTreeList = vTree;
    }
    public void run() {
        XmlRpcClientConfigImpl config =
            new XmlRpcClientConfigImpl();
        XmlRpcClient client=new XmlRpcClient();
        try{
            config.setServerURL(new URL(server));
        }catch(Exception e){}
        client.setConfig(config);
        try{
            result = client.execute(funcName,
                structTreeList, valueTreeList);
            asyncNotify(result);
        }catch(Exception e){}
    }
}
```

6.6 Conclusion

Recently, stability and reliability have become a problem that affect many WSN applications. With the improvement in hardware technology, hardware faults have decreased visibly. More errors of system are caused by software. Our experience revealed that unit testing can find up to 70–80 % bugs in a software. In this paper, combining ERPC with web service, we proposed a TinyOS-based Automated Unit Testing Framework (AUTF) for WSN. First, user uploads source code through a web service. At the time of compiling, the server stub of ERPC, web service, and test case are generated automatically for corresponding function unit. The AUTF automatically runs test case through dynamic and asynchronous web service invocation, and then converts web service invocations into remote function invocations from ERPC Client to ERPC Server. Finally, automated unit testing is realized on sensor nodes under the AUTF.

Acknowledgments This paper was supported by National Key Technology R&D Program under Grant No. 2011ZX03005-001, Breakthroughs Project in Key Areas of Guang dong and Hong Kong under Grant No. 2011A011302002, and Youth Fund Program of The fifth Electronics Research Institute of the Ministry of Industry and Information Technology under Grant No. 12Q08.

References

1. Limin S, Jianzhong L, Yu C, Hongsong Z (2005) Wireless sensor networks. Tsinghua University Press, Beijing
2. Karl H, Willig A (2005) Protocols and architectures for wireless sensor networks. Wiley, New York
3. Qing S, Jin H, Suxian Z (2008) Software process quality management. Science Press
4. Levis P, Madden S, Polastre J, Szewczyk R, Whitehouse K, Woo A, Gay D, Hill J, Welsh M, Brewer E, Culler D (2005) TinyOS: an operating system for sensor networks. In: Ambient intelligence. Springer, Berlin Heidelberg
5. Dunkels A, Gronvall B, Voigt I (2004) Contiki—a lightweight and flexible operating system for tiny networked sensors. In: 29th annual IEEE international conference on local computer networks, IEEE Press
6. Z-Stack - ZigBee Protocol Stack. http://www.ti.com.cn/tool/cn/z-stack
7. Ma JY, Zhou XS, Zhang Y, Li SN, Li ZG (2012) Debugging sensor networks: A survey. Jisuanji Xuebao (Chin J Comput) 35(3):405–422
8. Werner-Allen G, Swieskowski P, Welsh M (2005) MoteLab: a wireless sensor network testbed. In: Proceedings of the 4th international symposium on information processing in sensor networks (IPSN), ACM
9. Ertin E, Arora A, Ramnath R, Nesterenko M, Naik V, Bapat S, Kulathumani V, Sridharan M, Zhan H, Cao H (2006) Kansei: a testbed for sensing at scale. In: Proceedings of the 5th international symposium on information processing in sensor networks (IPSN), ACM
10. Tianlin L, Fangmin L (2009) Wireless sensor networks test platform. Wuhan University of Technology, Wuhan
11. How TUnit Works. http://docs.tinyos.net/tinywiki/index.php/How_TUnit_Works
12. Whitehouse K, Tolle G, Taneja J, Sharp C, Kim S, Jeong J, Hui J, Dutta P, Culler D (2006) Marionette: using RPC for interactive development and debugging of wireless embedded networks. In: Proceedings of the 5th international symposium on information processing in sensor networks (IPSN), ACM
13. Okola M, Whitehouse K (2010) Unit testing for wireless sensor networks. In: Proceedings of the 2010 ICSE workshop on software engineering for sensor network applications, ACM
14. Birrell AD, Nelson BJ (1984) Implementing remote procedure calls. ACM Trans Comput Syst (TOCS) 2:39–59
15. Bershad BN, Anderson TE, Lazowska ED, Levy HM (1990) Lightweight remote procedure call. ACM Trans Comput Syst (TOCS) 8:37–55
16. Erl T (2004) Service-oriented architecture: a field guide to integrating XML and web services. Prentice Hall PTR, New Jersey
17. Okola M, Whitehouse K (2010) Automatically exporting TinyOS modules as web services. In: The first international workshop on networks of cooperating objects (CONET)
18. Leitner P, Rosenberg F, Dustdar S (2009) Daios: efficient dynamic web service invocation. IEEE Internet Comput. 13:72–80

Chapter 7
Performance Evaluation of WSNs-Based Link Quality Estimation Metrics for Industrial Environments

Guangchao Gao, Heng Zhang and Li Li

Abstract Industrial wireless sensor network (WSN) is operating under severe conditions of electromagnetic interference (EMI) and multi-path interferences. Few existing simulation tools have dealt fairly with modeling EMI in industry site. Therefore, most of them are unable to meet the requirements of WSN simulation experiments. In this paper, the industrial environment is firstly categorized into different topographies, and the definition based upon the specific physical characteristics of the local surroundings is given to reflect both large-scale fading and multipath interference. The propagation model is well expressed by the one-slope path-loss model. The excellent agreement with a lognormal distribution is also obtained. Then the simulation environment is set up based on the on-site data in OPNET simulator. Finally, in order to obtain the best routing metric under different industrial environments, four commonly used link-quality metrics in WSNs: ETX, Hop Count, PRR and WMEWMA are investigated. The simulation results show that ETX is the most optimal routing metric on the overall performance. To the best of our knowledge, this paper is the first work to compare the performance between these link quality based metrics with networks of different qualities under different industrial conditions.

Keywords Routing metric · LQI estimation · OPNET

G. Gao · H. Zhang · L. Li (✉)
School of Computer and Information Science, Southwest University, Chongqing 400715, China
e-mail: ggch2010@swu.edu.cn

H. Zhang
e-mail: dahaizhangheng@swu.edu.cn

L. Li
e-mail: lily@swu.edu.cn

X. Wang et al. (eds.), *Advanced Technologies in Ad Hoc and Sensor Networks*, Lecture Notes in Electrical Engineering 295, DOI: 10.1007/978-3-642-54174-2_7,

7.1 Introduction

Today, WSN have emerged as a growing research area where most research scholars and industrialists focus on. Especially, wireless technology continues to see active use in industrial applications [1–3]. By using electromagnetic waves as their transmission medium, wireless systems avoid the limitations of wired networks and offer competitive advantages in terms of cost, flexibility, and ease of use. WSNs can be used in industrial mobile robots, real-time inventory management and process and equipment monitoring [4–8], etc.

It is known that deploying wireless networks in industrial environments comes with issues related to their quality of service (QoS) in terms of reliability and timeliness, especially when operating under severe conditions of electromagnetic interference (EMI) and multi-path interferences [9–11]. Because the cost of wireless sensor nodes is high, and the application environments differ in many aspects. So these reasons make it difficult to study on the problems of performance evaluation and verification for WSN protocols. Consequently, simulation can be used to provide useful recommendations on the configuration of the network and the important initialization parameters setting. Accurate modeling method of electromagnetism is the core function in the simulation of industrial wireless network. However, there are few researches concerned about it.

To support the end-to-end communication in industrial wireless networks, routing algorithms play a significant role including good paths and forwarding nodes between sources and their destinations. Currently, there are a lot of researches have done to study the routing metric construction method [12–15]. However, few attempts have been made until now to evaluate the performance of routing metric in real environment properties. Consequently, the contribution of this work may include: Firstly, main industrial environment characteristics and potential applications of WSNs in industrial have been introduced along with the related technical challenges; Then, on the basis of the analysis of the industrial environment, we describes wireless channel model, and we propose a method to accurate model the electromagnetism interference in industrial environments. Finally, according to the on-site test data, we have simulated the performance of several routing metric which are outstanding presently in the different industrial environment.

The rest of the paper begins with the description of the development of WSN-based industrial applications in Sect. 7.2. Section 7.3 describes the channel model, along with the on-site data of different types of industrial environment. We introduce the performance evaluation and provide an intuitive summary of the results in Sect. 7.4. Finally, Sect. 7.5 concludes this paper.

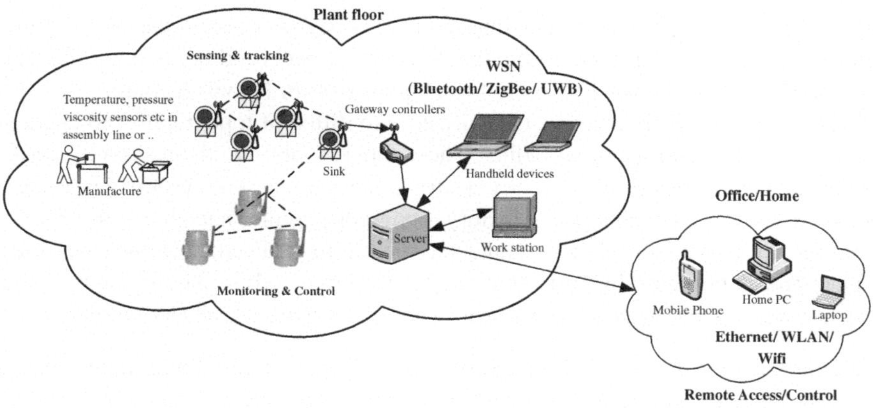

Fig. 7.1 The overview of delivery system

7.2 An Overview of WSN-Based Industrial Applications

Nowadays, wireless technology is used more and more to obtain the real-time control data in manufacture. Normally, it has several types of network to content the various application under the industrial environment. The typical network structure of industrial environment is shown in Fig. 7.1. The whole network can be connected through plant floor by Internet and the user can monitor the whole network in the Office. To assure the reliability of the system, the wired network is mostly used in real-time control systems and maintenance systems. The wireless sensor network which uses short-distance wireless communication can collect the relevant data, such as platform, device and industrial environment. After all data is delivered to monitoring center through the server, the engineers can access the data at any time through the devices which are connected to Internet.

7.3 Accurate Simulation Method of Industrial Environment

7.3.1 The Classification of Industrial Buildings

The modern industrial factory buildings have similar physical properties. The buildings were made of concrete floors and metal ceilings supported by steel truss work. And all factory walls were made of thick, precast concrete. What's more, industrial inventory in all facilities consists of the largest part of similar metal machinery [16]. Therefore, the experiments conducted by this article would be applicable to industrial environments sharing these physical characteristics. Before deploying WSNs in industrial applications, it should be aware that the coverage areas and reliability of data in the industrial environment may suffer due to noises, co-channel interferences, multipath and other interferers. The signal strength can be severely affected in factory environments due to reflections of walls and floors, interferences from ISM bands and noise generated from equipments or heavy machines.

The industrial environment is categorized into different topographies to characterize large-scale and temporal fading. These topographies are defined on the basis of physical properties of the indoor industrial environment. Observations indicated that there are four general types of factory topography common to all manufacturing facilities. Classifying the physical characteristics of the radio channel into categories is an approach proposed by Turin et al. used in the urban radio environment [17] and it is one we use here. Categories of factory topography are defined as follows:

(1) Line of Sight Path with Light Surrounding Clutter: Paths found along major aisles that are surrounded by relatively empty storage areas or low density work areas (such as a machine shop) where most scatterers are lower than the height of the transmitter and receiver antenna. Line of sight between transmitting and receiving antenna exists at every point along the path.
(2) Obstructed Path with Light Surrounding Clutter: Such paths exist when a LOS path is blocked by inventory or machinery that is approximately the same height as the receiving antenna. Such a radio path would typically exist across a machine shop or a manual assembly area.
(3) Obstructed Path-Heavy Surrounding Clutter: Such paths exist throughout areas of the factory where the skyline is busy, such as within a metal foundry, across an automated assembly line area, or between aisles in a stocked warehouse.
(4) All LSF topographies: The integration of the above three topographies.

7.3.2 Channel Model

As it is known, WSN use wireless channel as the communication media. When an electromagnetic signal propagates, it may be diffracted, reflected, and scattered [7]. All these effects have two significant effects on the signal strength. First, the signal decreases exponentially with respect to communication distance. Second, for a given communication distance, the signal strength is random and log-normally distributed about the mean distance dependent value [7, 18]. The industrial indoor has the characterization of large scale of fading and temporal fading.

The empirical model for path loss and temporal fading can be used to predict the wireless range of an IEEE 802.11b/g system with the validated statistically by measurements of the received signal strength or RSS reported by WLAN monitoring software [19]. The successful use of wireless communications in industrial environments requires an understanding of the propagation characteristics in these environments. To this end, various works have been done in literature in the field of empirical modelling of the indoor industrial channel: industrial large-scale, small-scale and temporal fading have been investigated through narrow-band measurements in [17, 20]. Through these works, we can find that, the log-normal shadowing path

Table 7.1 Log-normal shadowing model parameters

Frequency (MHz)	Propagation environment	PL (d_0) (dB)	n	σ(dB)
2400	LOS	63.57	2.40	4.79
2400	OBS (light clutter)	63.57	2.77	5.42
2400	OBS (heavy clutter)	63.57	4.29	8.42
2400	All LSF topographies	63.57	3.44	8.63

loss model is used to evaluate the industrial environment wireless channel. In the following, a one-slope path-loss model is well expressed and lognormal distribution is also obtained to express the shadowing fading.

$$\mathrm{PL}(d) = \mathrm{PL}(d_0) + 10n \bullet \log(d/d_0) + X_\sigma$$

$\mathrm{PL}(d_0)$ is the path loss in dB at an arbitrarily chosen reference distance d_0 in m, and n is the dimensionless pathloss exponent. σ is the shadowing deviation, and X_σ provides the randomness for the Log-normal channel modeling.

Several propagation experiments were performed within each topographical category within each factory. The experiments were designed to determine path loss, temporal fading, shadowing, and small-scale spatial fading characteristics. The experiments data is shown in Table 7.1 [20].

7.4 Performance Evaluations

7.4.1 Simulation Environment

In this study, we have evaluated the state-of-the-art four different routing metrics, which are Hop Count, PRR, WMEWMA and ETX. In the four metrics, the ETX and fourbit is the active monitoring, which nodes periodically send probe packets to test the link stability, and this scheme will cause additional network load.

To further evaluate the performance of routing metrics, we used OPNET modeler version 14.5 with the parameters presented in Table 7.1. The Collection Tree Protocol (CTP) is adopted as routing protocol, The CTP is a routing and data collection protocol building and maintaining a tree towards the sink node based on link quality estimation. We first evaluate the impact of the parameters values in a grid network of 7*7 nodes, the sink being located at the center of the simulation area. The experiments of parameter settings are: sending power for 0.003 W, channel transmission rate is 1 Mbps, data packet size is 1024 bytes, Channel error proximity is 0.01 %, packet reception threshold is −86 dBm.

Fig. 7.2 Double nodes communication network

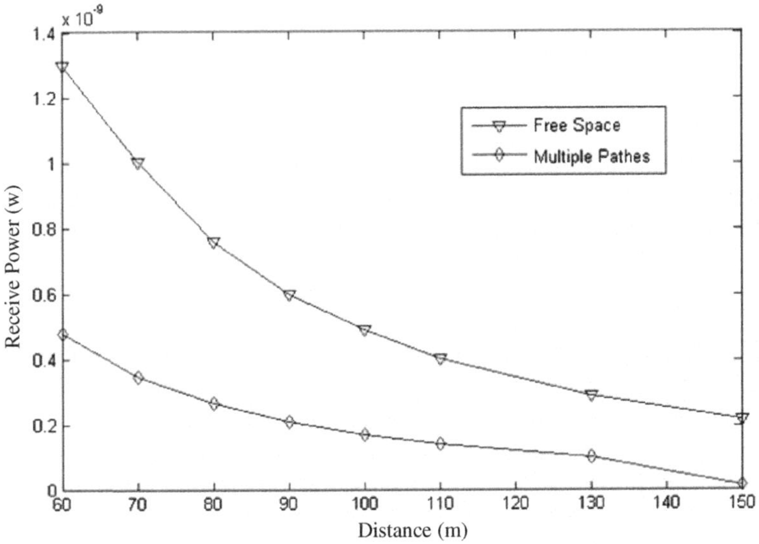

Fig. 7.3 Double nodes communication network

7.4.2 Performance of Channel Fading

The Experiment of network topology is shown in Fig. 7.2. Consider the scenario of two nodes one-way communication. Node A sends data to node B. We will investigate the receiving power of node B with changed distance between A and B. The result of experiments is shown in Fig. 7.3.

In Fig. 7.3, x-coordinate abscissa denotes the communication distance, and y-coordinate denotes receive power. Top line represents free space propagation model. In free space model, we can find that receive power is strict computation with the formula below.

$$L\text{pass} = 20 \lg f + 20 \lg d - 27.56\, dB$$

We can see from the formula above that receiving power is only influenced by distance and frequency. It does not conform to the actual situation. The lower line in Fig. 7.7 represents the improved model, on the basis of multipath and shadow effect. The receiving power obviously is less than the value of the free space model.

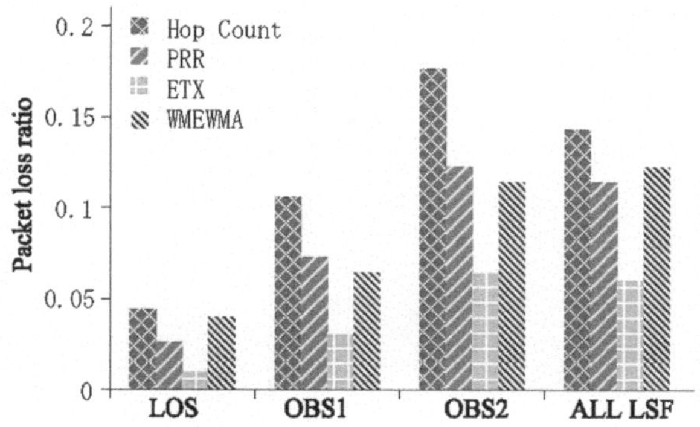

Fig. 7.4 The result of packet loss ratio

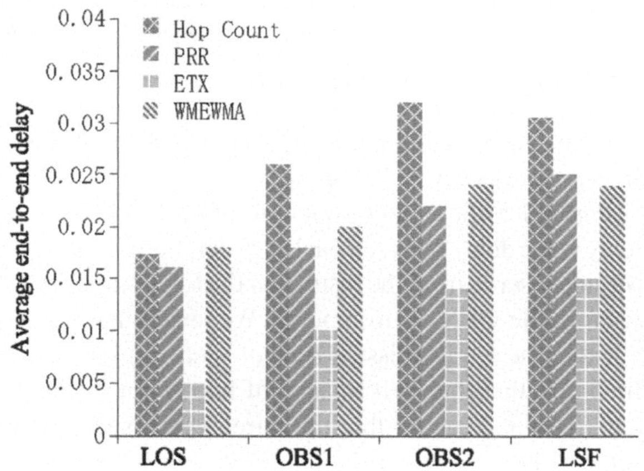

Fig. 7.5 The result of end-to-end delay

With increasing of space communication, the receiving power changes in small scale. That is because that signal attenuation and distance is in a linear relationship in actual situation.

7.4.3 Performance of Routing Metric

Figure 7.4 shows the performance result of packet loss ratio of ctp protocol using four routing metrics in different industrial environments. Each experiment is in the same experiment condition. In the Fig. 7.4, we can see that the ETX is optimus in packet delivery ratio under deferent industrial environments. On the other hand, the packet reception ratio (PRR) and WMEWMA are perform not very good. This is because the PRR and WMEWMA only evaluated for one-way link without considering the asymmetric link. While the hop count has the poor performance among the four metrics, the reason is the object of minimum hop tend to select the long distance

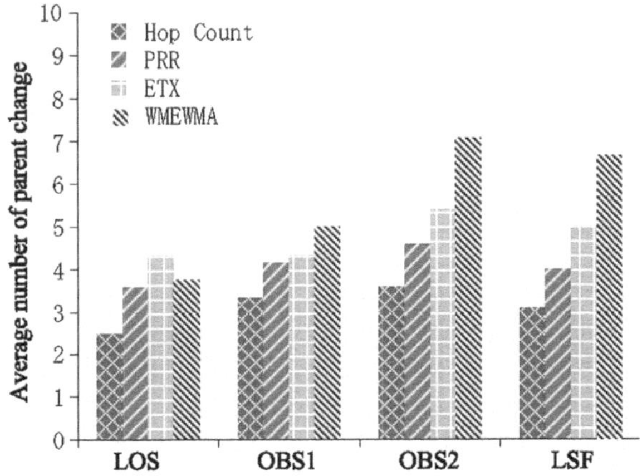

Fig. 7.6 The result of number of parent change

path, yet the communication quality of these pathes are often not very good and the links are very easy to be failure.

Figure 7.5 show the throughput of each scheme in end-to-end delay. End-to-end delay is the delay between packet generation and its reception by the final destination. We can observe from the result that the ETX outperform than other metrics, especially when in the OBS environments. We aware that this is much because that the ETX consider the retransmission time of successful packet delivery. And we can see from the figure that the performance of hop count is also acceptable in the line-of sight path. That is because that the wireless transmission environment is good, and the minimal hop count can be the good routing metric. Figure 7.6 compares LQEs in terms of average number of parent changes under different industrial environments. We observe that WMEWMA has the highest number of parent changes and Hop count has the lowest one. Regarding WMEWMA, this result is expected since WMEWMA has been shown to be very reactive to link changes. This results in faster and more frequent changes of parents as link quality degrades.

On the other hand, following the same reasoning, it would have been expected that PRR would have more parent changes than ETX. However, we observe the opposite. This is explained by the fact that PRR is inaccurate, thus it may select routes with bad links, leading to unstable routes that will be quickly broken, resulting in more frequent parent changes.

The general conclusion of this observation is that the accuracy of the estimator may have a controversial effect on the stability of the routes. In fact, an accurate estimator, such as ETX, may lead to stable paths, which minimizes the number of parent changes. Oppositely, an accurate estimator may also induce very frequent changes due to its excessive reactivity to quality changes, as it has been observed in Fig. 7.6, in particular when links are bad or moderate. It can also be seen that PRR and ETX provide a good compromise for these antagonist effects.

We compare the impact of link metrics on the performance of network throughput. Figure 7.7 shows the result. We can observe from the figure that Hop count metric has

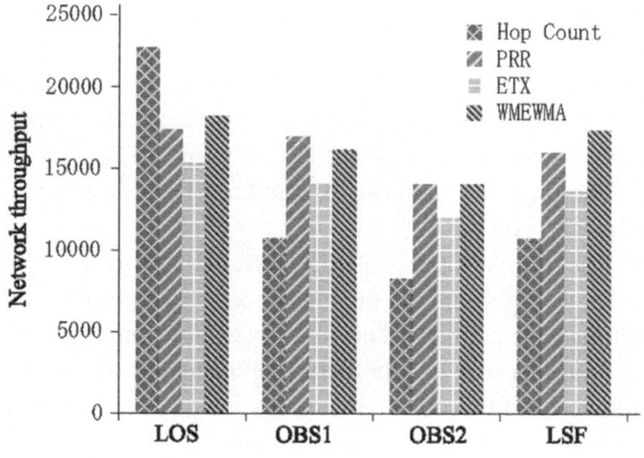

Fig. 7.7 The result of network throughput

very good performance in environment LOS. This is because when the link quality is in a acceptable situation e.g., in LOS environment, the hop count has the good performance. The reason is the hop count has the advantages of small overhead and fast speed of link construction. However, hop count shows the poor performance in the next two environments which may have the low link quality caused by EMI. In Fig. 7.7, we also observe that the ETX, PRR and WMEWMA have similar behaviors in network throughput. Although the ETX has the advantages of little time delay and low packet loss ratio, the ETX is the active based routing metric, it will consume a lot of bandwidth.

7.4.4 Summary

(1) As discussed in Sect. 7.4.1, ETX is more tolerant to losses when the link quality is complex and profound changes. So, ETX gives the better estimated value to links with a high delivery rates as compared to other metrics.
(2) The characteristic of Hop count leads to a smaller overhead for delivering a packet, while maintaining a high end to end delivery rate in LOS environment. This causes Hop count to change route when the link quality to the next hop drops to a very low level.
(3) WMEWMA and PRR exhibit the same preference in choosing paths with near perfect links. As a result, the average path length for WMEWMA and PRR is almost the same.
(4) ETX allows CTP to choose better neighbors when the neighbor table is full. As a routing protocol, CTP tends to choose nodes with high quality path to the sink. This results in a very stable routing topology for the entire network, comprising of directed links to the sink. However, paths from multiple nodes to the sink will intrinsically lead to congestion at some shared links. This may cause degradation in link quality and hence, lowers the path quality and end-to-end delivery rate, especially in high density networks.

7.5 Conclusion

In this paper the issue of accurately modeling of electromagnetism interference within complete network simulators was discussed. The initial implementations for each of models were first proposed followed by a discussion of their performance in terms of accuracy. A simulation method has been proposed to deal with each type of EMI in industrial sites.

The simulation of an IEEE 802.11b network is also presented as a case study in order to highlight the peculiarities of each strategy and the algorithm's impact on simulation accuracy of correlation of time and frequency. Our approach successfully simulated a real life industrial environment, taking into account various complications relevant to our application such as the interference of signals. Our simulation software can provide a reliable and flexible model of the notoriously unpredictable nature of industrial environments, forming a versatile platform from which further research can be conducted.

Acknowledgments The work is supported by National Natural Science Foundation of China (61170192), China-Canada joint research and development (R&D) projects and the Fundamental Research Funds for the Central Universities (2009DFA12100, No. XDJK2012C019).

References

1. U.S. Department of Energy, Office of Energy Efficiency and Renewable Energy (2002) Industrial wireless technology for the 21st century, Dec 2002
2. Akyildiz I, Su W, Sankarasubramaniam Y, Cayirci E (2002) Wireless sensor networks: a survey. Comput Netw 38(4):393–422
3. Low KS, Win NW, M J Er (2005) Wireless sensor network for industrial environment. In: International conference on computational intelligence for modeling, control and automation, November, 2005, pp 271–266
4. Moon TK, Kuc TY (2004) An integrated intelligent control architecture for mobile robot navigation within sensor network environment. In: Proceedings of IEEE/RSJ international conference on intelligent robots and system, Oct 2004, pp 565–570
5. Hochmuth P (2005) GM cuts the cords to cut costs. Mobility and Wireless, Article of Techworld
6. Technology@Intel Magazine (2005) Expanding usage models for wireless sensor networks Aug 2005, pp 4–5
7. U. S Department of Energy (2004) Sensors and automation: Eaton wireless sensor network for advanced energy management solutions, June 2004
8. Aakvaag N, Mathiesen M, Thonet G (2005) Timing and power issues in wireless sensor networks—An industrial test case. In: IEEE international conference on parallel processing workshops, Jun 2005, pp 419–426
9. Zamalloa MZ, Krishnamachari B (2007) An analysis of unreliability and asymmetry in low-power wireless links. ACM Trans Sens Netw 3(2):7. Article ID 1240227
10. Zhai H, Fang U (2006) Distributed flow control and medium access in multihop ad hoc networks. IEEE Trans Mobile Comput 5(11):1503–1514
11. Zuniga M, Krishnamachari B (2004) Analyzing the transitional region in low power wireless links. In: Proceedings of the SECON, Santa Clara, CA, USA, October 2004, pp 517–526
12. Gambiroza V, Sadeghi B, Knightly E (2004) End-to-End performance and fairness in multihop wireless Backhaul networks, In: ACM Mobicom, 2004

13. De Couto DSJ, Aguayo D, Bicket J, Morris R (2003) A high-throughput path metric for multi-hop wireless routing. In: ACM Mobicom, 2003
14. Draves R, Padhye J, Zill B (2004) Comparison of routing metrics for static multi-hop wireless networks. In: ACM SIGCOMM, 2004
15. Draves R, Padhye J, Zill B (2004) Routing in multi-radio, multi-hop wireless mesh networks. In: ACM Mobicom, 2004
16. Seidel SY, Rapport TS (1992) 914 MHz path loss prediction model for indoor wireless communication in multi floored buildings. IEEE Trans Antennas Propag 40(2):207–217
17. Rappaport TS, Mcgillem CD (1989) UHF fading in factories. IEEE J Sel Areas in Commun 7(1):40–48
18. De Couto DSJ, Aguayo D, Chambers BA, Morris R (2002) Performance of multihop wireless networks: Shortest path is not enough. In: Proceedings of the first workshop on hot topics in networks (HotNets-I), Oct 2002
19. Emmeric T et al (2009) Statistical validation of WLAN range calculated with propagation models for industrial environments by chipset-level received signal strength measurements. IET Sci Meas Tech 3(3):244–255
20. Tanghe E, Joseph W, Verloock L et al (2008) The industrial indoor channel: large-scale and temporal fading at 900, 2400, and 5200 MHz. IEEE Trans Wirel Commun 7(7):2740–2751

Chapter 8
Security Evaluation for Wireless Sensor Networks Based on Attack Test and Fuzzy Comprehensive Judgement

Xueying Zhang, Xiaotong Zhang, Yue Qi, Liwen He, Zuoxin Xiahou and Chunyuan Li

Abstract Security evaluation for wireless sensor networks analyzes the security performance of network, helps to find vulnerabilities and shortages, and provides advices to take effective security technologies, which is an important way to improve the security of the network proactively. Considering that the evaluation is a complex process and involves many factors, this paper proposes a security evaluation method for the wireless sensor networks based on fuzzy comprehensive judgement. First, we establish an evaluation index system based on proactive attack, which reflects the security performance objectively and truly. Second, we determine the index weights adopting the analytic hierarchy process method, which can reduce the subjectivity of weights decision. Third, we adopt the fuzzy comprehensive evaluation method to evaluate the security performance qualitatively and quantitatively. Finally, this method is proved to be reasonable and effective in simulation experiments.

Keywords Security evaluation · Wireless sensor networks · Fuzzy comprehensive evaluation · Attack test

---

The paper is sponsored by the National High Technology Research and Development Program of China (863 Program) (No. 2011AA0401), National Natural Science Foundation of China (No. 61003251, 61173150), Doctoral Fund of Ministry of Education of China (No. 20100006110015), the Fundamental Research Funds for the Central Universities (No. 2302012FRF-TP-12-085A) and Project for Research Center of Beijing Municipal Education Commission.

X. Zhang · X. Zhang (✉) · Y. Qi · L. He · Z. Xiahou · C. Li
University of Science and Technology Beijing, School of Computer and Communication Engineering, Beijing, China
e-mail: zhangxueying199011@163.com, zxt@ustb.edu.cn

X. Wang et al. (eds.), *Advanced Technologies in Ad Hoc and Sensor Networks*, Lecture Notes in Electrical Engineering 295, DOI: 10.1007/978-3-642-54174-2_8,

8.1 Introduction

With the characteristics of low-power consumption, low-cost, distributed and self-organizing, wireless sensor networks (WSN) are used for various fields such as environment alert systems, medical monitoring systems, military observation and intelligent transportation, and so on. However, because of its limited resources, the openness of communication, the non-controlled environment, it will face many threats such as information leakage, data tampering, replay attacks, node capture, denial of service [1]. To ensure that the network is secure and reliable, it is critical to take effective measures, including key management, encryption, authentication, and security evaluation. Security evaluation analyzes the security performance of network, which will help to find vulnerabilities and shortage, and provide advices to take effective security technologies [2]. Therefore, it plays an important role to improve the security of the network actively.

There are many methods used in security evaluation at present. Among them, the evaluation based on vulnerability scanning is charactered with active and non-destructive, but its performance depends on the integrity of the vulnerability library, and it will take up a lot of time [3]. The evaluation based on artificial neural network adopts a parallel and multi-objective method, which takes advantage of its good learning ability. It can avoid the interference of the subjective factors, but needs lots of training samples, which is not suitable for WSN [4]. The evaluation based on bayesian networks integrates the prior knowledge and samples knowledge, and utilizes probabilistic reasoning ability of bayesian network. It can reduce the subjective bias and noise effects, but it also needs lots of training samples. Meanwhile, the accuracy of evaluation may be decreased when the frequency of some factors is very low [5]. The evaluation based on network entropy puts forward the concept of network entropy from the view of the security characteristic change of network after it was attacked, which is usually applied to integrity evaluation of the network [6]. Fuzzy comprehensive evaluation is based on the fuzzy mathematics, and works well in dealing with the imprecise and fuzzy information. It transforms the qualitative evaluation with many factors into quantitative evaluation, and makes the result clear and systemic, which is worth using for reference [7–9]. But if we adopt this method, we need to determine the index weights subjectively. AHP overcomes the subjectivity of weights decision to some extent and improves the accuracy of evaluation [10]. Fuzzy comprehensive evaluation is a common method, but it didn't take the dynamic change of the attack state into consideration, and has not been combined with WSN well yet.

Therefore, this paper proposes a WSN security evaluation method which combines the fuzzy comprehensive evaluation with the characters of WSN. First, according to characters of wireless sensor networks, this paper builds an evaluation index system for the network security. Second, in the experimental platform, we use specific attacks test for different indexes, and get real and objective indexes data. Third, we use the analytic hierarchy process to determine the index weights, and adopt the

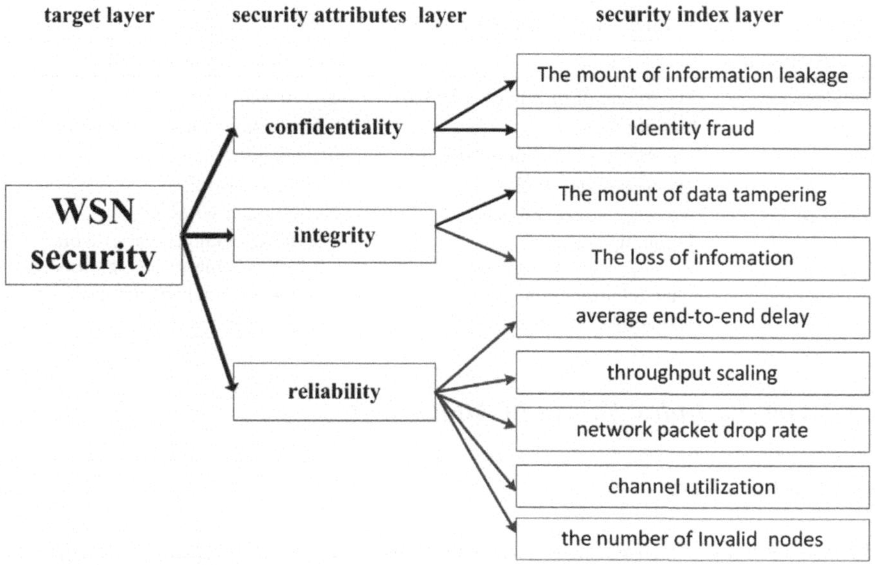

Fig. 8.1 The evaluation index system for the network security

fuzzy comprehensive evaluation method to evaluate the security performance of the network qualitatively and quantitatively.

This paper is followed by four sections. Chapter 2 puts forward the concrete process of the evaluation method. Chapter 3 shows an example of the WSN evaluation in experiment platform which demonstrates the feasibility and applicability of the method. Then, in Chap. 4, this method is proved to be reasonable and effective through the comparison of two cases. Finally, we get the conclusion in Chap. 5.

8.2 Security Evaluation for Wireless Sensor Networks

8.2.1 Build Evaluation Index System

According to characteristics of wireless sensor networks, this paper builds an evaluation index system for the network security. We put forward a top-down three-layer index model which contains target layer, security attributes layer, security index layer. Target layer is the security of the WSN, and security attributes layer is composed of confidentiality, integrity, and reliability. The value of indexes in the security index layer can be got in the attack test. Figure 8.1 illustrates the system.

Table 8.1 Attack test of the wireless sensor networks

Security attributes	Security index	Test with the following attack
Confidentiality	The mount of information leakage	Interception
	Identity fraud	Malicious node identity spoofing
Integrity	The mount of data tampering	Malicious data tampering
	The loss of infomation	Selective forwarding
Reliability	Average end-to-end delay Throughput scaling Network packet drop rate Channel utilization The number of invalid nodes	The spurious routing attacks, replay attacks, sinkhole attack, sybil attack, wormhole attack, flooding attack, channels interference and asynchronous attack

8.2.2 *Get the Index Values in Attack Test*

8.2.2.1 Attack Test

The common attacks in wireless sensor networks include interception, interruption, modification, malicious node spoofing, selective forwarding, the spurious routing attacks, replay attacks, sinkhole attack, sybil attack, wormhole attack, flooding attack, and asynchronous attack. In the experimental platform, we use specific attacks for different indexes, and get objective indexes value. Table 8.1 shows the network attacks for various indexes.

8.2.2.2 Index Value Normalization

Index value normalization is completed by comparing the attack state with the normal state, using the Formula (8.1).

$$x' = (x - \min)/(\max - \min),\ x \in [\min, \max] \tag{8.1}$$

x is the original value to be normalized; min is the index value in the normal state; max is the maximum of the index may be determined by practical or technical experience. After normalization, the values of all the indexes are in [0, 1], which makes the necessary preparations for subsequent calculations.

8.2.3 *Determine Weights for the Indexes*

Analytic hierarchy process (AHP) is an assessing method combines qualitative analysis with quantitive analysis, which can quantify multiple uncertainty and vagueness of decision-making process. In this paper, we adopt AHP methods to construct the judgment matrix among various layers, calculate the weight vector, check consistency, and determine the weights.

Table 8.2 1–9 ratio scale comparison

Nine-scale	Meaning
1	a_i is as important as a_j
3	a_i is a little bit important than a_j
5	a_i is obvious important than a_j
7	a_i is consuming important than a_j
9	a_i is extreme important than a_j

Table 8.3 The value of the random consistency index $RI(n)$

n	1	2	3	4	5	6	7	8	9	10	11
RI	0	0	0.58	0.90	1.12	1.24	1.32	1.41	1.45	1.49	1.51

8.2.3.1 Construct Judgment Matrix

We may need experts to help us in this step. According to the customary 1–9 ratio scale comparison table shown in Table 8.2, experts group take comparison between two factors with the same level, and construct the comparison judgment matrix, as following:

$$A = (a_{ij})_{n\times n} = \begin{bmatrix} a_{11} & a_{12} & \dots & a_{1n} \\ a_{21} & a_{22} & \dots & a_{2n} \\ \dots & \dots & \dots & \dots \\ a_{n1} & a_{n2} & \dots & a_{nn} \end{bmatrix} \tag{8.2}$$

where a_{ij} indicates relative importance between a_i and a_j, and $a_{ij} = 1/a_{ji}$. In Table 8.2, 2, 4, 6, 8 falling in between neighborhood judgment.

8.2.3.2 Calculate the Weight Vector and Check Consistency

First, we calculate maximum eigenvalue λ and maximum eigenvector W of the judgment matrix A.

Then, we calculate the consistency check index using the Formula (8.3)

$$CI = \frac{\lambda - n}{n - 1} \tag{8.3}$$

where n is the order of the matrix; The value of the random consistency index RI is shown in Table 8.3.

If $CR = CI/RI < 0.1$, the consistency of judgment matrix satisfies the request and W is the weight vector we want. Otherwise, the judgment matrix should be reconstructed.

8.2.4 Multilevel Fuzzy Comprehensive Evaluation Method

The WSN evaluation involves multiple indexes and factors, and the security attributes such as confidentiality, integrity, reliability have a certain degree of ambiguity. These factors must be integrated in order to make a reasonable assessment. Therefore, this article adopts multilevel fuzzy comprehensive evaluation method.

8.2.4.1 Build the Comments Set

This paper defines the comments set $V = \{\text{excellent, good, ok, bad, terrible}\}$.

8.2.4.2 Membership Matrix of the Index Layer

For an indexes X, we have got its normalization value x'. Now we calculate the membership of the index using the trapezoidal membership function as the following formulas (8.4)–(8.8):

$$u_1(x) = \begin{cases} 1 & x \le H_1 \\ \frac{x-H_1}{H_2-H_1} & H_1 < x \le H_2 \\ 0 & x > H_2 \end{cases} \tag{8.4}$$

$$u_2(x) = \begin{cases} 0 & x \le H_1 \\ 1 - u_1(x) & H_1 < x \le H_2 \\ 1 & H_2 < x \le H_3 \\ \frac{x-H_3}{H_4-H_3} & H_3 < x \le H_4 \\ 0 & x > H_4 \end{cases} \tag{8.5}$$

$$u_3(x) = \begin{cases} 0 & x \le H_3 \\ 1 - u_2(x) & H_3 < x \le H_4 \\ 1 & H_4 < x \le H_5 \\ \frac{x-H_5}{H_6-H_5} & H_5 < x \le H_6 \\ 0 & x > H_6 \end{cases} \tag{8.6}$$

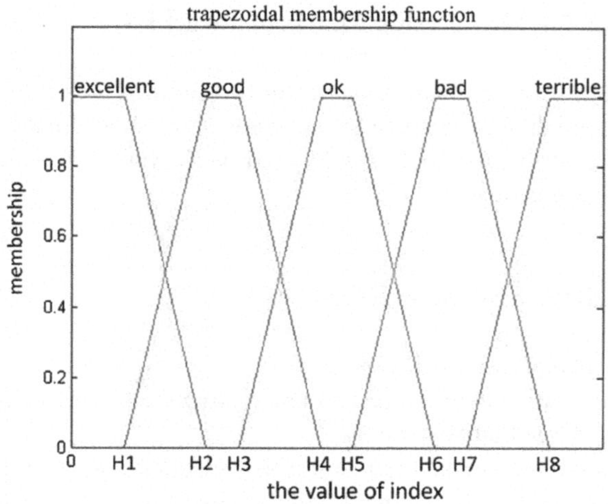

Fig. 8.2 Trapezoidal membership function

$$u_4(x) = \begin{cases} 0 & x \le H_5 \\ 1 - u_3(x) & H_5 < x \le H_6 \\ 1 & H_6 < x \le H_7 \\ \frac{x - H_7}{H_8 - H_7} & H_7 < x \le H_8 \\ 0 & x > H_8 \end{cases} \tag{8.7}$$

$$u_5(x) = \begin{cases} 0 & x \le H_7 \\ 1 - u_4(x) & H_7 < x \le H_8 \\ 1 & x > H_8 \end{cases} \tag{8.8}$$

where the central range of the five comment are(0, H_1), (H_2, H_3), (H_4, H_5), (H_6, H_7), (H_8, 1). Now we can get the membership matrix of the index $B_{ij} = [u_1(x'), u_2(x'), u_3(x'), u_4(x'), u_5(x')]$. Figure 8.2 illustrate the range of the five comments.

8.2.4.3 Membership Matrix of the Security Attributes Layer

For a security attribute, we get its fuzzy evaluation matrix $R_i = (B_{i1}, B_{i2}, B_{i3} \ldots\ldots B_{in})$, where B_{ij} is the membership matrix of the index which subordinates to this security attribute. The membership matrix of the security attribute $B_i = W_i \bullet R_i$, where W_i is the weight vector of the indexes which subordinates to this security attribute. In this paper, the characteristics of the evaluation are based on a comprehensive consideration of all factors. So we use $M(\bullet, \oplus)$ operator model.

8.2.4.4 Membership Matrix of Target Layer

The fuzzy evaluation matrix of the target $R = (B_1, B_2, \ldots, B_m)$, where B_i is the membership matrix of the attribute which subordinates to the target layer. The membership matrix of target $B = W \bullet R$, where W is the weight vector of the three security attribute.

8.2.4.5 Evaluation Result

(1) Qualitative analysis
According to the largest membership principle, we select the maximum as the result.
(2) Quantitative analysis
In order to make the results more clear, we quantify the result. We give percentile score for each comment as following: $90 \le c_1 < 100$ (excellent), $80 \le c_2 < 90$ (good), $70 \le c_3 < 80$ (ok), $60 \le c_4 < 70$ (bad), $50 \le c_5 < 60$ (terrible). So we get a score vector: $C = (c1, c2, c3, c4, c5) = (95, 85, 75, 65, 55)$. We can get the result score:

$$S = \frac{1}{\sum_{i=1}^{n} b_i} BC^T = \frac{\sum_{i=1}^{n} b_i c_i}{\sum_{i=1}^{n} b_i} \tag{8.9}$$

8.3 Simulation Analysis

In the experimental platform, we build a mesh–star network. Then, we launch attacks on it, and get the objective index value from the experiments. Next, we analyze the network security using the fuzzy comprehensive evaluation model.

8.3.1 Build an Evaluation Index System

This paper builds an evaluation index system as described in Fig. 8.1, and normalizes the index values. Table 8.4 illustrates the normalized values.

8.3.2 Determine the Weights Using AHP Method

We adopt AHP method to determine the weights. The weight vector of security attributes $W = (0.1919, 0.1744, 0.6337)$. As $CR < 0.1$, the consistency of judgment

Table 8.4 The normalized value and the weights of the indexes

Security attributes (weights)	Security index (weights)	Normalized value (case 1)	Normalized value (case 2)
Confidentiality (0.1919)	The mount of information leakage (0.5)	0.23	0.23
	Identity fraud (0.5)	0	0
Integrity (0.1744)	The mount of data tampering (0.6667)	0.63	0.15
	The loss of infomation (0.3333)	0.40	0.20
Reliability (0.6337)	Average end-to-end delay (0.2308)	0.31	0.17
	Throughput scaling (0.1640)	0.42	0.30
	Network packet drop rate (0.1240)	0.62	0.43
	Channel utilization (0.0937)	0.40	0.34
	The number of invalid nodes (0.3875)	0.33	0.30

matrix satisfies the request. We can also get the weight vector of the index layer. The weight vector of indexes for confidentiality $W_1 = (0.5, 0.5)$. The weight vector of indexes for integrity $W_2 = (0.6667, 0.3333)$. The weight vector of indexes for reliability $W_3 = (0.2308, 0.1640, 0.1240, 0.0937, 0.3875)$. The results are illustrated in Table 8.4.

8.3.3 Multilevel Fuzzy Comprehensive Evaluation

8.3.3.1 Membership Matrix of the Index Layer

Using the trapezoidal membership function, we get the membership matrix of every index and generate the fuzzy evaluation matrix of the security attribute layer.

The fuzzy evaluation matrix of confidentiality $R_1 = \begin{bmatrix} 0.155 & 0.845 & 0 & 0 & 0 \\ 1 & 0 & 0 & 0 & 0 \end{bmatrix}$.

The fuzzy evaluation matrix of integrity $R_2 = \begin{bmatrix} 0 & 0 & 0.355 & 0.645 & 0 \\ 0 & 0.45 & 0.55 & 0 & 0 \end{bmatrix}$.

The fuzzy evaluation matrix of reliability $R_3 = \begin{bmatrix} 0 & 1 & 0 & 0 & 0 \\ 0 & 0.32 & 0.68 & 0 & 0 \\ 0 & 0 & 0.42 & 0.58 & 0 \\ 0 & 0.45 & 0.55 & 0 & 0 \\ 0 & 0.905 & 0.095 & 0 & 0 \end{bmatrix}$.

8.3.3.2 Membership Matrix of the Security Attribute Layer

The membership matrix of the security attribute $B_i = W_i \bullet R_i$.

The membership matrix of confidentiality $B_1 = \begin{bmatrix} 0.5775 & 0.4225 & 0 & 0 & 0 \end{bmatrix}$.

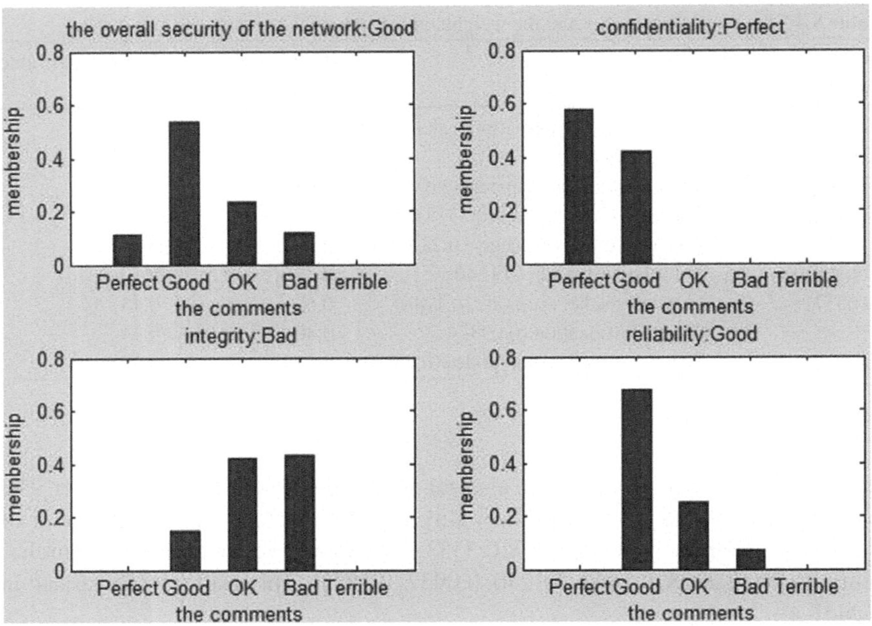

Fig. 8.3 Evaluation result in case 1

The membership matrix of integrity $B_2 = \begin{bmatrix} 0 \ 0.1500 \ 0.4200 \ 0.4300 \ 0 \end{bmatrix}$.

The membership matrix of the reliability $B_3 = \begin{bmatrix} 0 \ 0.6761 \ 0.2519 \ 0.0719 \ 0 \end{bmatrix}$.

8.3.3.3 Membership Matrix of Target Layer

The fuzzy evaluation matrix of target $R = (B_1, B_2, B_3)$, and the membership matrix of target $B = W \bullet R$. Now, we can get $B = \begin{bmatrix} 0.1108 \ 0.5357 \ 0.2329 \ 0.1206 \ 0 \end{bmatrix}$.

8.3.3.4 Case Study

(1) Qualitative analysis
As shown in Fig. 8.3, the first bar graph reflects the membership of the overall network security. According to the largest membership principle, we can see the state of the network is "good". The remaining three figures reflect the three security attributes including confidentiality, integrity, and reliability. The network is excellent in confidentiality, bad in integrity, and good in reliability.
(2) Quantitative analysis
By calculating, we can get the overall safety $S = 81.3677$. Meanwhile the confidentiality $S_1 = 90.775$, integrity $S_2 = 72.1996$, reliability $S_3 = 81.0421$.

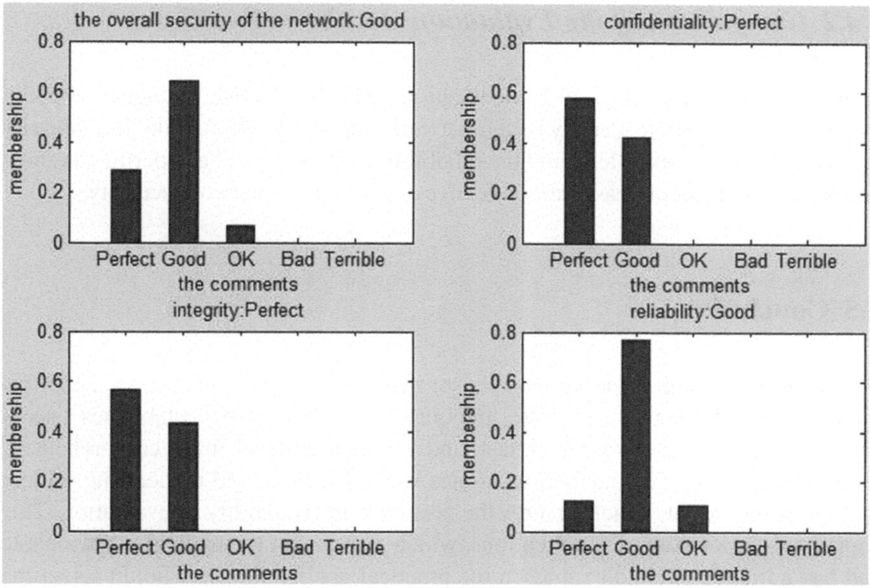

Fig. 8.4 Evaluation result in case 2

8.4 Analysis of the Evaluation Model

8.4.1 The Security Evaluation in Another Case

In the above case, the integrity of the network is bad. Therefore, we add the message authentication techniques in the second case. Meanwhile, we use self-adaptive frequency hopping technology to enhance the network reliability. According to the above method, we can get membership matrixes of security attributes.

The membership matrix of confidentiality $B_1{}' = \left[0.5775\ 0.4225\ 0\ 0\ 0\right]$.

The membership matrix of integrity $B_2{}' = \left[0.5667\ 0.4333\ 0\ 0\ 0\right]$.

The membership matrix of the reliability $B_3{}' = \left[0.1258\ 0.7668\ 0.1074\ 0\ 0\right]$.

The membership matrix of target $B' = \left[0.2894\ 0.6426\ 0.0680\ 0\ 0\right]$.

(1) Qualitative analysis

As shown in Fig. 8.4, the state of the network is "good". The network is excellent in confidentiality and integrity, and good in reliability.

(2) Quantitative analysis

By calculating, we can get the overall security $S = 87.2132$. Meanwhile the confidentiality $S_1{}' = 90.775$, integrity $S_2{}' = 90.6668$, reliability $S_2{}' = 85.1841$.

8.4.2 Comparison of the Evaluation Results in Two Cases

After we adopted the message authentication and self-adaptive frequency hopping technology, the overall security of the network improved. Meanwhile, the integrity turned from poor to excellent, and the reliability is enhanced. The experiments show that the method can be reasonably effective assessment of network security.

8.5 Conclusion

The security evaluation for wireless sensor networks is a complex process, and it is hard to be comprehensive, accurate, and quantitative. This paper establishes a security index system based on attack test, and adopts multilevel fuzzy comprehensive evaluation method. This method integrates various factors, and reduces the subjectivity to some extent, which ensures the accuracy and reliability of evaluation. This paper also shows example applications, which proves that the method is reasonable and has a high application value. In the practical application, we should select the appropriate model parameters to improve the performance of the method.

References

1. Du X, Chen HH (2008) Security in wireless sensor networks. IEEE Wirel Commun 15(4):60–66
2. Ashraf A, Hashmani M, Chowdhry BS, Mussadiq M, Gee Q, Rajput AQK (2008) Design and analysis of the security assessment framework for achieving discrete security values in wireless sensor networks. Paper presented at Canadian conference on electrical and computer engineering
3. Lv Z, Zhang J, Zhang J (2005) Research on network security leak scan and vulnerability analysis. Aeronaut Comput Tech 35(2):118–121
4. Zhao DM, Liu JX, Zhang ZH (2009) Method of risk evaluation of information security based on neural networks. Paper presented at 2009 international conference on machine learning and cybernetics, vol 2, July 2009
5. Jia W, Lian YF, Feng DG, Chen S (2008) Bayesian network approximate reasoning based method for network vulnerabilities evaluation. J Commun 29(10):191–198
6. Zhang Y, Xian M, Wang G (2004) A quantitative evaluation technique of attack effect of computer network based on network entropy. J Commun 25(11):158–165
7. Chen C, He Z, Jia J, Kuang J, Zhang Z (2012) Fuzzy evaluation algorithm for system effectiveness of wireless sensor networks. Paper presented at 2012 IEEE global high tech congress on electronics (GHTCE)
8. Liu D, Yang S (2009) An information system security risk assessment model based on fuzzy analytic hierarchy process. Paper presented at international conference on e-business and information system security, 2009 (EBISS '09)
9. Zhang R, Huang L, Xiao M (2010) Security evaluation for wireless network based on fuzzy-AHP with variable weight. Paper presented at 2010 second international conference on networks security wireless communications and trusted computing (NSWCTC)
10. Zhao D, Ma J, Wang Y (2007) Model of fuzzy risk assessment of the information system. J Commun 28(4):51–56

Chapter 9
High Accurate Time Synchronization Mechanism for WIA-PA Network

Yuwei Zhou, Qin Wang and Yadong Wan

Abstract Time synchronization has been a heated issue of industrial wireless network, and it is also the guarantee and premise of the stable and reliable operation of the entire wireless network. WIA-PA (wireless network for industrial automation-process automation) is an IEC-approved industrial wireless standard. Having taken the features and architecture of WIA-PA into consideration, we implement and analyze two approaches of time synchronization mechanisms of WIA-PA, test and validate these mechanisms under actual hardware platform and complex environment. The experimental results show that combining these two time synchronization mechanisms, the synchronization time between nodes can be controlled within 4 μs, which can achieve high precision and strong feasibility.

Keywords Wireless sensor network · WIA-PA · Time synchronization · Industrial wireless

---

This work has been performed under the National Natural Science Foundation of China (Grants No. 61003251 and No. 61172049) and Doctoral Fund of Ministry of Education of China (Grant No. 20100006110015).

Y. Zhou · Q. Wang · Y. Wan (✉)
Computer and Communication Engineering, University of Science and Technology Beijing, Beijing 100083, China
e-mail: yadong.wan@gmail.com

Y. Zhou
e-mail: zhouyuweiustb@gmail.com
http://www.ustb.edu.cn/

Q. Wang
e-mail: wangqin@ies.ustb.edu.cn

X. Wang et al. (eds.), *Advanced Technologies in Ad Hoc and Sensor Networks*, Lecture Notes in Electrical Engineering 295, DOI: 10.1007/978-3-642-54174-2_9,

9.1 Introduction

Industrial wireless technology is a kind of novel wireless communication technology which provides information exchange for low-rate, short-range-oriented devices. As the use of spread spectrum communication technology, multi-hop communication technology and mesh network technology, it has been equipped with a strong anti-jamming capability, ultra-low energy consumption, real-time communications and other technical features, and is suitable for harsh industrial environments [1]. With the rapid development of wireless communication technology, more and more technical standards, such as 802.11, RFID, Zigbee, Bluetooth, IEEE 802.15.4 and so on, have been applied to environmental monitoring and protection, health care, military, and other fields. Along with the development of these technologies, some international organizations have been actively promoting industrial wireless network technology standardization program in recent years, which mainly include WIA-PA (Wireless Networks for Industrial Automation-Process Automation) [2, 3], WirelessHART [4] and the ISA SP100 [5].

Theoretically, WIA-PA is one kind of network form of wireless sensor networks (WSN) which is applied in industrial process automation. Facing the complexity of the environment in the industry, WIA-PA network has high reliability and real-time optimization as primary objectives [6]. In order to ensure network reliability, WIA-PA network takes CSMA and TDMA hybrid access mode, using the multi-channel communication mechanisms and adaptive frequency hopping mechanisms, while all these mechanisms require accurate time synchronization. Each node is clocked by an internal or external crystal oscillator, for the reasons of cost restrict and crystal manufacturing process constraints, and that the crystal oscillator is vulnerable to environmental impacts, crystal aging, and other factors during operation, the frequency generated by each oscillator is impossible to be exactly the same. So the time error between the network nodes is inevitable.

WIA-PA standard has proposed network synchronization approaches, and has already provided two synchronization mechanisms. One synchronization mechanism is based on IEEE 802.15.4 beacon frame and the other is based on a frame called the synchronization frame in the DLL layer. The beacon frame-based synchronization mechanism can be coarse-grained synchronization, and only using the beacon frame based synchronization is not enough for WIA-PA network. As to the synchronization frame-based synchronization mechanism, there is no specification and implementation of the mechanism in the standards. Meanwhile, the synchronization implement solution and precision are closely related to WIA-PA network features such as superframe length, time slot length, and details of sending and receiving data between nodes. Therefore, the synchronization performance analysis and the impact on the network is an important problem to be addressed.

In this paper, to solve the existing problems in WIA-PA standard at this stage, we have designed and implemented the two synchronization mechanisms in WIA-PA standard and combined them with the features of the standard such as: network device type, network operation period, the network superframe structure, the slot length, device synchronization cycle, etc., the quantitative analysis of synchronization mechanisms is given.

9.2 Related Work

Currently, there are already some existing wireless standards abroad and domestic such as ISA100.11a standard, WirelessHART standard and WIA-PA standard. This section would briefly introduce and analysis the synchronization mechanisms implemented in the other two standards and several other classical synchronization algorithms for sensor networks.

WirelessHART standard is the extension of HART communication protocol which is designed for industrial environments. The real advantage of WirelessHART is time synchronized mesh networking protocol (TSMP) [7, 8]. Time synchronization mechanism consists of two steps. First select clock source, the network management will select numbers of nodes as clock source. Then each node passively or actively sync to clock source.

ISA100.11a standard proposed two mechanisms for time synchronization: one is the broadcast frame-based time synchronization mechanism and the other is acknowledgement frame-based time synchronization mechanism. Both synchronization mechanisms work together to build a unique ISA100.11a standard time synchronization mechanism.

Literature [9] introduced the time synchronization algorithm used in RBS protocol. RBS protocol does not synchronize the time between the sender and receiver, but synchronize time between the receiving nodes who received packets from the sender. According to statistical data, we can consider that the recipient receives the synchronization messages at the same time.

Literature [10] introduced the synchronization algorithm used in TPSN protocol. First of all, all nodes are leveled according to the hierarchy. And then each node will synchronize to higher level node, and finally synchronize all nodes to the root node.

Literature [11] introduced the synchronization algorithm used in DMTS protocol; the main idea is that the sender adds a timestamp t_0 in synchronization packets and send the packet, assuming that the preamble and the total length of the sync word are n bits, and we can calculate time of sending a single bit Δf according to the transmission rate. After the recipient receives the preamble and the sync word and record the local time t_1, record local time t_2, and the local time can be calculated as $t_0 + n * \Delta f + t_2 - t_1$.

The synchronization mechanisms and algorithms above have their own advantages and application situations, and are capable to meet the synchronization requirements under certain circumstances, but because of the restriction of WIA-PA protocol standard itself, these algorithms cannot be completely applied to WIA-PA. We need to implement the mechanisms according to the WIA-PA protocol.

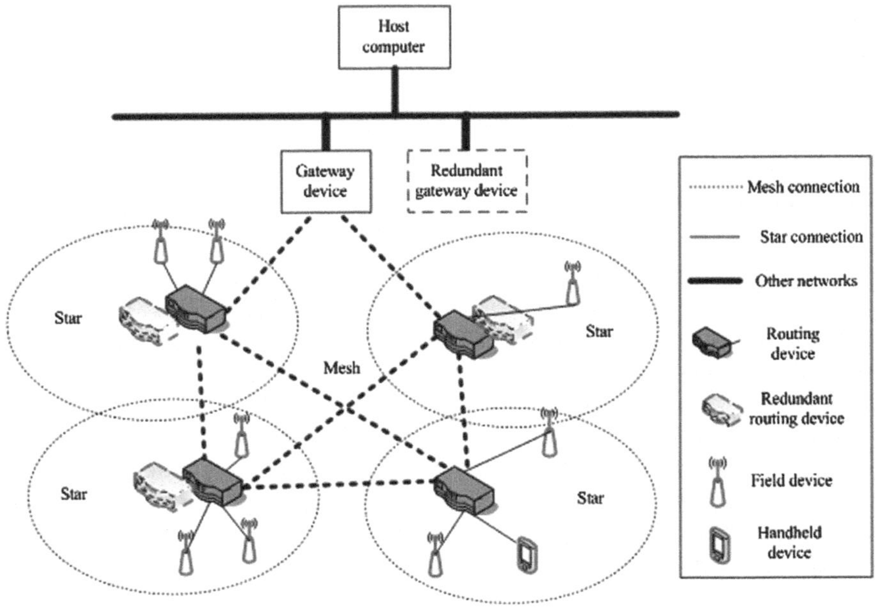

Fig. 9.1 WIA-PA network topology

9.3 WIA-PA Time Synchronization Mechanism

9.3.1 WIA-PA

WIA-PA is a kind of wireless network systems which is based on IEEE 802.15.4 short-range wireless communication standards, using free band in accordance with China radio management committee for industrial process measurement, monitoring and controlling, aiming at solving the interference to radio communications under the harsh environment and providing high reliability, real-time wireless communications services.

WIA-PA use a combination of star and Mesh network topology, the network topology is shown in Fig. 9.1. The first layer is the Mesh structure which is constituted by the gateway and router. Network manager and security manager for system management can be in a gateway or in the host computer. And the second layer is the star structure which is constituted by routing devices and field devices or handheld devices.

Compared with other industrial wireless technologies, WIA-PA has the following characteristics: (1) Star-Mesh topology makes the maintenance of the network topology more flexible. (2) CSMA/TDMA hybrid access mode supports for periodic and nonperiodic data communications. (3) Combination of a variety of measures to ensure communication reliability. (4) WIA-PA network's physical layer and data link sublayer are based on IEEE 802.15.4-2006 standards [12].

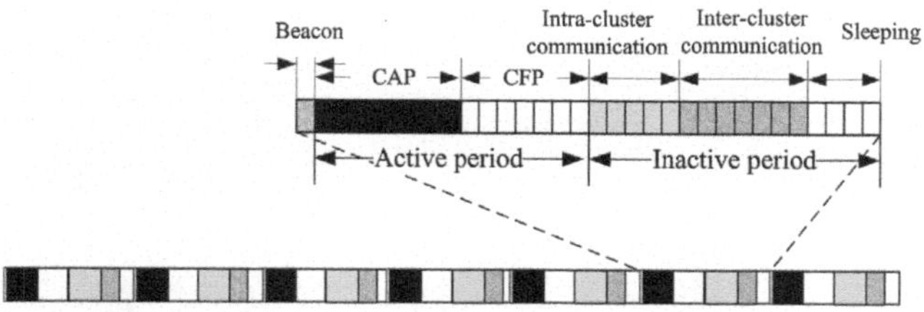

Fig. 9.2 WIA-PA superframe structure

WIA-PA data link layer is compatible with IEEE 802.15.4 superframe structure, the layer manages the slots through superframe. The superframe structure is shown in Fig. 9.2. WIA-PA superframe is divided into two parts: active and inactive period, and the CAP period in active period is used for CSMA communication while the inactive period for TDMA communication.

In order to ensure the reliability of TDMA communication, the devices in the network need to perform time synchronization. Devices in the WIA-PA network only need to synchronize with time source for relative time synchronization to distinguish sequence of events, as to whether time source should perform absolute time synchronization is not required in the standard.

9.3.2 Beacon Frame-Based Synchronization Mechanism

As is mentioned above, there are two types of time synchronization mechanisms in the WIA-PA standard. Beacon frame-based synchronization mechanism is mainly working when the device is joining the network. When the devices are joining the network of WIA-PA, it will periodically scan the beacon frame sent from the proxy router or gateway, and perform time synchronization through the beacon frame. As there is no time-related information in the beacon frame, such synchronization mechanism is coarse-grained synchronization.

As is shown in Fig. 9.3, according to the WIA-PA superframe structure, transmission of a beacon frame is fixed in the start of a superframe cycle, so $T_{start} = 0$. Ignoring the transmission delay, when a joining device scanned a beacon frame, it will consider that the current time is the start time of a superframe. Then the device will set the time as the relative time to the parent device which sends the frame and configures the local superframe from the data in the scanned beacon frame. Assuming that node A re-enters a superframe at the time of T_{start} and sends the beacon at the time of $T_{guardtime}$. Node B receives the preamble of the beacon frame and receives the whole packet after the time of $T_{receive}$, and then use time of $T_{process}$ to obtain the packets from the RFID and analysis beacon. Then the time that node B receives the beacon frame T_1 can be calculated by the formula (9.1) as follows:

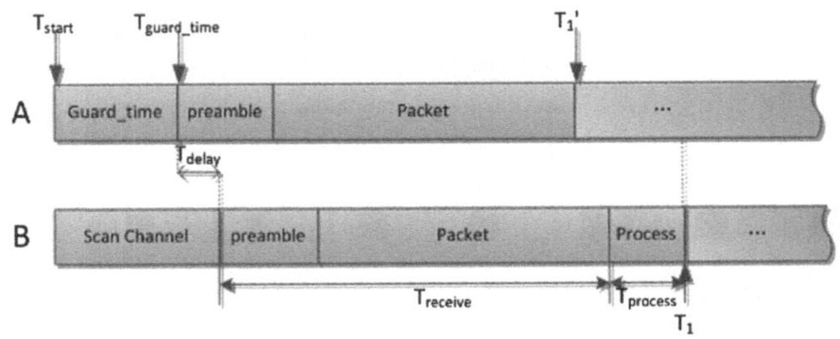

Fig. 9.3 Beacon frame-based synchronization mechanism

$$T_1 = T_{\text{guardtime}} + T_{\text{delay}} + T_{\text{receive}} + T_{\text{process}} \tag{9.1}$$

In the formula, $T_{\text{guardtime}}$ is a protection time that we artificially set before the packet transmission to prevent the failure of receiving the packet when synchronization error between the nodes is within the reasonable range. T_{delay} is the time consumed for the transmission of the packet. T_{receive} is the time used for the RF receiver of the device to receive the beacon frame. And T_{process} mainly includes two parts, one part is for the device to get the packet from the RFID buffer, and the other part is for the node to run the necessary programs.

Before the completion of the joining procedure, the router device would still receive the beacon frame each superframe cycle and perform time synchronization. When the joining procedure is complete, all the link resources and routing configurations that the network manager allocated will take effect. At the same time, the router will obtain its own CSMA channel, periodically send beacon frames, so it has been unable for the router to receive a beacon frame from gateway or his proxy router.

9.3.3 Synchronization Frame-Based Synchronization Mechanism

After the router finishes joining the network, the router cannot perform the time synchronization through beacon frame any more, while TDMA communication requires more precise time synchronization. WIA-PA standard design is an alternative synchronization mechanism of the synchronization frame-based synchronization mechanism.

WIA-PA standard synchronous frame format shown in Fig. 9.4, includes a byte of data link layer control subfield, a byte of command frame type, and five bytes of TimeValue field. So the synchronization frame payload is seven bytes. After the device has joined the network, the link resources and timeslots that network managers

Fig. 9.4 Format of synchronization frame

Octets: 1	1	5
DLSL frame control	Command frame identifier	Calibrated value of TimeValue

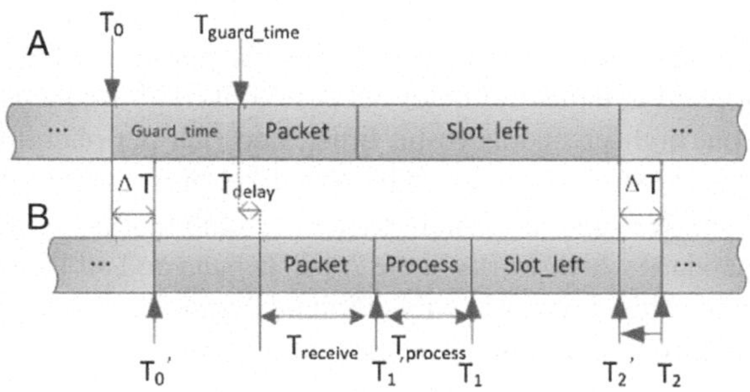

Fig. 9.5 Synchronization frame-based synchronization mechanism

allocated will become effective immediately, and each node will be assigned with a synchronization slot.

The procedure of synchronization frame-based synchronization mechanism is shown in Fig. 9.5. The parent node A triggers the synchronization timeslot at T_0, while Node B triggers synchronization timeslot at T_0'. Node A switches the RF module to sending state at time T_0, and sends the synchronization frame at time $T_{\text{guardtime}}$. Before the time of sending the frame, set the local time to the TimeValue field in the payload of the frame. Node B will consume T_{receive} time to receive the frame and in time T_1' gets the total frame, and then consumes T_{process} time to obtain the sync frame from the RF module buffer, gets the TimeValue T_0 from the payload, and reads the current time of the node T_1. Node B calculates the current synchronization error ΔT between the two nodes by comparing the local time T_0 and T_1, ΔT can be calculated by the formula (9.2):

$$\Delta T = T_0 - T_1 + T_{\text{delay}} + T_{\text{receive}} + T_{\text{process}} + T_{\text{guardtime}} \quad (9.2)$$

If $\Delta T > 0$, then the time of Node B is slower than the time of node A, the time of Node B will forward by T before the triggering of the next timeslot. As is shown in figure, the next slot boundary T_2 will be adjusted to T_2'. If $\Delta T < 0$, then the time of node B is faster than that in node A, the trigger time of next timeslot of node B will delay by ΔT. So that the time error will be corrected and the nodes will be synchronized.

The meaning of T_{delay}, T_{receive}, $T_{\text{guardtime}}$, T_{process} are the same that explained above.

Nodes will use the synchronization timeslot to receive the synchronization frame broadcasted from the parent node (gateway or proxy routing) to perform time synchronization. And the parent node will broadcast the synchronization frame once every superframe cycle, so nodes will perform time synchronization between its parent node by synchronization frame once every superframe.

9.4 Analysis and Test

9.4.1 Test Platform

The test platform designed in this chapter uses STM32 as CPU. STM32 is widely applied in industrial fields because of its high frequency, ultra-low power consumption, high-precision 16-bit timer, and rich peripheral interfaces, etc. We select RF chip AT86F231 produced by ATMEL Corporation as the RF module in the test platform. It is a kind of radio frequency chip that is compliant with the IEEE 802.15.4 standard which works in the 2.4 GHz band and all RF key components are integrated in the chip. It can not only efficiently complete RF transceiver tasks, but also greatly simplify the design of the hardware node.

9.4.2 Synchronization Accuracy and Performance Analysis

This section will focus on the precision and performance analysis of synchronization frame-based synchronization mechanism.

The test platform uses the 16-bit counter to implement the timing of slots in the node. STM32 itself uses 8 MHz external crystal oscillator as the node's clock source, and the clock frequency to the timer will reach 72 MHz through internal PLL clock multiplier. Therefore, the minimum counter precision can reach 72 MHz. The payload of the synchronization frame TimeValue is the internal counter register in the timer of STM32.

Now that the formula for the synchronization frame-based synchronization mechanism has been derived as formula (9.2), we will take quantitative analysis on the parameters in formula (9.2).

T_{delay}: As WIA-PA is mainly used in the situation of short distance wireless communication, electromagnetic wave propagation speed equals to the speed of light. So compared to accuracy of counter in the CPU, the packet propagation delay is negligible, that means, $T_{\text{delay}} = 0\,\mu\text{s}$.

T_{receive}: AT86f231 has provided eight basic interrupt operations in the hardware, while two of them are RXStart (received data starts interrupt) and TRXEnd (send or receive data ends interrupt). When the RF starts to receive a packet, RF enters RXStart interrupt and records the timer counter register value as T_{RXStart}. When the RF completely receives the packet, it enters TRXEnd interrupt, and records the timer counter register value T_{TRXEnd}. And then we have: $T_{\text{receive}} = T_{\text{TRXEnd}} - T_{\text{RXStart}}$.

T_{process}: Every time the RF module receives a packet, the processor needs to get the packet from the buffer of RF module through the SPI interface and the node's packet process time is mostly the SPI transmission time. As is shown in Fig. 9.6, the transmission time between the CPU and RF module is closely related to the length of the packet. The measured data shows a linear relationship between the two. The payload of the synchronization frame is seven bytes, plus the MAC layer header, the

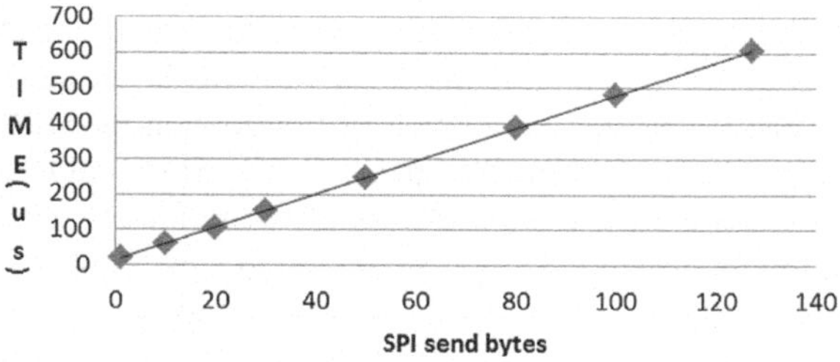

Fig. 9.6 SPI transmission time

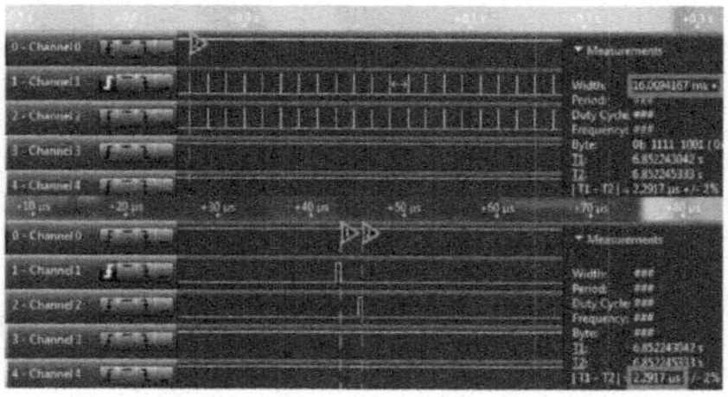

Fig. 9.7 Sample of synchronization result

total length of the synchronization frame is fixed to 22 bytes, the transmission time is about 107–108 μs. Therefore, make $T_{\text{process}} = 108\,\mu\text{s}$.

$T_{\text{guardtime}}$: It has been specified in the WIA-PA standard that the maximum synchronization error between the routing devices under the longest period of the superframe should not exceed 10 % of the length of basic timeslot, while the reason to set $T_{\text{guardtime}}$ is to prevent the packet loss because of reasonable synchronization error. So if the length of a slot is T_{slot}, then $T_{\text{guardtime}} = 10\,\% * T_{\text{slot}}$.

In order to test the synchronization results, we use logic analyzer as an auxiliary equipment. Logic analyzer is capable to acquire and display digital signals from the tested equipment. Every time the device starts a timeslot, STM32 would trigger a specified GPIO port to generate an instantaneous voltage transition, and then the logic analyzer can capture the voltage transition. By comparing and observing the generate time of voltage transition, we can get the synchronization error between nodes.

Figure 9.7 shows a sampling results by logic analyzer, this result is measured under the situation that the length of timeslot is 16 ms, while a superframe cycle contains 64 slots. The picture top of Fig. 9.7 shows that the length of a slot is 16 ms, the picture bottom of Fig. 9.7 shows that the synchronization error the moment the node just received synchronization frame, and the actual synchronization error after synchronization is about 2.3 μs. After multiple sets of data analysis, we can get the final synchronization accuracy which can be controlled within 4 μs.

The crystal frequency stability of this platform $\Delta f = 20\,\text{ppm}$, that is, within each one million cycles the maximum possible difference of the crystal frequency is 20 cycles, i.e., maximum error of one second is 20 μs. The maximum synchronization cycle of an equipment means the maximum working period when the device has not received the synchronization information and time update. If more than this time the device fails to receive clock update, the device may fail to communicate properly with the time source time because of the synchronization error. The MAC layer of WIA-PA is based on IEEE 802.15.4 MAC layer, the length of each timeslot within a superframe and the number of timeslots in the superframe is determined by two parameters called the BO and SO ($0 \leq SO \leq BO \leq 14$). SO determine WIA-PA slot length T_{slot}, BO and SO jointly determine the number of slots in a superframe slotnum. Synchronization frame-based synchronization mechanism can only sync once in a superframe period, so synchronization error is accumulated between two synchronization frames. Consider the worst case, all errors are accumulated in the same direction, then when the maximum synchronization error $T_{max} > T_{guardtime}$, the nodes may be unable to communicate properly, which can be the maximum synchronization cycle formula as is shown in formula (9.3):

$$N_{max} = T_{guardtime}/(T_{slot} * \Delta f) \tag{9.3}$$

According to WIA-PA standards, $T_{guardtime} = 10\,\% * T_{slot}$, and consider the worst case, $\Delta f = 20\,\text{ppm}$, then using formula (9.3), we can get $N_{max} = 5000$. According to the MAC layer defined in IEEE 802.15.4 standard, the total number of timeslots in a superframe $\text{Slotnum} = 16 * 2^{(BO-SO)}$. In order to ensure the normal operation of the network, Slotname must be less than N_{max}. So we can further get the restrictive relationship between BO and SO under this mechanism, as is shown below:

$$0 \leq SO \leq BO \leq 14 \;\&\&\; BO - SO \leq 8$$

As to power consumption, in addition to the necessary power consumption for processing and transmitting the synchronization packets, because of the protection mechanism, the RF module in the node needs to wait $T_{guardtime}$ time before receiving and sending data, and during this period the nodes must be in the receiving state, which will bring additional energy consumption. Therefore, additional synchronization energy will change linearly with the timeslot length. According to RF module datasheet, the RF module energy consumption in the receiving state $P_{rx} = 12.3\,\text{mA}$, while in the sending state $P_{tx} = 14\,\text{mA}$. Therefore, the additional energy consumption P_{extra} of the transmission nodes can be calculated by the formula shown in formula (9.4):

$$P_{extra} = P_{tx} * T_{guardtime} + P_{rx} * (T_{guardtime} - \Delta T) \tag{9.4}$$

For beacon frame-based synchronization mechanism, this mechanism is mainly used when the node is joining the network, coarse-grained synchronization mechanism is enough to ensure the normal data communication in the CSMA slot of WIA-PA superframe. And after the node joining the network, devices will not use CSMA slot to have further communication to his neighbors. The test and analysis method for beacon frame-based synchronization mechanism is same to that of synchronization frame-based synchronization mechanism, so details will not be described here, just give the conclusion: the synchronization accuracy of beacon frame-based synchronization mechanism can achieve $20\,\mu s$.

9.5 Conclusion

WIA-PA standard proposed two synchronization mechanisms, beacon frame-based synchronization mechanism and synchronization frame-based synchronization mechanism. Both methods have their own advantages and disadvantages. In order to ensure that the WIA-PA network works regularly and orderly, the two synchronization mechanisms should work together and collaboratively.

Beacon frame-based synchronization mechanism is mainly used when the node is joining the network. Devices can sync to its father node by scanning the beacon frames when the devices receive no information about the network and have no communication resources been allocated. No additional packet is needed during the synchronization procedure. Comparing to the synchronization frame-based synchronization mechanism, the synchronization error of this mechanism is relatively larger, but it is enough for CSMA communication. The synchronization accuracy of beacon frame-based synchronization mechanism can achieve less than 20 μs.

Synchronization frame-based synchronization mechanism is primarily used when the node complete joining the network. Father devices synchronize the child devices by periodically broadcast the local time value within the synchronization frame. The mechanism has high synchronization accuracy. However, this mechanism needs to occupy a special slot and extra link resources and this mechanism can only achieve point-to-point synchronization. Nodes can only synchronize with their parent node, and the synchronization error will accumulate with the increase of network topology layers. Synchronization frame-based synchronization mechanism can achieve less than 4 μs accuracy. In the premise of ensuring the normal and periodical packets communication, this synchronization mechanism limits the maximum synchronous cycle of the device, as well as the superframe structure. What is more, this mechanism will bring extra energy consumption to the device.

References

1. Xu Y, Zhang B (2010) Design and implementation of low power consumption networks based on WIA-PA. Comput Meas Control 18(9):2130–2132
2. China Industry Wireless Alliance (2008) Wireless industry automation specification for process automation. IEC Pass/62601
3. Industrial communication networks-field bus specifications-WIA-PA communication network and communication profile, IEC/PAS 62601, 31 Oct 2008
4. Wireless HART Communication, Standard, HART 7.0 Specifications, The HART, Communication Foundation(HCF), 7 Sept 2007
5. ISA100.11a Draft Standard release, ISA100.11a Working Group, 12 Sept 2008
6. Xu W, Liang W, Feng C (2011) Design and implementation of key technologies in WIA-PA networks. Appl Res Comput 28(6):2265–2270
7. Miao X (2012) Application and technology of WirelessHART wireless sensor networks. Autom Panorama 3:34–38
8. Technical overview of the time synchronized mesh protocol (TSMP). White Paper of Dust Networks. www.dustnetworks.com

9. Raghunathan V, Schurgers C, Park S, Srivastave MB (2002) Energy-aware wireless microsensor networks. IEEE Signal Process Mag 19:40–50
10. Xiaofeng X, Binyu Y, Fangyuan L (2010) Improved algorithm of TPSN for wireless sensor networks. Comput Meas Control 18(6):1465–1466
11. Qitao G, Xiaoping F, Qian C (2010) Research on time synchronization algorithm for wireless sensor networks. Electron Technol 47(7):5–7
12. Xiaoxiang P, Zhenjie L (2011) Hardware platform design based on WIA-PA networks. Instrum Tech Sens 4:60–62

Chapter 10
A Reliable Routing Algorithm Based on Link Quality Evaluation for Wireless Sensor Networks

Yongrui Chen, Weidong Yi, Binghua Wang and Fei Qin

Abstract Due to the lack of accurate evaluation of the transmission characteristics of the wireless communication links, routing algorithms in wireless sensor networks may result in poor network performance. This paper presents a link quality-based reliable routing algorithm. According to the link qualities evaluated by RSSI, neighbor nodes can be partitioned into different regions. First, based on a realistic link model and test results, an asymmetric link region is indicated, where selecting a neighboring node in this region as the next hop node will greatly degrade the transmission reliability. Based on this, the neighboring area of a node is divided into four regions, that is, connected, transitional, asymmetric, and disconnected regions. Then, route selection ways for different regions are presented, and different priorities of the nodes in different regions to be selected as the next hop node are assigned. Finally, the performance of the RSRE is tested in a real-world test-bed consists of 40 nodes, and compared with the Min-hop routing algorithm and the ETX based routing algorithm. The results show that the RSRE has better data transmission reliability, and hop distribution.

Keywords Routing protocol · Reliability · Link quality evaluation · Wireless sensor networks

---

Y. Chen (✉) · W. Yi · B. Wang · F. Qin
Department of Electronic and Communication Engineering,
University of Chinese Academy of Sciences, Beijing 10049, China
e-mail: chen_yong_rui@163.com

W. Yi
e-mail: weidongyi@ucas.ac.cn

B. Wang
e-mail: wangbinhua10@mails.gucas.ac.cn

F. Qin
e-mail: feiqin@ucas.ac.cn

X. Wang et al. (eds.), *Advanced Technologies in Ad Hoc and Sensor Networks*,
Lecture Notes in Electrical Engineering 295, DOI: 10.1007/978-3-642-54174-2_10,

10.1 Introduction

Recent studies in wireless sensor networks have shown that the transmission characteristic of underlying links can have a significant impact on routing protocols [1–3]. For example, some studies have revealed the existence of three distinct reception regions in wireless links [7]: connected, transitional, and disconnected. The transitional region is often quite significant in size, and is generally characterized by high variance in reception rates and asymmetric connectivity. Other studies also found that the low power wireless communication links display complex spatial and temporal correlations and characteristics. Therefore, the idealized perfect reception-within-range models used in common network simulation tools can be very misleading. Taking these factors into consideration, many routing protocols are designed to take advantage of the realistic link layer models. For instance, Couto et al. [1] proposed a new routing metric called ETX (expected transmission numbers), and show a better performance in terms of throughput than minimum hop count metric for the nodes in the transitional region. On the same line of the work, by evaluating link estimator and neighborhood table management, the author in [2] found that cost-based routing using a minimum expected transmission metric has a good performance.

There are many factors to be considered for designing an efficient routing protocol in wireless sensor networks, such as energy efficiency, network lifetime, throughput, and delay. The optimization goals for a routing protocol are depended mostly on the definition of routing metric. Among them, reliability may not be the most important, but has to be guaranteed. If the transmission links are unstable, the retransmission or reroute are typically very energy and time-consuming.

Based on this consideration, RSRE applies different routing selection methods for different links. For reliable links in connected region where almost 99 % perception rate can be guaranteed, a cost-based routing is adopted, to optimize some specified designing goals, using a route metric such as minimum hop, largest residual energy or combination of these (in this paper, min-hop is adopted to minimize transmission energy and delay). For unreliable links in transition region, where perception rate varies from 1 to 99 %, the most reliable link is chosen, to give transmission reliability with top priority, by using a link quality estimator based on the average measured RSSI. Also, the reliable links are given a higher priority than unreliable links.

To evaluating link quality, typically there are two kinds of link monitoring: (i) active link monitoring and (ii) passive link monitoring. In active link monitoring, a node monitors the links to its neighbors by sending probe packets. Probe packets can be sent either by broadcast [1], or by unicast [4]. Probe packets are generally sent at a certain rate, which yields a trade-off between energy efficiency (low rates) and accuracy (high rates). To reduce overhead and improve energy efficiency, probes may be multiplexed by any other control messages, such as shake hand messages in MAC layer, ACKs in link layer, synchronization messages, or routing advertisements.

Passive link monitoring is another widely used method due to its energy efficiency compared to active link monitoring. Unlike active link monitoring, passive link monitoring exploits existing traffic without incurring additional communication overhead. In fact, a node listens to transmitted packets, even if these packets are not addressed to it (overhearing) [5]. However, passive monitoring incurs the overhead of probing idle links [4]. In addition, when the network operates at low data rate or unbalanced traffic, passive link monitoring may lead to the lack of up-to-date link measurements. Consequently, it leads to inaccurate link quality estimation.

In RSRE, we use a hybrid link monitoring method which combines the benefits of active monitoring and passive monitoring. Probes (beacons) are broadcast to evaluating the link quality in a coarse-grain and as the criterion for region division. The average Required Number of Packet Transmissions (RNP) is maintained for the active link to detect if the link is in bad quality and needs route repair.

Since the link quality of broadcast probes are very different with unicast data packets, we found when active link monitoring is adopted, there is another unique region besides the three regions mentioned above, which we called asymmetric region. Unlike the asymmetric link in transitional region, which is transient asymmetric and caused by hardware asymmetry and radio irregularity, most links in asymmetric region are persist asymmetric and caused by the asymmetry of probe broadcast and packets unicast. If a sender uses a link in this region for data transmission, the performance of routing algorithm will degrade seriously. Based on this finding, RSRE divides the spatial area into four regions: connected, transitional, asymmetric, and disconnected. We analyze the boundary of asymmetric region and testified in tested experiments. To guarantee reliable transmission, the links in asymmetric region have to be excluded.

In summary, the contributions of this paper are as follows:

1. Indicating the existence of asymmetric region when active link monitoring is used;
2. Presenting a routing strategy using different routing selection criteria according to different regions;
3. Validating the effectiveness of RSRE by experiments in a 40 nodes real-world testbed.

10.2 Related Works

Traditional routing protocols in wireless sensor networks often evaluated through simulations that make simplifying assumptions about the link layer, such as the binary perfect-reception-within-range model. Several empirical studies [6–8] have questioned the validity of these assumptions. In [6] it is shown that the dynamics of even the simplest flooding mechanism and the topology of data gathering trees constructed in dense sensor networks can be significantly affected due to the asymmetric and occasional long-distance links caused by nodes present in the transitional region. In [9] also, it is argued that the routing structures formed taking into account unreliable links can be very different from the structures formed based on a simplistic model.

Since the behavior of links have great impact on the performance of routing algorithms, some studies proposed new link models based on empirical data. In [7], it is shown that the transmission range is divided by three regions, each with an irregular shape, dynamic bounds (changing over the time), and specific features. These regions are: (i) connected region, where links are often of good quality, stable, and symmetric, (ii) transitional region, where links are of intermediate quality (in long-term assessment), unstable, not correlated with distance, and often asymmetric, and (iii) disconnected region, where links have poor quality and are inadequate for communication. Particularly, the transitional region was the subject of several empirical studies because links within this region are extremely unreliable and even unpredictable. These intermediate quality links, referred also as intermediate links, are commonly defined as links having an average PRR between 10 and 90 %.

For the link qualities in the transitional region, it was demonstrated in several works that link quality is not correlated with distance. Some researches found that link quality is not isotropic, and some revealed sensor nodes that are geographically close to each other may have high spatial correlation in PRRs (packet receive rates). Some studies explores link quality variation over time. It is shown in [10] that links with very low or very high average PRRs are more stable than links with moderate average PRRs. The temporal variation of link quality is due to changes in the environment characteristics. Several studies analyzed the asymmetry of low-power links. It is found that asymmetric links are mainly located at the transitional region and not correlated with distance. Link asymmetry may or may not be persistent. Hardware asymmetry and radio irregularity constitute the major causes of link asymmetry.

Empirical observations on low-power links raised the need for link quality estimator (LQE) as a fundamental building block for routing protocols. RSSI measurement is a LQE which can be directly read from the radio transceiver (e.g., the CC2420). Their advantage is that they do not require any additional computation. Empirical studies such as [11] proved the existence of a RSSI value (−87 dBm [11]) above which the PRR is consistently high (99 % [11]), that is, belonging to the connected region. Below this threshold, a shift in the RSSI as small as 2 dBm can change a good link to a bad one and vice versa, which means that the link is in the transitional or disconnected region [12]. Also, RSSI was shown very stable (standard deviation less than 1 dBm) over a short time span (2 s), thereby a single RSSI reading (over a packet reception) is sufficient to determine if the link is in the transitional region or not [12]. Therefore, RSSI can provide a quick and accurate estimate of whether a link is of very good quality (connected region).

The Window Mean with Exponentially Weighted Moving Average (WMEWMA) [13] is a receiver-side LQE based on passive monitoring. It smoothes PRR estimates using the EWMA filter, which provides more stable but sufficiently agile estimation compared to PRR.

The Kalman-filter-based link quality estimator (KLE) [15] was proposed to overcome the poor reactivity of average-based LQEs, including PRR. Upon packet reception, RSS (Received Signal Strength) is extracted and injected to a Kalman filter, which produces an estimation of the SNR. Using a precalibrated PRR-SNR curve at the receiver, the approximated SNR is mapped to an approximated PRR, which represents the KLE link quality estimate.

The Required Number of Packet Transmissions (RNP) [16] is a sender-side estimator that counts the average number of packet transmissions/retransmissions required before successful reception. It can be computed as the number of transmitted and retransmitted packets during an estimation window, divided by the number of successfully received packets, minus 1 (to exclude the first packet transmission). RNP assumes an ARQ (Automatic Repeat Request) protocol at the link-layer level, that is, a node will repeat the transmission of a packet until it is correctly received.

The Expected Transmission Count (ETX) [1] is a receiver-side estimator that uses active monitoring. ETX is the inverse of the product of the forward delivery ratio and the backward delivery ratio, which takes into account link asymmetry. Couto et al. [1] showed that routing protocols based on the ETX metric provide high-throughput routes on multihop wireless networks. However, Wang et al. [14] found that ETX based on passive monitoring fails in overloaded (congested) networks, since a large number of nodes are not able to compute the ETX because they do not receive packets.

Four-bit [17] LQE is designed to be used by routing protocols and provides four bits of information, compiled from different layers. Four-bit assesses link quality as an approximation of the packet retransmissions count by combining two metrics (RNP and WMEWMA), through the EWMA filter. Fonseca et al. [17] found that CTP based on four-bit provides better performance (e.g., packet delivery) than the original version of CTP and MultiHopLQI. The L-NT and L-ETX are two sender-side LQEs that approximate the RNP [10]. They are referred as data-driven LQEs because they are based on feedback from unicast data packets. MetricMap is proposed by Wang et al. [14] as an alternative LQE for MintRoute, a hierarchical routing protocol, when the original LQE ETX fails to select routes. The Weighted Regression Estimator (WRE) is proposed in [3]. The author argued that the received signal strength is correlated with distance. Hence, WRE derives a complex regression function based on an input vector that contains a set of node locations together with their link quality known in advance. The Fuzzy Link Quality Estimator (F-LQE) [18] is a receiver-side estimator. In contrast to existing LQEs, which only assess one single link property thus providing a partial view of the link, F-LQE estimates link quality on the basis of four-link properties in order to provide a holistic characterization of the link, namely Smoothed Packet Reception Ratio (SPRR), link stability factor (SF), link Asymmetry Level (ASL), and channel Average Signal-to-Noise Ratio (ASNR). The DoUble Cost Field HYbrid (DUCHY) [19] is a routing metric that allows to select routes with short hops and high quality links. DUCHY is based on two LQEs. The first is receiver-side and uses active monitoring (based on beacon traffic). It is called Channel State Information (CSI). The second estimator is the RNP.

10.3 Link Quality Analysis

10.3.1 Analysis of Transitional Region

Due to the unique characteristic of each environment, most radio propagation models use a combination of analytical and empirical methods. One of the most common radio propagation models is the log-normal shadowing path loss model [13]. The model is given by:

$$\mathrm{PL}(d)(\mathrm{dB}) = \mathrm{PL}(d_0) + 10n\log(d/d_0) + X_\sigma \tag{10.1}$$

where d is the transmitter-receiver distance, d_0 a reference distance, n the path loss exponent (rate at which signal decays), and X_σ a zero-mean Gaussian RV (in dB) with standard deviation σ (shadowing effects). In the most general case, σ is a random process that is a function of time, but, since we are not assuming dynamic environments, we model it as a constant random variable over time for a particular link.

Assuming the transmission power is a constant, then the receiving power is:

$$\mathrm{Pr}(d)(\mathrm{dBm}) = P_\mathrm{t}(\mathrm{dBm}) - \mathrm{PL}(d)(\mathrm{dB}) \tag{10.2}$$

where P_t is the transmission power.

Then, the signal-to-noise ratio (SNR) of the receiving node can be expressed as:

$$\gamma(d) = \mathrm{Pr}(d) - P_\mathrm{n} = P_\mathrm{t} - \mathrm{PL}(d) - P_\mathrm{n} \tag{10.3}$$

where P_n is the noise floor of the transceiver. For example, $P_\mathrm{n} = -105\,\mathrm{dBm}$ when CC2420 is considered.

The bit-error-rate (BER) is expressed as:

$$P_\mathrm{e} = \beta_\mathrm{M}(\gamma) \tag{10.4}$$

where $\beta_\mathrm{M}(\gamma)$ is the relationship between the BER and SNR, which is decided by the modulation and encoding schemes of specific transceiver. Then, the packet receive rate (PRR) can be expressed as:

$$\phi(\gamma) = (1 - \beta_\mathrm{M}(\gamma))^{8f} \tag{10.5}$$

where $\phi(\gamma)$ is the PRR, and f is the frame size of the packet.

Typically, transmission range is divided into three regions according to the PRR. When PRR is between ϕ_u to 1, it's connected region; when PRR is between ϕ_i to ϕ_u, it's transitional region; when PRR is below ϕ_i, it's disconnected region. Denote the SNR in the outer boundary of connected region is γ_u, and the SNR in the outer boundary of transitional region is γ_l, it can be derived from (10.5):

$$\gamma_{\rm u} = \beta_{\rm M}^{-1}(1 - \phi_{\rm u}^{1/8f}) \tag{10.6}$$

$$\gamma_{\rm l} = \beta_{\rm M}^{-1}(1 - \phi_{\rm l}^{1/8f}) \tag{10.7}$$

For CC2420 transceiver, which use DSSS-O-QPSK as its modulation scheme, the PER function $\beta_{\rm M}(\gamma)$ is Q function, which can be approximated as:

$$\beta_{\rm M}(\gamma) = Q(\sqrt{(K-1)/3N + N_0/(2E_{\rm b})})^{-1} \tag{10.8}$$

where K is the user number, N is the spreading gain of DSSS. Assume that no interference caused by concurrent transmission, K is set to 1. For CC2420, $N = 8$. Then (10.8) can be simplified as:

$$\beta_{\rm M}(\gamma) = Q(\sqrt{(2E_{\rm b})/N_0}) \tag{10.9}$$

RSSI is measured by the hardware and can be read from the register of CC2420. SNR is RSSI subtract the noise floor, which is -105 dBm for CC2420. SNR and E_b/N_0 has the following relationship:

$$\gamma = \frac{E_b}{N_0}\frac{R}{B_N} \tag{10.10}$$

where R is the data rate, B_N is the system bandwidth. For CC2420, the data rate is 250 Kbps and the bandwidth is 384 KHz when the roll-off factor of the raised cosine filter is 0.5. There is:

$$\beta_{\rm M}(\gamma) = Q(\sqrt{3\gamma}) \tag{10.11}$$

Since CC2420 use NRZ encoding schemes, the PRR can be derived as:

$$\phi(\gamma) = (1 - \beta_{\rm M}(\gamma))^{8f} = (1 - Q(\sqrt{3\gamma}))^{8f} \tag{10.12}$$

Then the boundary conditions of the connected region and transitional region are:

$$\gamma_{\rm u} = 3[Q^{-1}(1 - 0.9^{1/8f})]^2 \tag{10.13}$$

$$\gamma_{\rm l} = 3[Q^{-1}(1 - 0.1^{1/8f})]^2 \tag{10.14}$$

10.3.2 Analysis of Asymmetric Region

In many routing algorithms, there are two types of packets. One is control packets, the length is fixed and short, used to evaluate the links to neighboring nodes and exchange messages with neighboring nodes. Another is data packets, the length is related with typical application and not fixed, and long. Since control packets carries less information than data packets, and to avoid collision, congestion and overhead caused by long control packets. The length of control packets is always much less than data packets. Since the length of these two types of packets are different. Therefore, in the routing algorithms, there is a asymmetric links, where the reception rate of control packets is larger than the PRR of the data packet in the same condition. To explain easily, we use the data gathering application as an example to explain the mechanism of asymmetric region. In data gathering sensor networks, multiple nodes consist a dynamic tree topology, called data gathering tree. Each node in the tree sends its packets to the sink through one or multiple hops. For multi-hop transmission, first a node has to send its packets to its next-hop node, that is, the parent node in the data gathering tree. Then the routing algorithm is how to select the parent node in the tree. By periodically listening the beacons from its neighboring nodes, a node select the optimal node in its neighborhood as its parent node. In a real-bed experiment, we found that if parent node are in the a particular region of the child node, then the child node can receive the parents beacon with a certain probability, however, parent node can not receive the data packets from child node. The reason is that the length of data packets and probes are not the same length. From the analysis in the above section, we can find under the same SNR condition, the longer the packet, the lower the reception rate. This is because the PRR is the accumulation of the bit error, the longer the frame length, the more easily the packet is dropped.

From the analysis of the above section, we can find that when the SNR is below a certain threshold, the receiving node can hardly hear the packets from the sending node. However, since the probes is short than data packets, it can still receive the probes from the sending node. When a node cannot receive a probe, its SNR γ_a satisfies:

$$\phi(\gamma_a) = (1 - \beta_M(\gamma_a))^{8f_2} = \phi_l \tag{10.15}$$

where f_2 is the length of probes, and ϕ_l is 0.1. therefore:

$$\gamma_a = 3[Q^{-1}(1 - 0.1^{1/8f_2})]^2 \tag{10.16}$$

Thus, adding the asymmetric region, the boundary condition of the four regions are:

$\gamma > \gamma_u$: connected region;

$\gamma_l \leq \gamma \leq \gamma_u$: transitional region;

$\gamma_a \leq \gamma \leq \gamma_l$: asymmetric region;

$\gamma < \gamma_a$: disconnected region;

where $\gamma_u = 3[Q^{-1}(1 - 0.9^{1/8f_1})]^2 \gamma_l = 3[Q^{-1}(1 - 0.1^{1/8f_1})]^2 \gamma_a = 3[Q^{-1}(1 - 0.1^{1/8f_2})]^2$ and f_1 and f_2 are the length of data packets and probes respectively. $\beta_M(\gamma)$ is decided by the specific transceiver, and the analysis CC2420 is mentioned in the above section.

It is worth to point out that, though asymmetric links also exist in transitional region, they are mostly transition, hardly to anticipate, and caused by the asymmetry of hardware and radio irregularity. However, the links in asymmetric region are mostly persist, and caused by asymmetry of data packets and probes. The effect of asymmetric region has much more effect on routing algorithm than asymmetric links in transitional region, since a parent node in the asymmetric region can hardly receive a packet from its child node.

10.4 Routing Algorithm

Based on the characteristic of links, a reliable routing algorithm is presented based on the link quality evaluation. First, all neighboring nodes are divided into three regions according to link quality: connected, transitional and asymmetric (disconnect region nodes can not be detected). Then, the parent node will be selected by different routing selection criteria according to different regions.

10.4.1 Link Quality Evaluation

Link quality evaluation is conducted by combining the passive LQE though RNP and active LQE through RSSI measurements.

For parent selection, since the links are all idle, link quality is evaluated by measuring the periodically broadcasted probes. Take CC2420 as example, the CC2420 transceiver can check the channel and measure the RSSI of the received packets. The receive power of CC2420 is in the range (−100 dBm, 0 dBm), and the value read from the register of RSSI.RSSI_VAL is in the range (−55, 45). Using CC2420, by analysis mentioned above and experiment in real test-bed, we found when the length of data packets and probes are 50 and 10 bytes respectively, the RSSI range of each region is: [−72 dBm, 0 dBm] for connected region, [−85 dBm, −72 dBm] for transitional region, [−100 dBm, −85 dBm] for asymmetric region and below −100 dBm for disconnected region.

The link quality is evaluated by averaging the RSSI measurements of probes with the window size as w. When the neighboring node is in the connected region, w is set to 1, since the RSSI of connected region is very stable, thus one read of RSSI is enough for route selection, making the routing algorithm more responsive for the routing selection. For the link in transitional region, w is set to 8, since the RSSI of transitional region is fluctuating and sensitive to the link condition. Therefore, a larger window is more appropriate for a long-term evaluation and the link quality assessment is more stable.

When the parent is selected, the link to the parent is active and a RNP counter is maintained. The retransmission number, drop packets number and total transmission number are all computed, and the value of RNP counter is computed as follows:

$$\text{RNP} = (N_{\text{re}} + N_{\text{dr}})/N_{\text{tot}} \tag{10.17}$$

where N_{re} is the retransmission number in a time window, N_{dr} is the number of drop packets in the time window, and N_{tot} is the total transmission number in the time window.

RNP is a metric to evaluate the expected retransmission number and drop number of a particular link and shows an effective link quality assessment for long-term link monitoring. When RNP is dropped below a threshold, or consequent packet drops are repeated for a certain times, the node will judge the current link is in bad quality, and reselect a new parent.

10.4.2 Loop Route Avoidance

Although using link quality as the routing metric can improve the reliability of data transmission, an apparent problem is that loop routes may appear since RSSI is not cumulative. Therefore, a loop route avoidance mechanism is needed. First, a node can not select any one of its children as its parent. Otherwise, it will form a parent-child loop. Then, for the loop route consists three or more than three nodes, we can prove that if each node only considers the nodes with a limitation of hop number as its parent node, then loop route will be avoided.

Lemma: Denote the hop number of node i as h_i, if i select a node j as its parent node, and $h_j < h_i + 2$, then loop route with at least three nodes will not appear.

Proof: if node i and node j forms a loop, and the number of nodes in the loop is no less than 3, then another node k exists, which k is is child node, and at the same time k is js parent or ancestor. That is, $h_k = h_i + 1$, and $h_j \geq h_k + 1$, therefore $h_j \geq h_i + 1$, contradiction. Hence, i and j can not form a loop route.

10.4.3 Routing Algorithm

The routing algorithm can be described as follows.

Step 1: Each node (denoted as node i with its hop number to the sink as h_i) in the network maintains a neighborhood table, which records each neighboring nodes link quality (by averaging measured RSSI from probes), route cost (e.g. ETX or hops, broadcasted in the probes), and hop number to the sink (to avoid loop route). A children table is also maintained.

Step 2: The neighborhood table is divide into three sub-tables containing nodes in three regions: connected, transitional and asymmetric, according to the comparison of link quality and the pre-calculated region threshold described in (10.13), (10.14) and (10.16).

Step 3: If the connected region table is not empty, the node with the minimum route cost in this table is chosen as the parent node. Go to the end.

Step 4: Otherwise, if the transitional region table is note empty, the node with the largest link quality (average RSSI measurements) is chosen as the parent node, if this node is not in the children table and its hop number is less than $h_i + 2$. Go to the end.

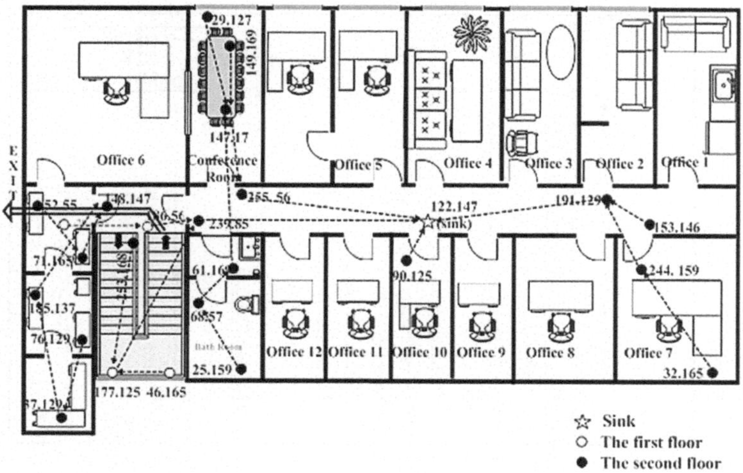

Fig. 10.1 The layout of the experiment

Step 5: Otherwise, no parent node will be selected, since the nodes in the asymmetric region have poor performance in data transmission. Go to the end.

10.5 Performance Evaluation

10.5.1 Experiment Scenario

RSRE is tested in a real-world testbed. The hardware platform is Tmote sky nodes, which supplied by two AA batteries. The software platform is contiki operating system. In MAC layer, simple CSMA and backoff mechanism is adopted to avoid the collision of data packets. The sensor nodes send data packets to the sink in a rate of 1 packet per minute.

The experiment is conducted in an office-floor building. There are many obstacles inside and outside the room, and many people move here and there. Therefore, the link conditions are complex and time varying. The deployment of nodes are shown in Fig. 10.1. Including the sink, 40 nodes are distributed in the room, corridor and stairway of the building, transmitting data packets through multi-hop ways. Two classic routing algorithm are implemented in the nodes for comparison: minimum-hop routing algorithm and ETX-based routing algorithm. The experiment time for each protocol is 6 h, and the sink collected a total of 14400 packets.

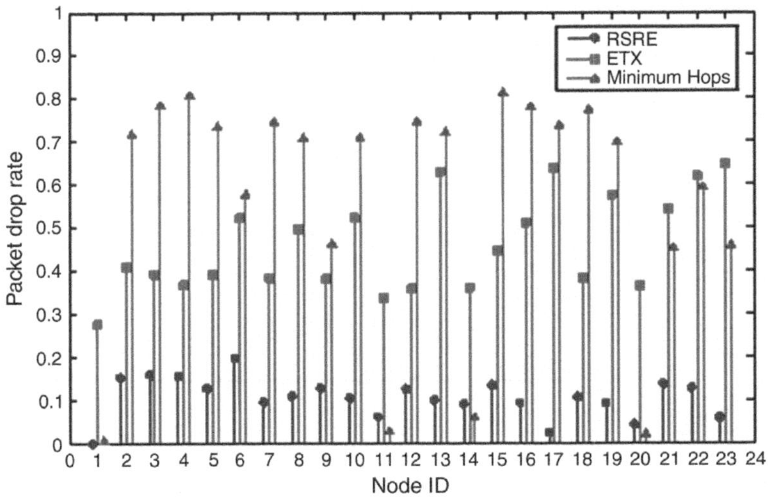

Fig. 10.2 Packet Drop Rate (PDR) of each node for the three algorithms

10.5.2 Experiment Results

From Fig. 10.2, we can see the RSRE has the least drop packet rate for every node, which is below 0.2 %. The reason is that RSRE take the link quality into consideration, and adaptively select hops or average RSSI as the route metric according to the link quality, which will guarantee the selected route is a reliable route. ETX is also based on link quality evaluation, however, ETX is only useful to active links that data packets are sent through it, moreover, ETX is a evaluation based on the average of historic values, which will fluctuated by the time. Therefore, ETX can not be quickly reflect the current link conditions, and thus can not guarantee the most reliable route. As for the min-hop route algorithm, the packet drop rate is even above 0.8, the transmission is very unreliable.

Figure 10.3 displays the hop distribution of the three routing algorithms. From Fig 10.3 we can see RSRE forms a four hops tree topology, most nodes are three hops. ETX-based routing algorithm forms a 5-hops tree topology, most nodes mostly are 4 or 3 hops. Min-hop routing algorithm forms a 3-hops tree topology, most nodes are 2 hops. As we know, data packet transmission in each hop contributes extra energy consumption and transferring delay. Therefore, a network with the less average hops number, is able to save more energy and reduce end-to-end delay. In this case, RSRE is better than ETX-based algorithm, but worse than min-hop algorithm. That is, RSRE achieves a trade-off between hop distribution and reliability, since different routing selection strategy is applied in different regions.

Figure 10.4 show the average packet loss rate for nodes when different hop numbers are considered. From Fig. 10.4 we can see that the packet loss rate increases as the hops grow. The reason is a packet may be dropped due to transmission unrelia-

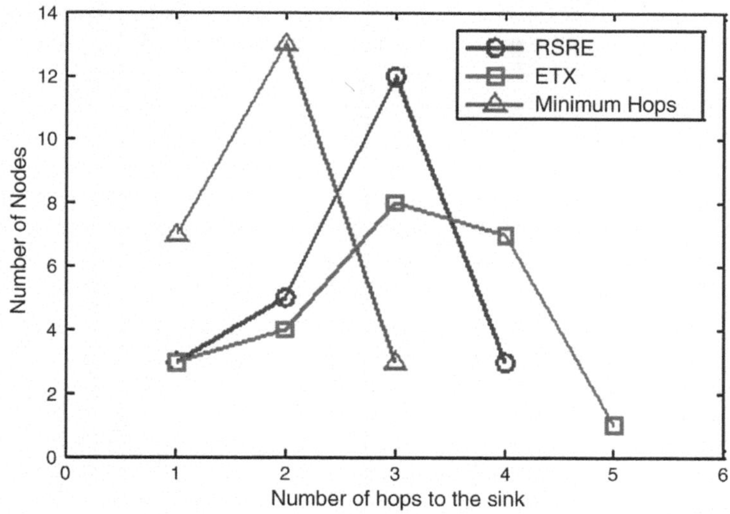

Fig. 10.3 The distribution of hop number for the three algorithms

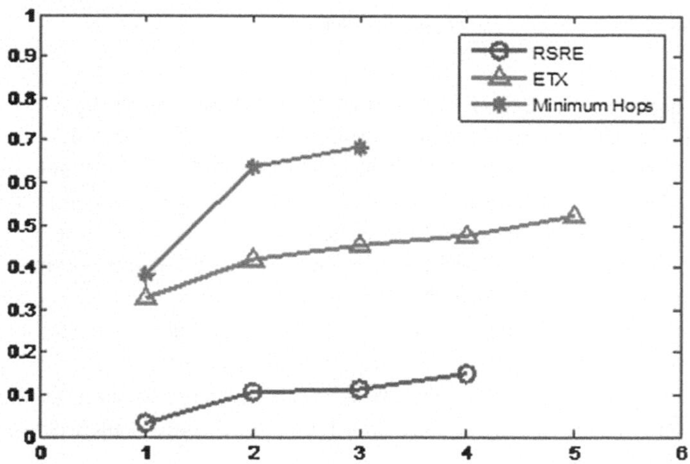

Fig. 10.4 Average packet loss rate for nodes with different hop numbers

bility in each hop of route. RSRE has a much lower packet loss rate than Min-hop and ETX-based routing algorithm, for nodes with all hops. That is, RSRE is the most reliable routing algorithm among all algorithms.

Figure 10.5 displays the number of each node changes parent node during the experiment. As we can see, ETX-base routing changes topology very frequently, and min-hop routing is most stable. RSREs topology is more inconstant than min-hop based routing but less changeful than ETX-based routing. Since link quality is variable, the ETX value for each link is fluctuating; therefore, a node may frequently change its parent. For example, node A is the parent of node B, since the link between

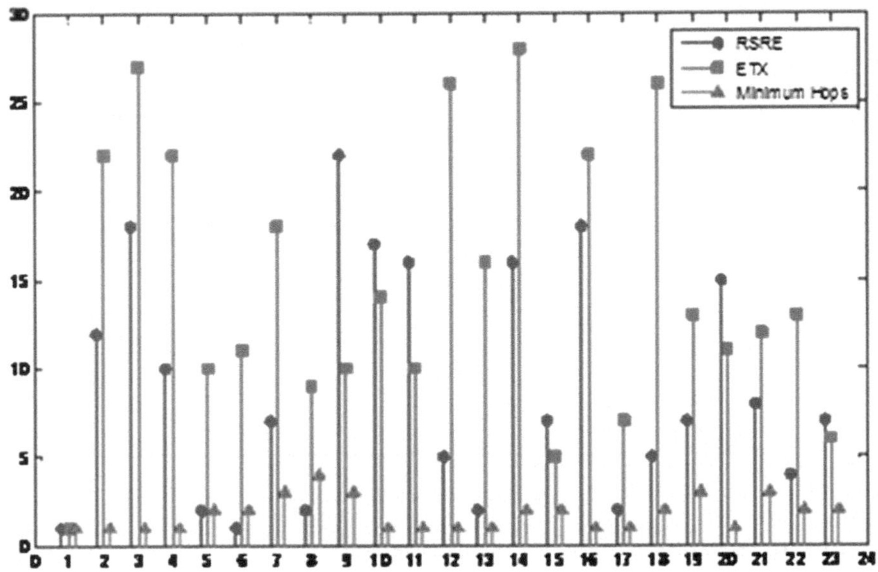

Fig. 10.5 Number of changing parent node for each node

A and B is not very good, the ETX value of the A-B link is dropping along the time. After the ETX value dropped below a certain value, e.g, the ETX value of B-C link, node B will select C as its parent. Then, since B-C link is not reliable either, the ETX of B-C link also decreases. After a while, B will select A as parent again. This procedure will continue, and B will sway between A and C time to time. However, for RSRE, this phenomenon will not happen. Though RSSI measurement is more sensitive to link quality variance, the RSSI of links in connected region is stable, and RSSI of links in transitional region is smoothed by averaging. As we know, frequent re-reroute may be very energy consuming, and make the network unstable, especially in a low power sensor networks. That is, RSRE is more stable in topology variation and insensible to link quality change than ETX-based route, which is important to energy conserving protocols.

10.6 Conclusion

In this paper, a reliable routing algorithm based on link quality evaluation is presented. First, we analyze that for the active link quality evaluation in routing algorithm, an asymmetric region exists which will seriously degrade the routing performance. In this region, a node can receive the neighboring nodes probes but can hardly receive data packets. Based on this analysis, we divide the neighboring area of the node into four regions: connected, transitional, asymmetric and disconnected. The boundary conditions of each region are also investigated.

Then, according to the characteristic of different regions, different routing selection method are proposed in RSRE: in the connected region, the neighboring node with the least route cost is selected as the next-hop node; in the transitional region, the neighboring node with the best link quality is selected as the next-hop node; the neighboring node in the asymmetric region is excluded to be the next-hop node. The node in the connected region always has higher priority than the node in the transitional region to be selected as the next-hop node. An active LQE by averaging RSSI measurements and a passive LQE by maintaining RNP count for active links are combined for a comprehensive and responsive link quality evaluation. Loop route avoidance mechanism is also proposed.

Finally, the performance of RSRE is testified in a 40 nodes test-bed deployed in an office-floor environment, compared with minimum hop route and ETX-based route. The experiment results showed that RSRE outperforms the other two routing algorithms in transmission reliability, reducing the packet drop rate by almost 1/2 compared with ETX-based route and nearly 1/4 compared with min-hop route. RSRE has a better hop distribution compared with ETX-based route, and also more stable in topology variation in condition of frequent link quality changes.

References

1. Couto DSJD, Aguayo D, Bicket J, Morris R (2003) A high-throughput path metric for multihop wireless routing. In: Proceedings of the 9th annual international conference on mobile computing and networking (MobiCom 03). ACM, New York, pp 134–146
2. Kotz D, Newport C, Elliott C (2003) The mistaken axioms of wireless-network research. Technical Report TR2003-467, Department of Computer Science, Dartmouth College
3. Xu Y, Lee WC (2006) Exploring spatial correlation for link quality estimation in wireless sensor networks. In: Proceedings of the 4th annual IEEE international conference on pervasive computing and communication (PERCOM 06). IEEE Computer Society, Los Alamitos, pp 200–211
4. Kim KH, Shin KG (2006) On accurate measurement of link quality in multi-hop wireless mesh networks. In: Proceedings of the 12th annual international conference on mobile computing and networking (MobiCom 06). ACM, New York, pp 38–49
5. Lal D, Manjeshwar A, Herrmann F (2003) Measurement and characterization of link quality metrics in energy constrained wireless sensor networks. In: Proceedings of the IEEE global telecommunications conference (Globecom 03). IEEE, San Francisco, pp 446–452
6. Ganesan D, Krishnamachari B, Woo A, Culler D, Estrin D, Wicker S (2002) Complex behavior at scale: an experimental study of low-power wireless sensor networks. UCLA CS Technical Report UCLA/CSD-TR 02-0013
7. Zhao J, Govindan R (2003) Understanding packet delivery performance in dense wireless sensor networks. In: Sensys 03
8. Woo A, Tong T, Culler D (2003) Taming the underlying issues for reliable multi-hop routing in sensor networks. In: SenSys 03
9. Woo A, Culler D (2003) Evaluation of efficient link reliability estimators for low-power wireless networks. Technical report UCB/CSD-03-1270, EECS Department, University of California, Berkeley
10. Zhang H, Sang L, Arora A (2010) Comparison of data-driven link estimation methods in low-power wireless networks. IEEE Trans Mob Comput 9:1634–1648

11. Liu H, Li J, Xie Z, Lin S, Whitehouse K, Stankovic JA, Siu D (2010) Automatic and robust breadcrumb system deployment for indoor fire-fighter applications. In: Proceedings of the 8th international conference on mobile systems, applications, and services (MobiSys 10). ACM, New York, pp 21–34
12. Srinivasan K, Dutta P, Tavakoli A, Levis P (2006) Understanding the causes of packet delivery success and failure in dense wireless sensor networks. In: Proceedings of the 4th international conference on embedded networked sensor systems (SenSys 06). ACM, New York, pp 419–420
13. Srinivasan K, Dutta P, Tavakoli A, Levis P (2010) An empirical study of low-power wireless. ACM Trans Sens Netw 6:1–49
14. Wang Y, Martonosi M, Peh L-S (2007) Predicting link quality using supervised learning in wireless sensor networks. ACM SIGMOBILE Mob Comput Commun Rev 11(3):71–83
15. Senel M, Chintalapudi K, Lal D, Keshavarzian A, Coyle EJ (2007) A Kalman filter based link quality estimation scheme for wireless sensor networks. In: Proceedings of the IEEE global telecommunications conference (GLOBECOM 07). IEEE, New York, pp 875–880
16. Cerpa A, Wong JL, Potkonjak M, Estrin D (2005) Temporal properties of low power wireless links: modeling and implications on multihop routing. In: Proceedings of the 6th international symposium on mobile ad hoc networking and computing (MobiHoc 05). ACM, New York, pp 414–425
17. Fonseca R, Gnawali O, Jamieson K, Levis P (2007) Four bit wireless link estimation. In: Proceedings of the 6th international workshop on hot topics in networks (HotNets VI), ACM SIGCOMM
18. Baccour N, Koubaa A, Youssef H, Ben Jamaa M, Do Rosario D, Alves M, Becker LB (2010) F-LQE: a fuzzy link quality estimator for wireless sensor networks. In: Proceedings of the 7th European conference on wireless sensor networks (EWSN10). Springer, New York, 240–255
19. Puccinelli D, Haenggi M (2008) DUCHY: double cost field hybrid link estimation for low-power wireless sensor networks. In; Proceedings of the 5th workshop on embedded networked sensors (Hot EmNets08). ACM, New York

Chapter 11
A Probabilistic Group Reverse k-Nearest-Neighbor Query in Sensor Networks

Jinghua Zhu, Bin Kan, Yong Liu, Tong Wang, Liqiang Pan and Dan Liu

Abstract Group Reverse k-Nearest-Neighbor Query (GRkNN) can be used to evaluate the influences of query objects and has wide applications. Traditional algorithms are not able to obtain satisfied results for uncertain sensor network data. A probabilistic threshold group reverse k nearest neighbor query (PT-GRkNN) is proposed in this paper. PT-GRkNN exploits filter-refine framework, geometric filter and probability filter are proposed to shrink the search space, the final query results can be obtained by refine phase. Experimental results show that the efficiency of PT-GRkNN is better than the existing methods.

Keywords Sensor network · Uncertain data · Reverse nearest neighbor

---

J. Zhu · B. Kan · Y. Liu (✉) · T. Wang · L. Pan · D. Liu
School of Computer Science and Technology, Heilongjiang University,
Harbin 150080, China

J. Zhu
Key Laboratory of Database and Parallel Computing of Heilongjiang Province,
Harbin 150080, China
e-mail: liuyong001@hlju.edu.cn

T. Wang
College of Information and Communication Engineering, Harbin Engineering University,
Harbin 150001, China

L. Pan
School of Computer and Science Technology, Harbin School of Technology,
Harbin 150001, China

D. Liu
School of Computer Science, Northeast Forestry University, Harbin 150001, China

X. Wang et al. (eds.), *Advanced Technologies in Ad Hoc and Sensor Networks*,
Lecture Notes in Electrical Engineering 295, DOI: 10.1007/978-3-642-54174-2_11,

11.1 Introduction

Given a set of query points Q, a group reverse k nearest neighbor (GRkNN) query returns objects having $q \in Q$ as their k nearest neighbor. GRkNN can be used to evaluate the influences of the query objects and plays an important role in the field of location-based service, decision support, and marketing. Take the geology disaster rescue as an example: the rescue team members who carry detection equipment are responsible for searching for victims separately. The team commander issues GRkNN query; the query returns victims who have the team members as their k nearest neighbors.

The naive method to solve the GRkNN query is to compute the RkNN of each object in the same group, and then union their RkNN into result set. In fact, the RkNN of the nearby objects have a large overlap. So the naive method has low efficiency for it does not consider the location similarity among the group.

In wireless sensor networks, due to the sense accuracy limitation, periodically sampling, transmission delay, and privacy protection, there are a lot of noises in data. Compared with the traditional certain data, GRkNN query on uncertain data has a lot of new challenges.

The possible locations of each uncertain data obey a probability distribution and can be approximately represented by a circle. Whether an object belongs to the GRkNN query result or not is no longer a certain event, instead it is a random event with some probability. Unfortunately, GRkNN query on traditional certain data dose not suit for uncertain data objects.

In this paper, we propose a Probability Threshold Group Reverse k Nearest Neighbor query (PT-GRkNN), it returns the uncertain objects whose probability of being the reverse k nearest neighbors of a group query objects above a given threshold. PT-GRkNN exploits the filter-refine framework: geometric filter and probability filter are designed to obtain the candidate objects. The final query results can be attained by the verification phase. We compare our algorithm with the naive algorithm. The experimental evaluation shows that our approach is able to significantly outperform the naive approach.

The contributions of this paper are as follows.

1. We propose a new query type named PT-GRkNN which can solve the decision support and influence set problems. 2. We propose geometric filter, probability filter, and verification methods, and integrate them into the PT-GRkNN algorithm. 3. We experimentally show that our proposed algorithm performs better than the naive solutions under various settings.

The rest of this paper is organized as follows: First we review the related work in Sect. 11.2, and then we give preliminaries and formally define the problem of PT-GRkNN queries in Sect. 11.3. In Sect. 11.4, we first describe the framework for PT-GRkNN query processing and then introduce the PT-GRkNN algorithm in detail. All proposed techniques are experimentally evaluated in Sect. 11.5. Finally Sect. 11.6 concludes the paper.

11.2 Related Work

Korn [1] proposed 60° region pruning algorithm which divides the space into six regions equally and searches each of the regions without any index. Tao et al. [2] proposed TPL that uses the property of perpendicular bisectors to prune the search space. TPL uses many combinations of k bisectors which is expensive. To overcome this issue, Wu et al. [3] proposed an algorithm called FINCH. Instead of using bisectors to prune the objects, they use a convex polygon that approximates the unpruned area. Cheema et al. [4] introduce the concept of influence zone which is the area such that every point inside this area is the RkNN of q and every point outside this area is not the RkNN. Wu et al. [5] proposed CRkNN algorithm to continuously monitor RkNN queries. They proposed a new continuous verification method called CRange-k and applied it to solve the continuous verification. Song et al. [6] proposed Group RkNN (GRkNN) in spatial database and designed the corresponding algorithms. There are currently two approaches for answering PRNN queries. (1)The approach from Chen et al. [7] which is designed for PRNN queries on uncertain objects represented by continuous probability density functions (PDFs) and (2) the approach from Cheema et al. [8] which works for the discrete case only.

Inspired by the concept of minimum enclosing circle [6] and the probability pruning method proposed in [7], we give the definition of PT-GRkNN query in this paper.

11.3 Preliminaries

11.3.1 Uncertain Data Model

A probabilistic database D is given by a set of uncertain objects $D = \{U1, U2, \ldots, Un\}$ with d uncertain attributes [9]. An uncertain object U_i is represented by a set of d-dimensional points $\{u_1, u_2, \ldots, u_m\}$ reflecting all possible instances of U_i. Each instance u_j is assigned with a probability $p\{u_j\}$ denoting the probability that U_i appears at U_j. The probability distributions of each two objects are pairwise independent and the events of occurrence of all instances are mutually exclusive, so $\mathrm{P(Ui)} = \sum_{j=1}^{m} u_j = 1$. A possible world $\mathrm{W} = \{u_1, u_2, \ldots, u_n\}$ is a set of instances containing one instance from each object and occurring with an appearance probability of $\mathrm{P(W)} = \coprod_{i=1}^{n} \mathrm{p}(u_i)$. Let Ω denote the set of all possible worlds, then $\sum_{w \in \Omega} \mathrm{P(W)} = 1$.

11.3.2 Query Definitions

Definition 1 (GRkNN). *Given a set of data points* $\mathrm{P} \subset R^d$ *and a set of query points* $Q = q_1, q_2, \ldots, q_n$, *GRkNN query returns all points p which have* $q_i (1 \le i \le n)$

as their k nearest neighbor formally: GRkNN(Q,P) = {p|p ∈ P ∧ (∃q ∈ Q ∧ q ∈ kNN(p, P))}.

Definition 2 (PT-GRkNN). *Given a set of data points* P ⊂ R^d *and a set of query points* $Q = q_1, q_2, \ldots, q_n$, *given a threshold probability t, PT-GRkNN query returns all points p whose GRkNN probability larger than t, formally:* PT-GRkNN(Q,U,t) = {Ui|Ui ∈ U ∧ $P(GRkNN(Q, Ui) \geq t\}$.

$P(GRkNN(Q, Ui) = \sum_{w \in \Omega} P(W) \cdot I_{GRkNN(Q,W)}(Ui)$. $P(GRkNN(Q, Ui) = \sum_{w \in \Omega} P(W) \cdot I_{GRkNN(Q,W)}(Ui)$ is an indicator function, if Ui is the GRkNN of Q in possible world W, then the function value is one, otherwise it is zero.

11.4 PT-GRkNN Query Algorithm

11.4.1 Algorithm Framework

PT-GRkNN exploits the filter-refine framework: geometric filter and probability filter are designed to obtain the candidate objects. The final query results can be attained by the verification phase. In geometric filter, we do not consider the probability distribution of uncertain data; the location distribution of each uncertain object can be represented by a circle approximately. We first compute a minimum cover circle which has the minimum area among the circles that contain all the query points. Then we incrementally shrink the space of uncertain data objects. The probability filter phase can reduce the number of candidate objects further. We compute the boundary of GRkNN probability. Those objects whose GRkNN probability upper bound are less than the given probability threshold cannot be query results. In verification phase, we compute the exact GRkNN probability for each candidate object, and return those objects whose GRkNN probability is larger than the given threshold.

11.4.2 Geometric Filter

Definition 3 (Minimum Cover Circle). *The circle which has the minimum area among the circles that contains all points is called Minimum Cover Circle (MCC). In* Fig. 11.1, *Circle O is the MCC of* $\{q_1, q_2, q_3, q_4\}$.

Definition 4 (O-halfplane). $\perp(O, O1)$ *is a perpendicular bisector between point O and* $O1$, *the half-plane contains O is called O-halfplane, and the other half-plane is called* $\overline{\text{O}}$- halfplane.

We can observe from Fig. 11.1 that those points in $O1—halfplane$ cannot be the reverse nearest neighbors of the points in minimum cover circle O, because they

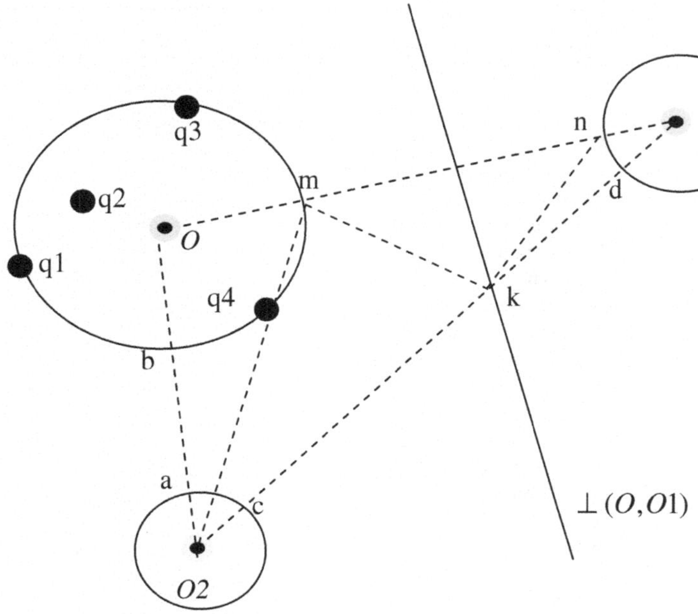

Fig. 11.1 Minimum cover circle and uncertain data circle

Fig. 11.2 Geometric filter

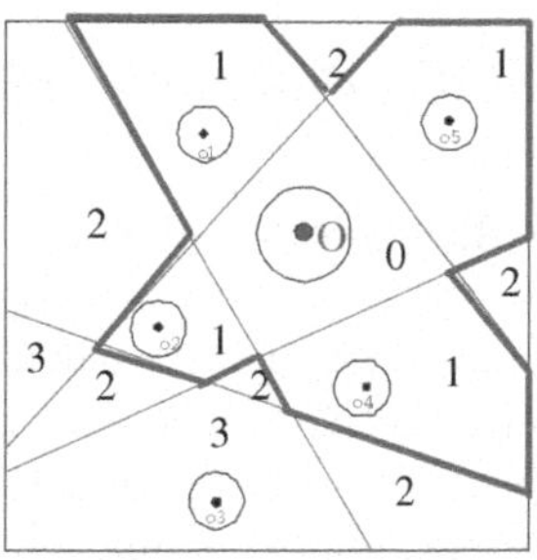

are nearer to uncertain data object $O1$ than to query circle O. Due to space limits, we omit the proof here. We design geometric filter based on this observation. We first compute the minimum cover circle of the query objects, the whole space is divided into several polygons by the perpendicular lines between uncertain object circle center and the query circle center. As shown in Fig. 11.2, the level of each polygon represents how many it locates in. If the level of polygon is less than k, that means it locates in at most $k - 1$ and it may contain the query result. Hence, the combination of all the polygons whose level less than k is the candidate, as shown in Fig. 11.2. The detail of geometric filter is shown in Algorithm 1.

Algorithm 1: G_Filter

Input: (1) Uncertain data object index tree $R*$, (2) Minimum cover circle of query objects O, and (3) k
Output: S_{cnd}

$V := \emptyset$, $S_{cnd} := \emptyset$, $SR :=$ the whole space;
H is the heap of $R*$ // increase order of the distance to circle O
while H is not empty **do**
  fetch entry (e, key) from H;
  if $e.MBR \cap SR! = \emptyset$ **then**
    if e is index node **then**
      for each node m in e **do**
        if $m.MBR \cap SR! = \emptyset$ **then**
          Insert $(m, mindist(m, O))$ into H;
    if e is leaf **then**
      for each data object n in e **do**
        if $n.MBR \cap SR! = \emptyset$ **then**
          Insert $(n, mindist(n, O))$ into H;
  if $e.MBR \cap SR = \emptyset$ **then**
    $S_{cnd} := S_{cnd} \cup e$;
    if $(key > 0)$ **then**
      $V := V \cup e$;
      Shrink the search space V;
Return S_{cnd};

11.4.3 Probability Filter

After geometric filter, the uncertain objects whose level larger than k can be pruned. However, in order to compute the GRkNN probability of each candidate, we need to enumerate all the possible worlds. So we design the probability filter to further prune the candidates. The main idea is similar to reference [8]. We use R^* tree to index the possible world of candidate objects. The index node is represented by $(key, prob)$. We exploit the dominate relation between the candidates to prune them further.

Definition 5 (Dominate Relation). *Given uncertain object* A and B, Q *is Query object,* $A \prec_Q B iff \sum_{i=1}^{d} \max(MaxDist(B_i, A_i)^2 - MinDist(B_i, Q_i)^2) < 0$.

Definition 6 (Dominate Probability). $P(A \prec_Q B) = \sum_{a_i \in A} \sum_{b_j \in B} \sum_{q_k \in Q} I(a_i, b_j, q_k) \cdot P(a_i) \cdot P(b_j) \cdot P(q_k)$. $I(a_i, b_j, q_k)$ *is an indicator functioni f the instance* a_i *of* A *dominates the instance* b_j *of* B *for query point* q_k*, then its value is 1, otherwise 0.*

In order to save the cost, we can compute the lower and upper bounds for the dominate probability approximately: $P_{LB}(GRNN(Q, U) = \sum_{Uj \in S_{cnd} \setminus U} P_{LB}(U \prec_Q U_j)$.
$P_{UB}(GRNN(Q, U) = \sum_{Uj \in S_{cnd} \setminus U} P_{UB}(U \prec_Q U_j)$

Those objects whose GRkNN probability upper bound less than the given probability threshold cannot be query results. Algorithm 2 shows the details.

Algorithm 2: P_Filter

Input: (1) Query points set Q, (2) Uncertain Objects $B \in S_{cnd}$, (3) k, and (4) t
Output: $\{-1, 1, 0\}$
$P_{LB}(GRkNN(Q, B)) = 0$, $P_{UB}(GRkNN(Q, B)) = 0$;
for all $Q' \in Q, B' \in B$ **do**
  $P_{LB}(GRkNN(Q', B')) = 1$, $P_{UB}(GRkNN(Q', B')) = 1$;
  for each $A \in S_{cnd} \setminus B$ **do**
    $P_{LB}(A_{\prec Q'} B') = 0$, $P_{UB}(A_{\prec Q'} B) = 1$;
    for each $A' \in A$ **do**
      if $A'_{\prec Q'} B'$ **then**
        $P_{LB}(A_{\prec Q'} B') = P_{LB}(A_{\prec Q'} B') + P(A')$
      if $Q'_{\prec A'} B'$ **then**
        $P_{UB}(A_{\prec Q'} B') = P_{UB}(A_{\prec Q'} B') + P(A')$
    Compute $P_{LB}(GRkNN(Q', B'))$ and $P_{UB}(GRkNN(Q', B'))$;
  $P_{LB}(GRkNN(Q, B)) = P_{LB}(GRkNN(Q, B)) + P(Q') \cdot P(B') \cdot P_{LB}(GRkNN(Q', B'))$;
  $P_{UB}(GRkNN(Q, B)) = P_{UB}(GRkNN(Q, B)) + P(Q') \cdot P(B') \cdot P_{UB}(GRkNN(Q', B'))$;
if $P_{LB}(GRkNN(Q, B)) > t$ **then**
  return 1; //B is result
if $P_{UB}(GRkNN(Q, B)) < t$ **then**
  return -1; //B is not result
return 0; //need to be verified

11.5 Experiments

In this section, we evaluate the performance of our algorithms for PT-GRkNN queries. We compare the geometric filter with TPL, and compare the probability filter with the naive method which we call baseline algorithm.

11.5.1 Performance of Geometric Filter

TPL algorithm is designed for single query point. It can be extended to a group query points by executing multiple RkNN query for each query point. However, this method does not consider the similar location among the same group. As seen from Fig. 11.3, G_Filter has a better prune capability than TPL. As seen from Fig. 11.4, with the increase of k in the query, G_Filter prunes much more uncertain objects than TPL.

11.5.2 Performance of Probability Filter

We compare our probability filter(P_Filter) with the baseline method which enumerates all the possible world to compute the GRkNN probability of uncertain objects.

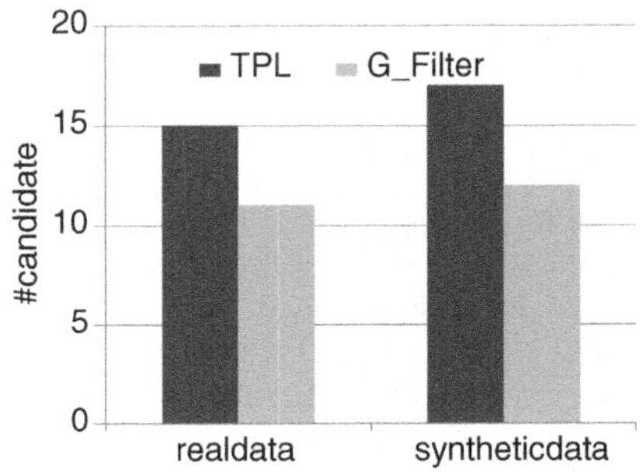

Fig. 11.3 The filter capacity of G_Filter and TPL

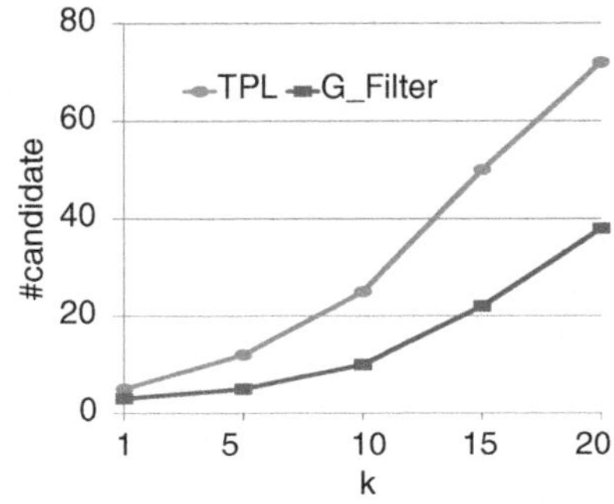

Fig. 11.4 Candidate size change with k

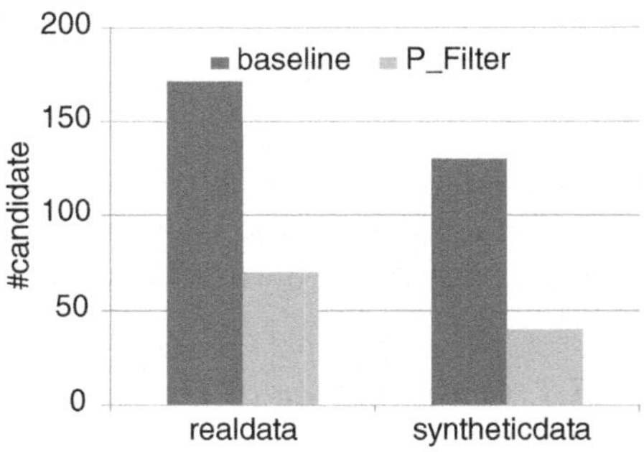

Fig. 11.5 The filter capacity of P_Filter and baseline

P_Filter method explores the dominant relation between objects and computes the probability bound, in this way it can prune much more candidates as early as possible. As seen from Fig. 11.5, P_Filter has a better prune capability than baseline method. We can see from Fig. 11.6 that the value of threshold affects the candidate size of baseline algorithm, but has not any affect on P_Filter algorithm.

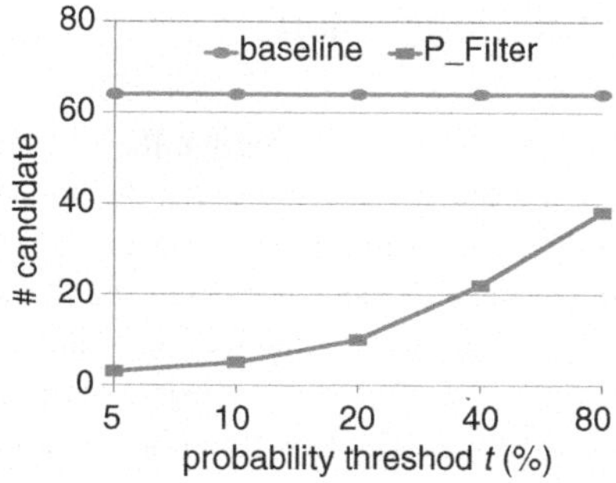

Fig. 11.6 Candidate size changes with t

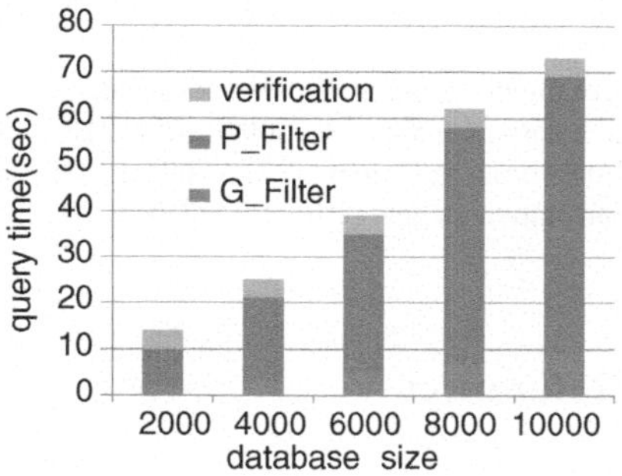

Fig. 11.7 Time cost changes with database size

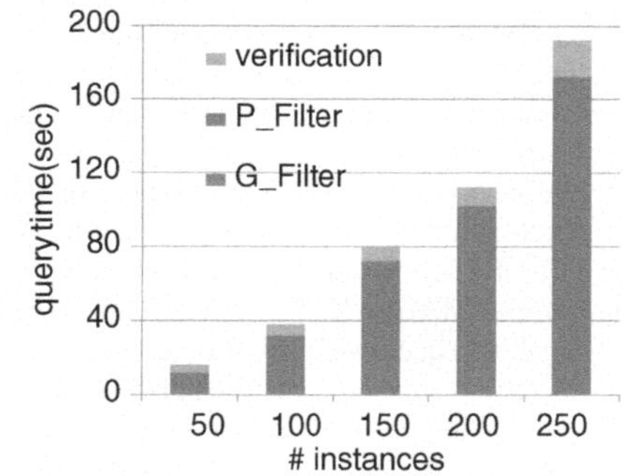

Fig. 11.8 Time cost changes with number of possible world instance

11.5.3 Time Cost

In this section, we compare the time cost of the three phases: geometric filter, probability filter, and verification. As seen in Fig. 11.7, the total query time includes three parts and it increases with the database scale. The probability filter occupied the most time. As seen from Fig. 11.8, when the instance of one uncertain object increases, the query time will increase accordingly and P_Filter also is very time-consuming. The G_Filter has little affect on the total query time in this case.

11.6 Conclusion

In this paper, we introduce the semantic of group k nearest reverse neighbor query for uncertain data and gives the solution for it. GRkNN can be used to evaluate the influences of the query objects and plays an important role in the field of location-based service, decision support and marketing. Traditional algorithms are not able to obtain satisfied results for uncertain sensor network data. The PT-GRkNN algorithm proposed in this paper consists of three phases: geometric filter, probability filter, and verification. The first two phases can prune a large amount of uncertain data objects and get the candidate set. The experimental evaluation shows that our approach is able to significantly outperform the naive approach.

Acknowledgments This work was supported in part by the National Science Foundation for Young Scholars of China (61100048, 61102105), Scientific Research Fund of Heilongjiang Provincial Education Department (12531476, 12531498), and the Research Foundation of Harbin for Youth Innovative Talents (2011RFQXG028, 2012RFQXG096).

References

1. Korn F, Muthukrishnan S (2000) Influence sets based on reverse nearest neighbor queries[C]. In: Proceedings of the 2000 ACM SIGMOD international conference on management of data (SIGMOD), Providence, Rhode Island, pp 201–212
2. Tao Y, Papadias D, Lian X (2004) Reverse KNN search in arbitrary dimensionality[C]. In: Proceedings of the 30th international conference on very large data bases (VLDB), Toronto, Canada, pp 744–755
3. Wu W, Yang F, Chan CY et al (2008) FINCH evaluating reverse K-nearest-neighbor queries on location data[C]. In: Proceedings of the 34th international conference on very large data bases (VLDB), Auckland, New Zealand, pp 1056–1067
4. Cheema MA, Lin X, Zhang W et al (2011) Influence zone: efficiently processing reverse K nearest neighbors queries[C]. In: Proceedings of the 27th international conference on data engineering (ICDE), Hannover, Germany, pp 577–588
5. Wu W, Yang F, Chan CY et al (2008) Continuous reverse K-nearest-neighbor monitoring[C]. In: Proceedings of the 9th international conference on mobile data management (MDM), Beijing, China, pp 132–139
6. Song XY, Yu CC (2010) GRkNN: group reverse nearest neighbor query in spatial space[J]. J Chin Comput 33(12):2229–2238
7. Lian X, Chen L (2009) Efficient processing of probabilistic reverse nearest neighbor queries over uncertain data. VLDB J 18(3):787–808
8. Cheema MA, Lin X, Wang W, Zhang W, Pei J (2010) Probabilistic reverse nearest neighbor queries on uncertain data. IEEE Trans Knowl Data Eng 22(4):550–564
9. Sarma AD, Benjelloun O, Halevy A et al (2006) Working models for uncertain Data[C]. In: Proceedings of the 22nd international conference on data engineering (ICDE), Atlanta, Georgia, USA, pp 1–7

Chapter 12
A Study on the Wireless Sensor Networks MAC Protocol for Aircraft Health Monitoring

Wei Zhou and Bo Jing

Abstract For the communication requirements of real-time, reliability, and energy consumption of clustering wireless sensor networks for aircraft health monitoring, we propose a modified TDMA protocol and relevant cluster members time synchronization mechanism. This modified TDMA protocol can adaptively allocate time slots based on sensor nodes data traffic, so that it can improve using efficiency of the channel, reduce the loss rate of the packets, time delay, and energy consumption. Along with the time slots, cluster head broadcasts its time stamps to sensor nodes, which adopt linear regression analysis method to estimate the relative drift and relative deviation of the clock. Then the sensor node could adjust its clock according to the head.

Keywords Aircraft · Media access control · Time synchronization · Wireless sensor networks · Time division multiple access

12.1 Related Work

Traditional aircraft sensing system has problems such as lead complexity, heavy weight, difficult and costly maintenance, line failure prone to occur, slow speed of information processing, etc. [1, 2]. Wireless sensor networks (WSNs) are composed of a large number of wireless sensor nodes according to specific communication protocols, which can compute and achieve wireless communications. Because WSNs lay emphasis on wireless communications, distributed data monitoring and processing, and wireless sensor nodes are portable and easy to be installed, all of which provide a very good way to collect data for aircraft health monitoring system [3, 4].

W. Zhou (✉) · B. Jing
PHM Lab, Aeronautic and Astronautic Engineering College,
Air Force Engineering University, Xian 710038, China
e-mail: zwz-kgy@139.com

X. Wang et al. (eds.), *Advanced Technologies in Ad Hoc and Sensor Networks*,
Lecture Notes in Electrical Engineering 295, DOI: 10.1007/978-3-642-54174-2_12,

In recent years, the research on wireless sensor networks application in aircraft condition monitoring has attracted increasing attention worldwide. In 2007, the United States Air Force started the "Advanced Subminiature Telemetry (ASMT)" program in Florida at Eglin Air Force Base aimed to develop a wireless sensor networks system for aircraft ground and flight test monitoring [5]. In 2009, "Fly-by-Wireless (FBW)" Alliance announced to largely fund four research projects to apply wireless sensor networks for Aircraft Health Monitoring; this task was assumed respectively by the NASA Langley Research Center, the State Science and Technology Commission, and the Canadian Mcgill University, etc. [6]. Up till now, a number of scholars have conducted researches to solve this kind of problem through experiments. The literature [7–9] proves that using the 2.2–2.45 GHz band, wireless sensor networks does not affect the normal work of aircraft airborne electronic equipment, neither does other airborne electronic equipment affect the normal work of wireless sensor networks. In [1] is presented a hierarchical wireless sensor networks topology for aircraft condition monitoring. Simulation results shows that this wireless sensor networks system can run efficiently. In [10–12] is constructed a WSNs prototype system for monitoring aircraft cabin environment. The preliminary experimental results show that this system can effectively monitor the aircraft cabin physical environment in real-time. Reference [13] put forward a kind of aircraft corrosion monitoring system based on low power and wireless embedded self-healing wireless sensor networks, which can efficiently handle the uncertainties in the process of fault diagnosis and prediction by the multi-sensor technology. Wireless sensor networks are used in aircraft control and health monitoring system in [14], which summarizes the previous research results in this field. In [15] is presented a kind of WSN node design method for aircraft structure health monitoring; this node adopts the thermoelectric power generation principle of power supply, and its feasibility is verified in a laboratory environment.

The study of conventional wireless sensor networks is mainly focused on the energy consumption, self-organization network topology, and positioning issues, and the wireless sensor networks for aircraft health monitoring have the following features: First, the sensor nodes in the physical space are not arbitrarily arranged, but in the vicinity of monitoring sites according to the arrangement plan of sensor layout scheme, therefore, large parts, even all, of the sensor nodes locations in the monitoring network are specified and fixed. Second, due to its relatively fixed position, the key sensor nodes of monitoring network in part or in whole can be continuously supplied by the power module, so the energy consumption is not the most difficult problem. Third, the core task of the monitoring network is to accurately acquire the data of each system state of aircraft in real-time, which requires the data of the local, or even all, correlation sensor nodes in aircraft have the same time course. Thus, wireless sensor networks for aircraft health monitoring need to focus on solving the problem of real-time, synchronization, and reliability of data monitoring issues [16], which needs the design of suitable network communication protocol. At present, most of the wireless sensor networks for aircraft health monitoring apply ZigBee communication protocol [2, 10–12]. This contest-based MAC protocol has high energy consumption, latency, and other shortcomings. In [17], the MAC layer of wireless sensor networks for health monitoring of rotorcraft structure applies the time division multiple access (TDMA) protocol, which has severe packet loss, large delay, etc.

Taking into consideration the peculiarity of the aircraft structure and airborne systems, wireless sensor networks for aircraft health monitoring should use the cluster network structure. After the nodes of the cluster-based wireless sensor networks self-organize to be clusters, the cluster member nodes send sensory data to the cluster head; this also needs to share the channel resources. Therefore, to solve the communication synchronization, reliability, energy consumption, and other issues of cluster-based wireless sensor networks, this paper proposes an improved TDMA protocol, and puts forward the corresponding one-way broadcast-based cluster clock synchronization algorithms. Unlike the typical scheduling MAC protocol, this improved TDMA protocol has obvious advantages in packet loss rate, time delay, and power consumption, which can better meet the communication needs of wireless sensor networks for aircraft health monitoring.

12.2 WSNs Architecture for Aircraft Health Monitoring

Since subsystems such as the engine, fuel, and cockpit environment in existing general aircraft are distributed in their respective regions, the aircraft body can be divided into left and right wing, vertical tail, left and right horizontal stabilizer, landing gear, front, middle, and rear sections of main body. Thus using clustering network architecture is more appropriate for the WSNs of aircraft health monitoring, with each subsystem or respective region of aircraft forming one or more clusters, and cluster head and sensor nodes in each cluster using star topology, which also coincidently meets the low delay and accurate synchronization network design requirements. Aircraft data acquisition system based on clustering WSNs is shown in Fig. 12.1.

Different types of nodes in the cluster network have different functions and tasks, so sensor nodes are distributed in the key points of the monitoring system in order to collect aircraft system status information in real-time, then send preliminarily managed information to the cluster head in accordance with the corresponding MAC protocol. Due to the limited number of key monitoring sites of the aircraft system, the sensor nodes which are corresponding to the key monitoring points can be numbered so as to locate and eliminate the fault in time. The distribution of the cluster head should first consider how the airborne power can continuously supply it. As not all sensor nodes can be supplied by the lasting airborne power, the cluster head should be laid in the appropriate position of the cluster, so that the sensor nodes that use limited power supply (battery) can the reduce energy consumption generated by the data transmission. The cluster head will send data to the Sink directly or through the neighboring cluster heads after the data are aggregated. On top of the whole network, Sink is a bridge to connect wireless sensor networks with other networks. Aircraft status information pooled by Sink, on the one hand, is sent to the cockpit through

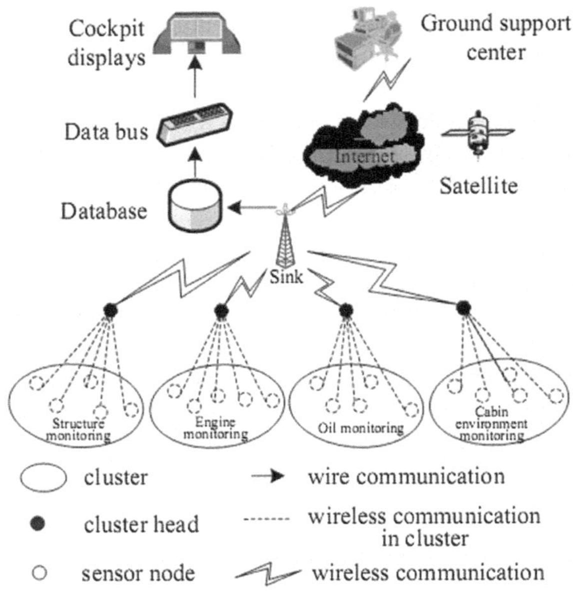

Fig. 12.1 Aircraft data collection system based on clustering WSNs

the airborne data bus to provide real-time reference for the flight crew; on the other hand, it is sent via data chain to the ground support center to provide data support for achieving aircraft prognostics and health management (PHM).

12.3 Modified TDMA Protocol Design

Since the sensor nodes of each cluster in cluster-based WSNs need to share channel resources when sending sensory data to the cluster head, the appropriate MAC protocol and clock synchronization mechanisms in the cluster are needed to solve the problem of synchronization, reliability, and energy consumption in intra-cluster communication.

12.3.1 TDMA Protocol

TDMA protocol is a kind of simple and efficient scheduling MAC protocol, which shares channel resources through time slot allocation. The node communicates with other nodes via the pre-assigned time slot scheduling. The adjacent nodes send and receive data at different time slots so as to avoid collision with neighbor nodes [18]. Meanwhile, the TDMA protocol also has an obvious disadvantage: all nodes have the same length slot allocation in a time slot, some nodes have a small flow, but still

Table 12.1 Time slot allocates request frame format of sensor node

Start-bit—Node ID—Requests slot length—Check-bit—Stop-bit

occupy the same length slot with the nodes that have a larger flow, and this may result in packet loss. This causes waste and unfair usage of channel resource, and increases the transmission delay of data.

12.3.2 Modified TDMA Protocol

For the demerit of TDMA protocol, this paper presents TDMA protocol using different time slots that can adaptively adjust the time slot length according to the node flow, thereby improving the channel utilization, reducing packet loss, and transmission delay. Time slot allocation process of the modified TDMA protocol is as follows:

Suppose the wireless channel capacity of a cluster is B; there are n working sensor nodes in the cluster, respectively S_1, S_2,..., S_n, and the occupation time slot length in the j round is $T_{1,j}$, $T_{2,j}$,..., $T_{n,j}$, wherein the time slot for sending a request to the cluster head is τ; sensor nodes data packet generation rate is k_1, k_2,..., k_n. Initially, each sensor node has the same length of time slot, i.e., $T_{1,1} = T_{2,1} = \cdots = L = T_{n,1}$.

During the j-th round, the to be transmitted message length of S_i in the message queue is L_i, packet length that does not include time slot assignment request transmit in time slot $T_{i,j}$ is $B(T_{i,j} - \tau)$. After that, the length of remaining messages in the message queue of sensor node is S_i is $L_i - B(T_{i,j} - \tau)$.

If $L_i - B(T_{i,j} - \tau) > 0$, that sensor node S_i does not send all data packets in the time slot of the current round, the next round of desired allocated time slot length is:

$$T_{i,j+1} = \left[L_i - B(T_{i,j} - \tau) + k_i T_{i,j}\right]/B + \tau \tag{12.1}$$

If $L_i - B(T_{i,j} - \tau) \leq 0$, the sensor node S_i sends all data packets in the time slot of the current round, the next round of desired allocated time slot length is:

$$T_{i,j+1} = k_i T_{i,j}/B + \tau \tag{12.2}$$

Request frame format that sensor nodes send time slot allocation request to the cluster head is shown in Table 12.1.

When the request slot length $T_{i,j+1}$ of sensor node is too large, it may cause the request time slots length of all sensor nodes to exceed the monitoring period, i.e., $\sum_{i=1}^{n} T_{i,j+1} > T_{\text{detection}}$. Therefore, the cluster head needs to rule on the request slot; the decision process is shown in Fig. 12.2.

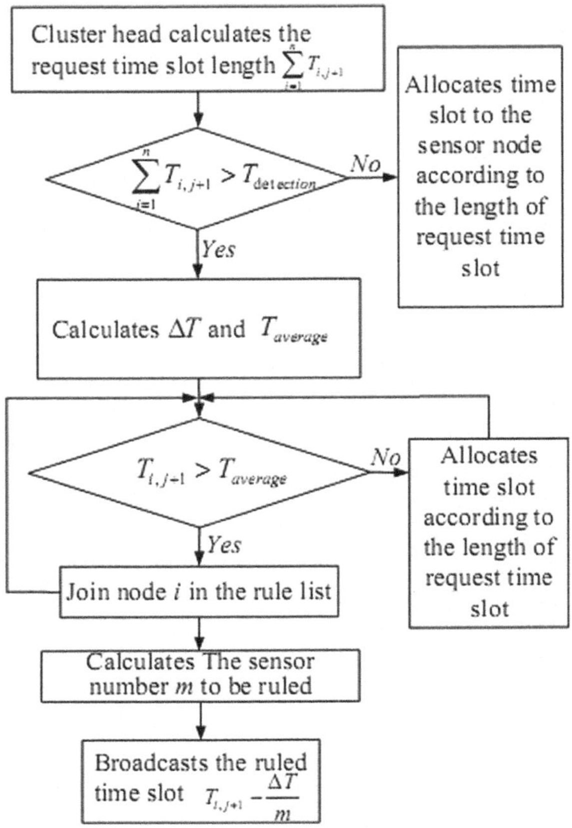

Fig. 12.2 Process of cluster head confirms the time slot of sensor node

The cluster head first calculates the request time slot length, and if $\sum_{i=1}^{n} T_{i,j+1} \leq T_{\text{detection}}$, then it allocates time slot to the sensor node according to the length of request time slot; otherwise, it calculates the difference $\Delta T = \sum_{i=1}^{n} T_{i,j+1} - T_{\text{detection}}$ between the length of the time slot request and the monitoring period, and the average request time slot length $T_{\text{average}} = \frac{1}{n} \sum_{i=1}^{n} T_{i,j+1}$.

If the request time slot length of m sensor nodes ($m \leq n$) is greater than T_{average}, then the actual time slot length distributed to node is $T_{i,j+1} - \frac{\Delta T}{m}$.

By ruling request time slots, it not only guarantees the channel resources be fully utilized, but also makes the time of a complete monitoring process to not exceed the user-defined monitoring period. Figure 12.3 is the work sequence diagram of cluster head and sensor nodes when the modified TDMA protocol completes a sensing data transmission. Sensor Node 1 transmits time slot allocation request to the cluster heads in time τ, and then goes to sleeping state. After a time of $(n - 1)\tau$, Sensor

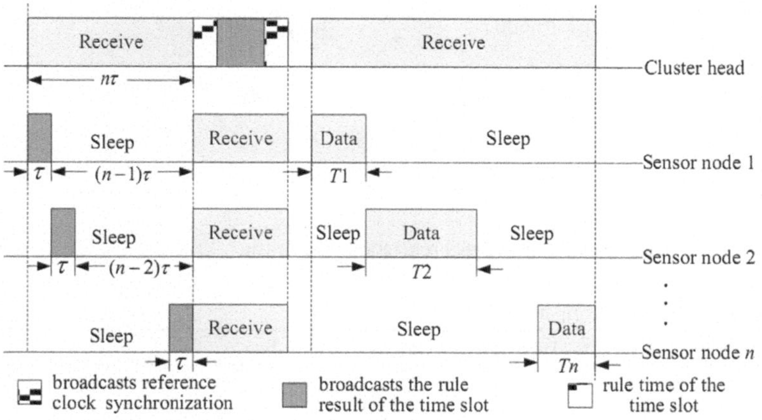

Fig. 12.3 Sequence diagram of modified TDMA protocol

Node 1 is in the receiving state, receiving the time slot allocation decisions from the cluster head. As can be seen, in the requests time slots sent from sensor nodes to the cluster head consider the impact of packet generation rate on time slot allocation, thus avoiding the network performance degradation caused by the TDMA protocol to all sensor nodes equal time slots allocation.

12.3.3 Clock Synchronization Algorithm for Modified TDMA Protocol

In wireless sensor networks, clock synchronization means all or parts of the network nodes have the same clock reference, i.e., different nodes maintain the same clock, or a node can convert other clock for local clock. Because there is drift in the local hardware clock of sensor node, deviation exists in the time course of the local task execution; this error will directly affect the performance of MAC protocol and routing protocol, which leads to a performance degradation of the whole aircraft health monitoring system.

A. **Clock Model**

1. Constant Rate Model

For a constant rate model, $r(t)$ is a constant, $r(t) = \frac{\mathrm{d}C(t)}{\mathrm{d}t}$, namely the crystal frequency is constant. When the required clock synchronization accuracy is far lower than the frequency change, the assumptions of the model are reasonable.

2. Drift Bounded Model

Typically, the clock rate $r(t)$ fluctuates within a certain range, i.e., $1 - \rho \leq r(t) \leq 1 + \rho$, where ρ is the clock drift. For low-cost network nodes, $\rho \in [1, 100]$ ppm.

3. Drift Change Bounded Model

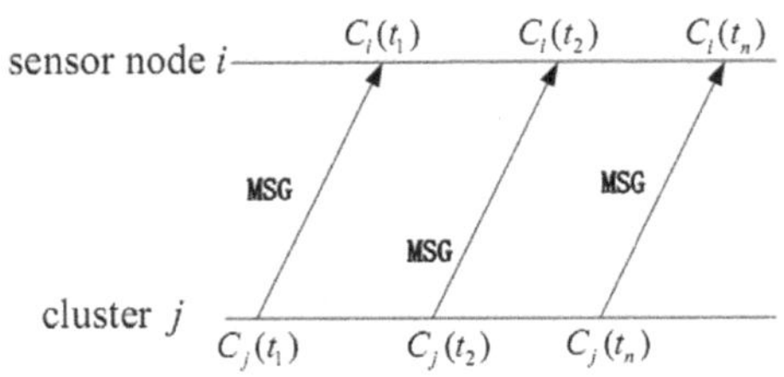

Fig. 12.4 Principle diagram of time synchronization mechanism

Table 12.2 MSG frame format

Start-bit—Cluster head ID—Time stamp—Check-bit—stop-bit

The clock drift change $\xi(t) = \frac{d\rho(t)}{dt}$ is bounded, i.e., $-\xi_{\max} \leq \xi(t) \leq \xi_{\max}$.

$C_i(t)$ and $C_j(t)$ mean the local time value for nodes i and j at time t. If the constant rate model is adopted , $C_i(t)$ can be expressed as $C_i(t) = a_i t + b_i$. As can be seen, the clock of any two nodes i and j, meet the following linear relationship:

$$C_i(t) = \frac{a_i}{a_j} C_j(t) + (b_i - \frac{a_i}{a_j} b_j) \tag{12.3}$$

In Formula (12.3), order $\frac{a_i}{a_j} = a_{ij}$ and $b_i - \frac{a_i}{a_j} b_j = b_{ij}$, where a_{ij} and b_{ij} are relative drift and relative offset of the two node clock. When $C_i(t) = C_j(t)$, the clock of the two nodes keeps synchronous, i.e., $a_{ij} = 1$ and $b_{ij} = 0$. Therefore, for the occasion of the constant rate model, clock synchronization is to estimate (a_{ij}, b_{ij}), and then adjust the local clock of the node, so that two nodes in the new time will keep the clock synchronized.

B. **Clock Synchronization Algorithm**

In the modified TDMA protocol proposed, when sensor nodes are in the conversion state of work, what is needed is to keep clock synchronization between cluster head and sensor nodes within the cluster. For this point, we put forward a kind of cluster clock synchronization algorithm based on one-way broadcast. Specific thinking path is like this: when a cluster head broadcasts time slot verdict to the nodes in the cluster, if the clock drift or clock shifts between the sensor node clock and the cluster head reference clock does not satisfy the synchronization accuracy requirements, the time stamp of the cluster head broadcasts to the sensor nodes in the cluster so that the sensor node can maintain clock synchronization with the cluster head, thus achieving the synchronization between the sensor nodes; the principle is shown in Fig. 12.4.

The cluster head broadcasts synchronization message MSG to the sensor nodes in the cluster; MSG frame format is shown in Table 12.2, containing a time stamp $C_j(t)$ of a cluster head.

After sensor nodes receive time synchronization message, with its time information $C_i(t)$, the linear regression method is used to estimate relative drift a_{ij} and relative offset b_{ij}, thereby adjusting the local clock accordingly. The cluster head can repeatedly broadcast MSG to the sensor nodes in the cluster at different times to improve the precision of regression analysis.

C. **Regression Analysis of Relative Drift and Relative Offset**

Considering the distance between cluster head and its cluster nodes is small in the space of the wireless sensor networks for aircraft health monitoring, the time that synchronous message MSG transmits between the cluster head and its cluster nodes can be ignored. Due to the linear correlation of $C_i(t)$ and $C_j(t)$, if point set $\{(C_i(t_k), C_j(t_k))|1 < k \leq n\}$ is available, then the linear regression method can be used to estimate a_{ij} and b_{ij}.

Assuming that the cluster head number is *j*, when the one-way broadcast synchronization occurs, the cluster head broadcasts synchronous message that contains a time stamp $C_j(t_k)$ to the cluster nodes in the cluster at time $C_j(t_k)$. When Sensor Node *i* in the cluster receives the synchronous message in time $C_i(t_k)$, Sensor Node *i* gets the reference clock and its own clock attribute pairs $(C_i(t_k), C_j(t_k))$.

When estimating a_{ij} and b_{ij} by using the linear regression method to meet the principle of minimum variance, Formula (12.4) takes the minimum.

$$R = \sum_{k=1}^{n} [a_{ij} C_j(t_k) + b_{ij} - C_i(t_k)]^2 \tag{12.4}$$

Therefore, the partial derivative of R about a_{ij} and b_{ij} should be 0:

$$\frac{\partial R}{\partial a_{ij}} = 2\left[a\sum_{k=1}^{n} C_j^2(t_k) + b\sum_{k=1}^{n} C_j(t_k) - \sum_{k=1}^{n} C_j(t_k)C_i(t_k)\right] = 0 \tag{12.5}$$

$$\frac{\partial R}{\partial b_{ij}} = 2\left[a\sum_{k=1}^{n} C_j(t_k) + nb - \sum_{k=1}^{n} C_i(t_k)\right] = 0 \tag{12.6}$$

Formulas (12.5) and (12.6) can be rewritten into a matrix form:

$$\begin{bmatrix} \sum_{k=1}^{n} C_j^2(t_k) & \sum_{k=1}^{n} C_j(t_k) \\ \sum_{k=1}^{n} C_j(t_k) & n \end{bmatrix} \begin{bmatrix} a_{ij} \\ b_{ij} \end{bmatrix} = \begin{bmatrix} \sum_{k=1}^{n} C_j(t_k)C_i(t_k) \\ \sum_{k=1}^{n} C_i(t_k) \end{bmatrix} \tag{12.7}$$

Accordingly, we have:

$$a_{ij} = \frac{n\sum_{k=1}^{n} C_j(t_k)C_i(t_k) - \sum_{k=1}^{n} C_i(t_k)\sum_{k=1}^{n} C_j(t_k)}{n\sum_{k=1}^{n} C_j^2(t_k) - \left[\sum_{k=1}^{n} C_j(t_k)\right]^2} \tag{12.8}$$

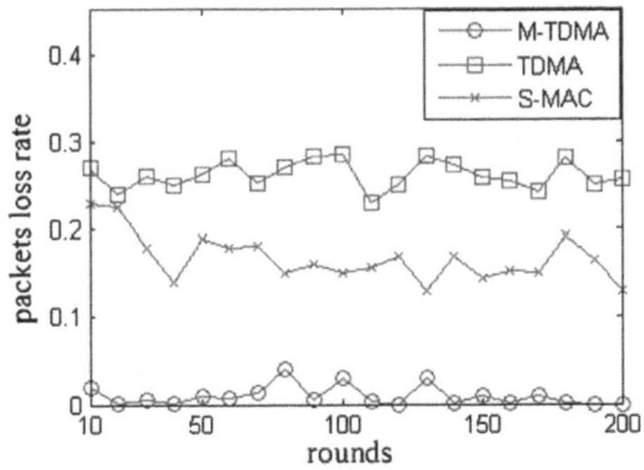

Fig. 12.5 Packets loss rate of the modified TDMA protocol

$$b_{ij} = \frac{\sum_{k=1}^{n} C_j^2(t_k)C_i(t_k) - \sum_{k=1}^{n} C_j(t_k)\sum_{k=1}^{n} C_j(t_k)C_i(t_k)}{n\sum_{k=1}^{n} C_j^2(t_k) - \left[\sum_{k=1}^{n} C_j(t_k)\right]^2} \tag{12.9}$$

The forms of a_{ij} and b_{ij} include four kinds of summation operation types: $\sum_{k=1}^{n} C_i(t_k)$, $\sum_{k=1}^{n} C_j(t_k)$, $\sum_{k=1}^{n} C_j(t_k)C_i(t_k)$ and $\sum_{k=1}^{n} C_j^2(t_k)$. Therefore, using linear regression method to estimate the relative drift and the relative offset just takes five storage units, four of which are used to preserve the intermediate results of summation arithmetic, and the last one is used to save the time stamp in the receive buffer of wireless chip.

12.4 Simulation of Modified TDMA Protocol

With the actual space of the aircraft system considered, the simulation parameters is set: the cluster radius $R = 20\,\text{m}$, communication radius $R_s = 15\,\text{m}$, the number of sensor nodes within the cluster is 16, the monitoring cycle is 1.5 s, the duty cycle of S-MAC protocol is 10 %, the flow of each sensor node is the random number in [0, 1] interval. Respectively compare the performance of S-MAC protocol, TDMA protocol, and the modified TDMA (M-TDMA) protocol in packet loss rate, time delay, and energy consumption, statistics once every 10 rounds and simulate a total of 200 rounds. The results are shown in Figs. 12.5, 12.6, and 12.7.

As can be seen from Fig. 12.5, the modified TDMA protocol is better than the performance of S-MAC protocol and TDMA protocol in the packet loss rate, because the modified TDMA protocol could adapt dynamic allocate time slots according to the flow of sensor nodes. But not all the packet loss rate reaches 0, which is mainly because in some rounds, all the request time slots lengths of the sensor nodes within a cluster are larger than the monitoring period, resulting in packet loss of some nodes.

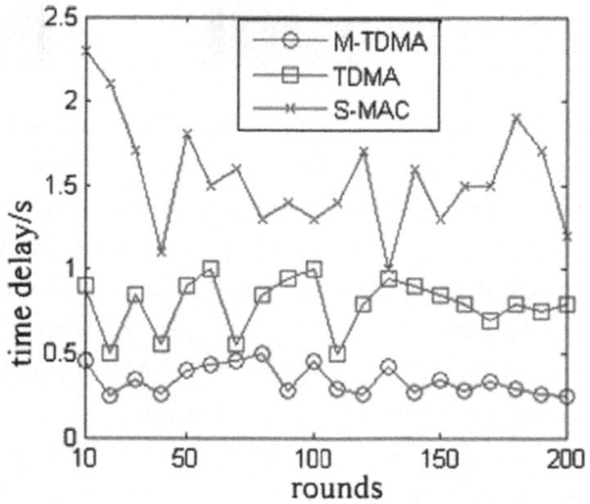

Fig. 12.6 Time delay of the modified TDMA protocol

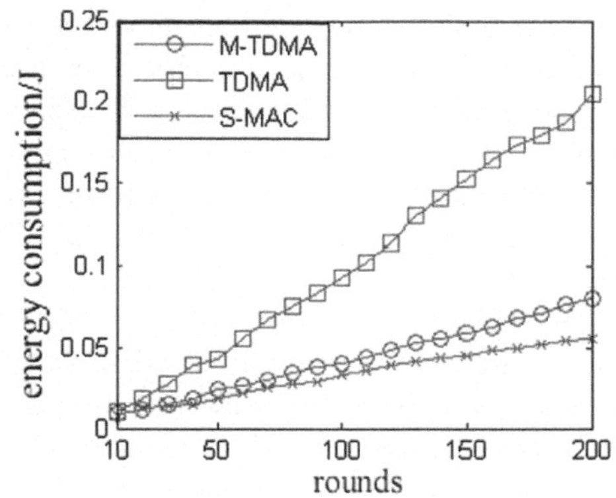

Fig. 12.7 Energy consumption of the modified TDMA protocol

As can be seen from Fig. 12.6, compared to the same time slots TDMA protocol allocates to all sensor nodes, this modified TDMA protocol reduces the time delay by introducing adaptive time slots allocation mechanism. The S-MAC protocol greatly increases the time delay by introducing periodic sensing sleep mechanism; the node cannot send and receive packets when sleeping.

Figure 12.7 shows that the S-MAC protocol has the minimum energy consumption owing to the energy-saving advantage of low duty cycle mechanism for S-MAC protocol, when the data generation rate is slow and the data network is idle, nodes can timely enter dormant state, thus greatly saving energy. According to the modified TDMA protocol in this paper, on the one hand, the sensor node communicates with the cluster head according to the pre-assigned time slot scheduling, neighboring nodes send, and receive data at different time slots to avoid the channel conflict with neighbor nodes; on the other hand, due to the traffic adaptive allocation mechanism, it reduces unnecessary idle listening. Therefore, compared with the TDMA protocol, this modified TDMA protocol has a significant reduction in energy consumption.

12.5 Summary

In this paper, clustering wireless sensor networks is used in aircraft health monitoring. Clustering wireless sensor networks for aircraft health monitoring must meet the communications requirement, such as real-time, reliability, and energy consumption. This paper proposes a scheduling MAC protocol of modified TDMA, which can adaptively allocate time slots based on sensor nodes traffic, so that it can improve usability of the channel, reduce the loss rate of the packets, time delay, and energy consumption. Compared with the S-MAC protocol, TMDA protocol, despite its energy consumption being greater than the S-MAC protocol, this modified TDMA protocol has advantages of packet loss rate and time delay, which can better meet the communications requirements for wireless sensor network of aircraft health monitoring. Given that scheduling MAC protocols require clock synchronization mechanism to coordinate the work state transitions between nodes, this paper proposes a kind of cluster clock synchronization algorithm based on one-way broadcast that is suitable for this modified TDMA protocol. When the cluster head broadcasts time slot decisions to the sensor nodes, it also broadcasts synchronous messages that contain the cluster head clock. After receiving this message, nodes use linear regression method to estimate the relative drift and the relative offset, and adjusts its clock according to this.

Wireless sensor networks provide a flexible, portable, and reliable distributed data acquisition mode for aircraft health monitoring, which needs to focus on the design of intelligent sensor nodes, network protocols, data processing technology, and other aspects of the research. About the network protocol, the next step is to focus on the communication protocol between each cluster head and between the cluster head and the Sink.

Acknowledgments This work is supported by the National Natural Science Foundation of China, under grant number 51201182 and the Aeronautical Science Foundation of China, under grant number 20101996012. The authors gratefully acknowledge the support.

References

1. Notay JK, Safdar GA (2011) A wireless sensor network based structural health monitoring system for an airplane. In: Proceedings of the 17th international conference on automation and computing, University of Huddersfield, pp 240–245
2. Ma J-C, Peng J-T, Zhang G-Q et al (2009) Wireless sensor network system for aircraft condition monitoring. Measur Control Technol 28(11):6–13
3. Sun L-M, Li JZ, Chen Y (2005) Wireless sensor networks. Tsinghua University Press, Beijing
4. Jing Bo, Zhang Jie, Sun Yong (2011) Smart network sensor and wireless sensor networks. National Defense Industry Press, Beijing
5. Musteric SA, Berdugo A (2007) The advanced subminiature telemetry system (ASMT): a wireless, network based, miniaturized instrumentation system. U. S. Air Force T and E Days. Florida, AIAA, pp 2007–1641

6. Aerospace safety enhancements at core of four projects selected by the CANEUS fly-by-wireless consortium, http://www.caneus.org/fbw/content.aspx?id=2009_home
7. Steven M, Albert B (2007) The advanced subminiature telemetry system (ASMT): a wireless, nonintrusive, network based, Instrumentation system. Test instrumentation workshop
8. Jianhua L, Demirkiran I, Yang T, Helfrick A (2009) Feasibility study of IEEE 802.15.4 for aerospace wireless sensor networks. In: Proceedings of digital avionics systems conference, IEEE/AIAA 28th, pp B3–10
9. Armstrong NL, Antar YMM (2008) Investigation of the electromagnetic interference threat posed by a wireless network inside a passenger aircraft. IEEE Trans Electromagn Compat 50(2): 277–284
10. Pook M, Loo SM, Kiepert J (2012) Monitoring of the aircraft cabin environment via a wireless sensor network. In: Proceedings of 42th international conference on environmental systems, AIAA 2012–3462, California
11. Kiepert J, Loo SM, Klein D et al (2011) Wireless sensor networks for aircraft cabin environmental sensing. In: Proceedings of 41st international conference on environmental systems, AIAA 2011–5108, Portland
12. Hall JA Jr, Loo SM, Stephenson D et al (2012) A portable wireless particulate sensor system for continuous real-time environmental monitoring. In: Proceedings of 42nd international conference on environmental systems, AIAA 2012–3441, California
13. Demo J, Steiner A, Friedersdorf F et al (2010) Development of a wireless miniaturized smart sensor network for aircraft corrosion monitoring. In: Proceedings of aerospace conference, 2010 IEEE, pp 1–9
14. Yedavalli RK, Belapurkar RK (2008) Application of wireless sensor networks to aircraft control and health management systems. J Control Theory Appl 9(1):28–33
15. Becker T, Kluge M, Schalk J et al (2009) Autonomous sensor nodes for aircraft structural health monitoring. Sens J 9(11):1589–1595
16. Yuan S-F, Qiu L, Wu J et al (2009) Challenge in structural health monitoring of large aircraft development. Aeronaut Manufact Technol 22:62–67
17. Arms SW, Townsend CP, Galbreath JH et al (2011) Flight testing of wireless sensing networks for rotorcraft structural health and usage management systems. In: Proceedings of 14th Australian International Aerospace Congress
18. Kredo K, Mohapatra P (2007) Medium access control in wireless sensor networks. Comput Netw 51:961–994

Chapter 13
A Multi-channel MAC Protocol with High Throughput for Wireless Sensor Networks

Bin Cheng, Linlin Ci, Chengping Tian and Taolin Guo

Abstract In wireless sensor networks (WSN), sensor nodes are hardly recharged, so energy efficiency is always considered to be of primary importance. However, recent applications, such as health monitoring applications, require high throughput to transmit sensing data in a burst. Aiming to maximize the integral throughput, we propose a new scheduled multi-channel MAC protocol, named LS-MAC, which adopts a fully distributed scheduling scheme through locally selecting and updating the time slot/channel pair information. Each node is assigned a time slot to transmit on a particular channel, making LS-MAC run efficiently without collisions under high-load traffic scenarios. Extensive simulation results on Glomosim platform show that LS-MAC exhibits higher bandwidth utilization and integral throughput without sacrificing energy efficiency.

Keywords MAC · Multi-channel · Wireless sensor networks · Glomosim

13.1 Introduction

Wireless sensor networks have been widely used in environmental monitoring, intrusion detection and location, medical detection, industrial service, and so on. As a kind of energy constrained self-organizing networks, especially in harsh deployment environments, the energy efficiency becomes the bottleneck of WSN. Existing MAC protocols, such as S-MAC [1], introduces the duty cycling technology to reduce energy consumption and prolong network lifetime. However, other performance metrics, such as end-to-end delivery latency, throughput, and fairness, undergo different degrees of negative impacts.

B. Cheng (✉) · L. Ci · C. Tian
School of Computer Science and Technology, Beijing Institute of Technology, Beijing, China
e-mail: chengbinpaper@gmail.com

T. Guo
Shandong University of Science and Technology, Shandong, China

X. Wang et al. (eds.), *Advanced Technologies in Ad Hoc and Sensor Networks*, Lecture Notes in Electrical Engineering 295, DOI: 10.1007/978-3-642-54174-2_13,

Every MAC protocol designed for WSN cannot perform well in every scenario. For example, constant change in environment can lead to many problems such as frequent sampling and burst transmission. For example, in order to timely detect abnormality of patients, nodes in health monitoring applications need to sample 500 times per second [2]. In addition, some multimedia WSNs equipped with an embedded camera and microphone used for sensing multimedia information require higher throughput than traditional applications.

Two major reasons leading to the decline of throughput: low bandwidth reuse rate caused by interference; waste of bandwidth caused by half-duplex wireless transceiver. Fortunately, multi-channel technology can well overcome the problem of adjacent parallel transmission interference in single-channel protocols.

In order to improve throughput performance in WSN, we propose a multi-channel MAC protocol, named LS-MAC. Traditional MAC protocols usually adopt global scheduling strategy, in which the sink is responsible for scheduling. It should get all the status information of each node among the entire network to uniformly decide how to allocate the time slot/channel, leading to high control cost. However, LS-MAC uses a fully distributed scheduling scheme, in which nodes exchange relevant information taken up by the slot/channel within the scope of sensing range, then each node can locally get a available free time slot/channel vector through bitwise operation. When the network topology changes, nodes can update their bitwise information to avoid potential conflicts.

This paper is organized as follows: Sect. 13.2 provides related researches on multi-channel MAC protocols; Sect. 13.3 gives the detailed design of LS-MAC; performance evaluations are given in Sect. 13.4; finally, Sect. 13.5 concludes the work.

13.2 Related Work

Recently, scholars have conducted numerous researches to improve WSN performance involving MAC protocol, routing protocol, and transmission protocol optimization. In the following, we will focus on analyzing and discussing current researches on multi-channel MAC protocols.

Based on whether to adopt one channel or multiple channels, MAC protocols can be divided into two categories: single-channel and multi-channel MAC protocols. In single-channel MAC protocols, for example, S-MAC [1], B-MAC [3], TMAC [4], and so on, control message exchange and real data transmission both perform on one common channel, which largely reduces channel utilization and throughput. To improve throughput, Z-MAC [5] introduced an efficient scheduling scheme in which a specified interval is allocated to each node to avoid conflicts. However, the improvement is very limited, owing to only one available channel for use.

With the emergence of wireless transmission module supporting for multiple bands, such as CC2420 wireless communication module used by MICAz and Telos nodes providing 16 independent channels with 5 MHz interval, multi-channel MAC protocols can realize parallel transmission over different channels, which greatly enhances the throughput. Thus multi-channel MAC protocols are one of the current research hotspots in the WSN field. Zhou et al. proposed a multi-channel MAC protocol, named MMSN [6], which divides time frame into several time slots based on traditional CSMA protocol, before real data transmission, neighboring nodes compete for channels fairly; then data transmission performs on designated time pots. The number of available channels is a bottleneck. When data transfer begins, each node must frequently switch between its own channel and the channel occupied by its destination to build connection. The frequent switching largely increases end-to-end delivery latency. In addition, MMSN selects one dedicated channel for broadcast transmission and sets aside a special field for broadcast operations. Although this makes broadcast transmission perform well, low channel utilization also restricts the throughput improvement. For the above shortcomings of MMSN, LS-MAC introduces a scheduling mechanism during medium access, making nodes complete data transmission in specified slots, which reduces energy waste caused by conflicts and retransmissions. On basis of the characteristics of data collection applications, Wu et al. proposed TMCP [7], aiming at the tree-based structure applications. It divides the whole tree-based network into several sub-trees, and assigns each subtree a orthogonal channel. Thus all sub-trees can transmit in parallel without interference. Although this kind of coarse-grained channel assignment method can overcome the shortage of available channels, the problem of inner-tree interference cannot be ignored, especially in large-scale dense networks. Moreover, TMMAC [8] and MCMAC [9] assign different channels to nodes within two hops to avoid potential interferences. However this node-level assignment mechanism brings more latency owing to frequent channel switching. LS-MAC proposes an adaptive distributed strategy to adjust node scheduling locally, reducing control overhead and improving network throughput.

13.3 LS-MAC Protocol Design

13.3.1 Basic Framework

In order to realize precise scheduling, the clock synchronization must first be solved. When the synchronization finishes, nodes choose appropriate slot/channel pairs according to certain optimization strategy to realize parallel transmission without conflicts. The basic framework of LS-MAC is shown in Fig. 13.1.

13.3.2 Initialization

After placing the nodes in a designated area, all nodes turn on radio, enter the initialization state, and this status continues until it successfully receives the required related control messages of clock synchronization. In addition, when battery replacement

Fig. 13.1 Basic framework of LS-MAC

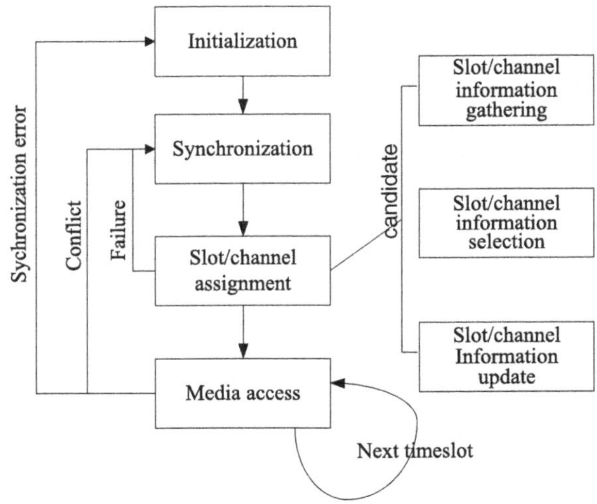

and resetting happen, leading networking topology changes. Nodes will turn back into the initialization state. Nodes in this state will continue to sample activities of channel so as to obtain relevant synchronization messages to complete clock synchronization. When a node receives relevant control messages, it turns into the clock synchronization state.

13.3.3 Synchronization

An unavoidable problem during synchronization is clock drift. Clock drift leads transceiver node cannot establish communication link. When the network runs for a long time, nodes will compare the number of current time slots and the time frame number contained in its received control message, in oder to check whether clock synchronization has error. If they are not equal, nodes will jump into initialization state for clock synchronization. Before starting and finishing data reception, LS-MAC sets protection interval of two crystal vibration cycle lengths, avoiding inconsistencies caused by clock drift , while ensuring that receiving nodes can be awakened before sending nodes transmit data packets, solving the problem of "early to sleep" caused by synchronization error.

13.3.4 Time Slot/Channel Pair Selection

When a node tends to join the network and transmit data, it must randomly select a time frame to get an available idle slot/channel pair, this random strategy can significantly reduce conflicts. Once the time frame is selected, nodes jump into slot/channel selection phase. The main task in this phase is to explore and collect information about time slot/channel pairs having been occupied in neighborhood, and this operation will last for the entire time frame. Each node can establish a conflict-free time slot/channel pair list and select an available slot/channel pair from the list. This section can be subdivided into three parts: discover, select, and update. Before presenting three operations, we give the definition of idle slot/channel pair. Before real-data transmission, nodes enter the idle time slot/channel exploratory stage in order to obtain free time slot/channel pair. The three essential judgment rules for idle slot/channel pair are as follows:

(1) Nodes cannot detect any data transfer activities and carriers: Data transfer activities and carriers both indicate that medium is busy in current time slot/channel. Hence, each node in the time slot/channel discovery phase should exclude all slot/channel pairs with data activities and carrier. This rule is similar to the RTS in CSMA/CA.
(2) Potential receiving nodes are not receiving data packets from other nodes: Potential receiving nodes should transmit control messages including the usage of time slot/channel, so nodes can exclude those time slot/channel pairs being occupied. This principle is similar to CTS in CSMA/CA.
(3) Neighbors within single-hop scope cannot hold the same time slot: For the half-duplex communication apparatus of transceiver, if adjacent nodes in exploring state occupy different channels in the same time slot, nodes cannot receive all the information of slot/channel pairs from two different channels at the same time, leading to the loss of information on slot/channel pair.

13.3.5 Selection and Update

In previous section, nodes can get the usage of slot/channel within two-hop range. How to select a idle slot/channel pair? LS-MAC follows a principle that, within a single hop range adjacent nodes cannot occupy the same time slot, regardless of whether their channels are the same or not. When in the slot/channel list exists two unoccupied time slot/channel pairs, nodes select the first idle slot/channel pair in our work. Of course, nodes can randomly select from the available time slot/channel pairs. The example shown in Fig. 13.2 can explain the whole process of "slot/channel" selection. We mark the new member as red node X, and the other eight nodes' vectors, used for recording the information of idle "slot/channel" pairs, have been given during the initialization phase. Divide each frame into five time slots (the number of time slots is equal to the maximum node density). We assume there exist two orthogonal channels, which are denoted as CH_1 and CH_2, respectively. So every node holds two bitwise vectors to record idle time slots on each channel. The value of each bit stands for whether the corresponding time slot is occupied. 1 means busy and 0 means idle. Through control messages exchange of time slot/channel pairs with adjacent nodes

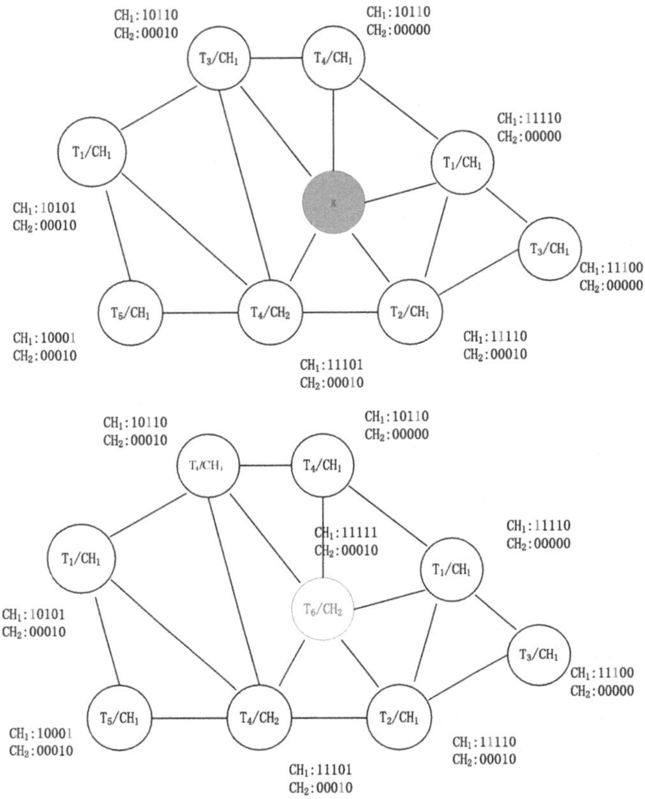

Fig. 13.2 The slot/channel pair selection of LS-MAC

A, B, C, D, E, node X can obtain through OR operation on CH_1, getting that no idle time slots are available except (CH_2,T_5). So red node selects (CH_2,T_5).

13.4 Performance Evaluation

In this section, we evaluate LS-MAC protocol on platform Glomosim. Performance evaluation mainly performs in terms of throughput, packet delivery rate, latency, and energy efficiency, compared with MMSN, TMCP, and CSMA (without RTS/CTS handshake) under the condition of varying the number of channels. All protocols run for 1,000 times, then calculate average results, and set a 95 % confidence interval. All source nodes initiate CBR (continuous bit rate) streams toward the sink and each node generates a packet each 2s (if nodes transmit more frequently, buffer overflow occurs). This scenario can be considered as a periodic data collection. In these simulations all raw data packets from sources are relayed toward the sink. The number of channels varies from 1 to 10, and 100 nodes are randomly deployed in $150 \times 150\,\text{m}^2$ area.

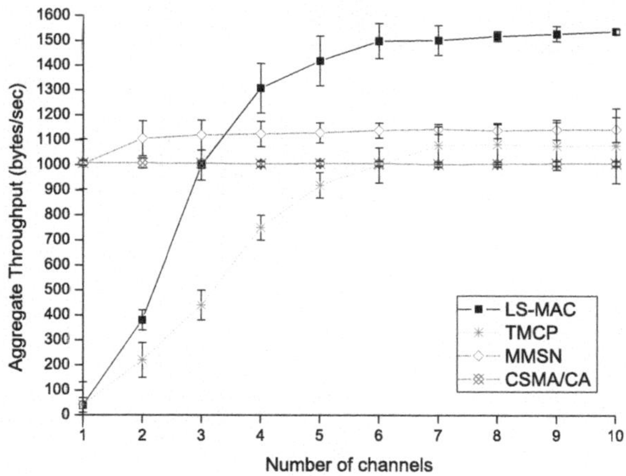

Fig. 13.3 Aggregate throughput versus number of channels

Figure 13.3 shows simulation results in terms of aggregate throughput when the number of available channels varies. In this experiment, we calculate the aggregate throughput by counting the number of packet bytes per second at the sink, and we divide each frame into 32 time slots. From the different curves presenting different protocols, we can easily get that CSMA cannot benefit from multiple channels, so its corresponding curve keeps flat when the number of available channels changes. Like [10], MMSN achieves higher throughput than other protocols except LS-MAC when the number surpasses 4. This can verify that distributed slot/channel assignment mechanism performs better. When the number of channels is lower than 7, TMCP runs at a higher speed of the increment in throughput than MMSN and CSMA/CA. This is because networking partition makes multiple data streams run in parallel, however, the network partition scheme has its own shortcoming of when the scale of network is fixed, the network reaches saturation even the number of available channels keeps increasing, as interference and conflicts among different sub-regions degrade performances. As shown in this experiment, the threshold is 7 and the curve of TMCP keeps flat.

Figure 13.4 shows simulation results in terms of packet delivery rate when the number of available channels varies from 1 to 10. To compare with MMSN, TMCP, and CSMA/CA, LS-MAC also does not conclude the acknowledgment mechanism of successful packet reception. As shown in Fig. 13.3, all the trend of three curves are similar to curves of aggregate throughput in Fig. 13.4. In LS-MAC, more channels make more nodes get free time slot/channel pairs, so the packet delivery rate increases rapidly. The MMSN curve nearly maintains flat around 64–73 % when the number of channels lies between 6 and 10, as it reaches saturation. The increasing speed of TMCP changes rapidly between 1 and 5, but decreases from 6 to 7, because more channels make the whole network be partitioned into more sections. The connectivity of network cannot be guaranteed.

The end-to-end delivery latency not only includes the data transmission time, but also contains time consumed during the control message exchange. As shown in

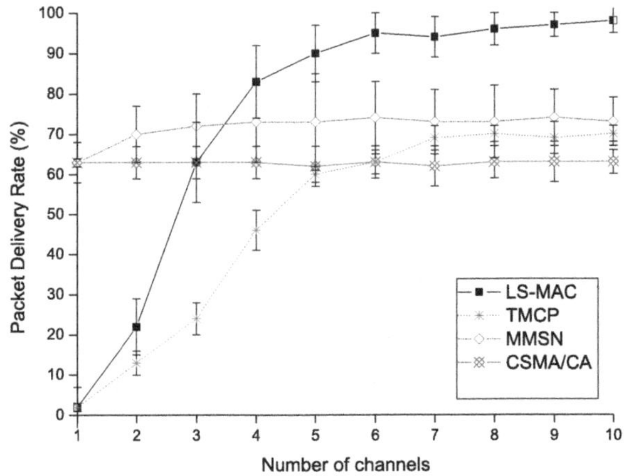

Fig. 13.4 Packet delivery rate versus number of channels

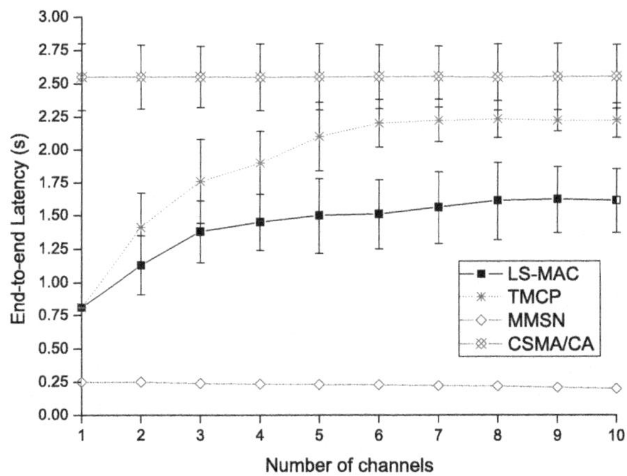

Fig. 13.5 Average end-to-end latency versus number of channels

Fig. 13.5, LS-MAC always performs significantly better than CSMA/CA and TMCP, but poorly than MMSN. The reason is that when one node wants to do transmission, it must wait for some extra time to get a free slot/channel pair. This mechanism brings more latency. As CSMA is a contention-based protocol, excessive contentions lead to exponential back off and retransmission operation. MMSN has much lower delay compared to the MC-LMAC protocol, owing to this higher latency is an inner characteristic of schedule-based protocols. If a node wants to transmit data, it has to wait for its assigned slot. The average delay from source nodes to the sink is equal to a frame. Figure 13.6 shows the energy efficiency when the number of channels changes. Energy efficiency is a major factor in wireless sensor networks. In this simulation, we evaluate the average energy consumption per byte of packet, which equals the value of the total energy consumed divided by the total number of successfully delivered packets. As the number of channels increases, both LS-LMAC and MMSN spend less energy than CSMA. LS-MAC largely outperforms the other three protocols in

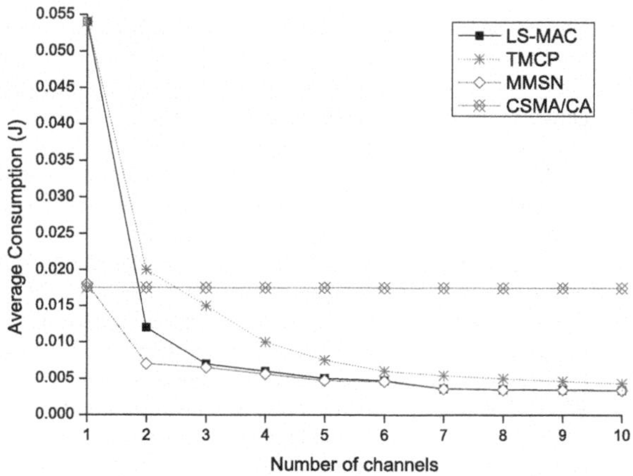

Fig. 13.6 Energy efficiency versus number of channels

terms of energy efficiency, owing to its distributed manner during slot/channel pair assignment, and less time on idle listening. But the number of attempts per transmission is much lower since some nodes cannot occupy an available free time slot due to interferences and contentions within common branches. Through extensive simulations, LS-MAC can better save energy in course of channel assignment under non-static application scenarios.

13.5 Conclusions

In order to improve the performance of WSN, this paper presents an efficient multi-channel MAC protocol named LS-MAC, which proposes a fully distributed scheduling mechanism to optimize the performance of traditional multi-channel MAC protocols. LS-MAC adopts the distributed strategy to efficiently complete conflict-free scheduling without sacrificing energy efficiency, and achieving higher packet rate and throughput. Compared with the three other classic MAC protocols, it can efficiently solve the interference and collision problems in dense wireless sensor networks, significantly improving network performance.

Acknowledgments Thanks to the support of the National Natural Science Foundation of China (61063042) and the Post-Doctoral Foundation of China (201104753).

References

1. Wei Y, Heidemann J, Estrin D (2002) An energy-efficient MAC protocol for wireless sensor networks. In: Paper presented at 21th annual joint conference of the IEEE Computer and Communications Societies, vol 3. IEEE Press, New York, pp 1567–1576

2. Chintalapudi K, Fu T, Paek J, Kothari N, Rangwala S, Caffrey J, Govindan R, Johnson E, Masri S (2006) monitoring civil structures with a wireless sensor network. Internet Comput 10:26–34
3. Van DT, Langendoen K (2003) An adaptive energy-efficient MAC protocol for wireless sensor networks. In: Paper presented at the 1st international conference on embedded networked sensor systems, ACM Press, Los Angeles, pp 171–180
4. Keshavarzian A, Lee H, Venkatraman L (2006) Wakeup scheduling in wireless sensor networks. In: Paper presented at the 7th ACM international symposium on mobile ad hoc networking and computing. ACM Press, Florence, pp 322–333
5. Rhee I, Warrier A, Aia M, Min J, Sichitiu ML (2008) Z-MAC: a hybrid MAC for wireless sensor networks. IEEE/ACM Trans Netw 16(3):511–524
6. Zhou G, Huang C, Yan T, He T, Stankovic JA, Abdelzaher TF (2006) MMSN: multi-frequency media access control for wireless sensor networks. In: Paper presented at 25th IEEE international conference on computer communications. vol 6. IEEE Press, Barcelona, pp 1–13
7. Wu Y, Stankovic JA, He T, Lin S (2008) Realistic and efficient multi-channel communications in wireless sensor networks. In: Paper presented at 27th Annual IEEE international conference on computer communications. IEEE Press, Phoenix, pp 1193–1201
8. Zhang JB, Zhou G, Huang CD, Son SH, Stankovic JA (2007) TMMAC: an energy efficient multi-channel MAC protocol for Ad Hoc networks. In: Paper presented at 2007 IEEE international conference on communications. IEEE Press, pp 3554–3561
9. So W, Walrand J, Mo J et al (2007) McMAC: A parallel rendezvous multi-channel MAC protocol. In: Paper presented at 2007 IEEE wireless communications and networking conference. IEEE Press, pp 334–339
10. Du S, Saha AK, Johnson DB (2007) RMAC: A Routing-enhanced duty-cycle MAC protocol for wireless sensor networks. In: Paper presented at 26th annual IEEE international conference on computer communications. IEEE Press, Alaska, pp 1193–1201

Chapter 14
A General Framework on Connected Sensor Cover in Homogenous Dense Sensor Networks

Jing Lv, Hongwei Du and Hejiao Huang

Abstract In a wireless sensor network, sensors are often deployed in a redundant and random fashion. However, if all the sensors are in an active mode, the lifetime of the network will be very limited. Selecting sufficient but not redundant sensors to meet the requirements (coverage and connectivity) of the network while the remaining sensors switch to the sleep mode is an efficient way to reduce energy consumption and prolong the lifetime of the network. This is called the connected sensor cover problem. In this paper, we present two approximation algorithms to construct such a connected sensor cover to satisfy the coverage and connectivity. Simulation results show that the performance of the two algorithms outperform the best-known algorithm.

Keywords Wireless sensor networks · Connected sensor cover · Coverage · Connectivity

14.1 Introduction

Recent technological advances in micromation with the emergence of small, low-power, limited data computation, and wireless communication capabilities sensors have brought researchers' strongly interested in wireless sensor networks [1]. A wireless sensor network [2–4] consists of a large number of such sensors for gathering some interested data such as light, temperature, or vibrations around its location. A wireless sensor network is usually deployed to accomplish some high-level sensing

J. Lv (✉) · H. Du · H. Huang
Harbin Institute of Technology Shenzhen Graduate School, Shenzhen, China
e-mail: jinglvbest@gmail.com

H. Du
e-mail: hwdu@hitsz.edu.cn

H. Huang
e-mail: hjhuang@hitsz.edu.cn

X. Wang et al. (eds.), *Advanced Technologies in Ad Hoc and Sensor Networks*, Lecture Notes in Electrical Engineering 295, DOI: 10.1007/978-3-642-54174-2_14,

tasks in military, environmental, health or home applications, etc., [5] and there is a periodical query, which from external source to require the network to report a summary of the sensed data. During the query execution, to some extent if only some needed and no more sensors participating in this activity will reduce energy consumption which includes communication and sensing energy costs.

In most actual deployment scenarios, sensors are usually deployed in a random and redundant fashion. Exploiting redundant deployment leads to the sensed data by neighbor sensors presenting highly spatiotemporal correlation. Thus, it's useful to design an effective algorithm or protocol to realize the above strategy for energy efficiency and data redundancy reduction.

In this paper, we consider to design effective algorithms which employ the above scheme as well as provide coverage and connectivity. The objective is to minimize the number of active sensors. We design two algorithms that are run by the network and result in some sensors constructing a topology which subjects to coverage and connectivity. Experiments show that our algorithms are energy efficient.

The rest of this paper is organized as follows. In Sect. 14.2, we review the related work on energy efficient coverage and connectivity issues. We introduce the homogenous network model and the problem formulation in Sect. 14.3. Then in Sect. 14.4, we present the two proposed algorithms for the homogenous network model. In Sect. 14.5, the simulation results are depicted to prove the performance of the proposed algorithms. Finally we conclude this paper.

14.2 Related Work

Many existing works have considered the problem of energy efficiency coverage and connectivity. Tian and Georganas [6] studied a special case that the sensors communication radius is at least twice of the sensing radius and they proved this is the sufficient condition to ensure that complete coverage preservation implies connectivity among active sensors if the original network topology is connected. Under this assumption, Stefan Funke et al. [7] first discussed algorithms for connected sensor cover, and then studied the connectivity of a sensor cover. Ajoy K. Datta et al. [8] proposed two solutions to the minimum connected sensor cover problem via forming a minimum connected dominating set. However, this rapid exhaustion of the active sensors may disconnect the network topology and leave the area uncovered. Hence, Javad Akbari Torkestani [9] proposed to make a good trade-off among the network connectivity, coverage, and lifetime, a proper number of sensors should be activated. Zongheng Zhou et al. [10] studied a small subset of active sensors which maintained a required degree of coverage and formed a connected network with a degree of fault tolerance.

Himanshu Gupta et al. [11] introduced the notion of a connected sensor that cover formally and designed a centralized approximation algorithm to construct a connected sensor cover within an $O(logn)$ factor of the optimal size, where n is the number of sensors in the network. Zou Sai et al. [12] put forward the concept of square grid partition and brought forth a distributed approximation algorithm for the problem of the minimal-connected coverage. A polynomial-time approximation under variable sensing radius and communication radius with performance ratio $O(logn)$ was designed by Zongheng Zhou et al. [13].

The connected set cover problem is well studied in the literature [14–16]. By obtaining connected set covers, the lifetime of the network is maximized. Similarly, Wei Shen et al. [17] proposed a heuristic approach to get more connected disjoint subsets of sensors for energy efficiency. In addition, they considered fault tolerance and required k-coverage. Tahiry Razafindralambo et al. [18] focused on connectivity preservation during the deployment of mobile wireless sensors according to different application requirements. To gain insight into the source of the complexity study, Weizhong Luo et al. [19] initiate a systematic parameterized complexity study of max-lifetime target coverage in wireless sensor networks. Given a set of targets and a set of sensors with two active phrases, Hongwei Du et al. [20] make efforts to find a sleep/active schedule of sensors to maximize the time period during which active sensors form a connected coverage set. Using the relationship between the group steiner tree and the connected sensor cover problem, Lidong Wu et al. [21] proposed an approximation algorithm to solve the connected sensor cover by employing linear programming.

14.3 Network Model and Problem Formulation

Consider a homogeneous static dense sensor network consisting of n sensors, where all the sensors have the same sensing radius, r and the same communication radius, R. We assume that the sensing range of a sensor is a disk with radius r, centered at a node. Let the undirected communication graph $G = (V, E(R))$ denote the induced communication graph in the network, where the sensors act as vertices and an edge exists between any two sensors if the Euclidean distance between them is less than R.

Connected Sensor Cover (CSC): Consider a sensor network consisting of n sensors randomly deployed in a sensing area A, denoted by $S = \{s_1, s_2, \ldots, s_n\}$. Let the sensing range covered by sensor s_i be A_i. Given a query Q over a region A_Q in the sensing area, where $A_Q \subseteq A$, a set of sensors $M = \{s_{i1}, s_{i2}, \ldots, s_{im}\}$ is said to be a connected sensor cover if the following two conditions hold:

(1) $A_Q \subseteq A_{i1} \cup A_{i2} \cup A_{i3} \cup \ldots \cup A_{im}$; (2) The induced communication graph on M is connected.

The CSC problem is to find the minimum size of a connected sensor cover. This problem is proved to be NP-hard, as the less general problem of covering points using line segments is already known to be NP-hard [22].

14.4 Algorithm Description

Definition 1. Intersection: Given a coverage region A, a point q is called an intersection if: (1) If $q \in A$, q is an intersection of the sensing cycles of any two sensors u and v; (2) If q is on A 's boundary, q is an intersection of sensor v if $|qv| = r$.

Combining cover set algorithm (CCSA) is described in the following:

CCSA. Compute a CSC with connected benefit.

Input: n sensors, sensing radius of sensors r and communication radius of sensors R

Output: A vertex set C

1 compute all intersections $I = \{t_1, t_2, \ldots, t_m\}, m \geqslant 1$

2 for each sensor s_i , compute $G_i = \{t \in I | t \text{ is covered by } s_i\}, 1 \leqslant i \leqslant n$

3 $C = NULL$

4 while there are intersections uncovered by the sensing area of C do

4.1 add the sensor s_i with the maximum G_i size to C

4.2 remove intersections which have been covered by C from other G_i

5 while the network is not connected do

5.1 for each sensor not in C do

5.1.1 compute its connected benefit

5.2 choose the sensor with the maximum connected benefit into C

6 return C

In step 1, it chooses all intersections in the sensor network as target points. With those target points, the CSC problem with the target area is transformed to a CSC problem with target points.

In step 4, let C denote the set of sensors already selected by the algorithm at any stage. Initially, C is empty. If there exists intersection point uncovered by the sensing area of C, it adds the sensor with the maximum G_i size to C, and removes the intersection point in other G_i if it is already covered by at least one sensor in C. It repeats this procedure until there is no intersection point uncovered by the sensing area of C.

In step 5, it checks the connectivity of C, if the network is connected, the algorithm ends and the result is C. Otherwise, for each sensor not in C, it computes its CB, and chooses the sensor with the maximum CB and adds it to C. It repeats this procedure until the network induced by C is connected. C is the set of selected sensors.

Definition 2. Connected benefit (CB): For a disconnected network, if we add one other sensor to it, the reduction number of connected components is called the connected benefit of this sensor.

In Fig. 14.1, $C1$–$C4$ denotes four connected components. The union of sensing range of sensors in $C1$–$C4$ can cover the required region, but the graph induced by these sensors is disconnected. s_1–s_7 denotes 7 sensors not in C. We compute their CB, respectively. $CB(s_1) = 1, CB(s_2) = 0, CB(s_3) = 2, CB(s_4) = 0, CB(s_5) = 0,$ $CB(s_6) = 0$, $CB(s_7) = 0$. Hence, we add s_3 to C.

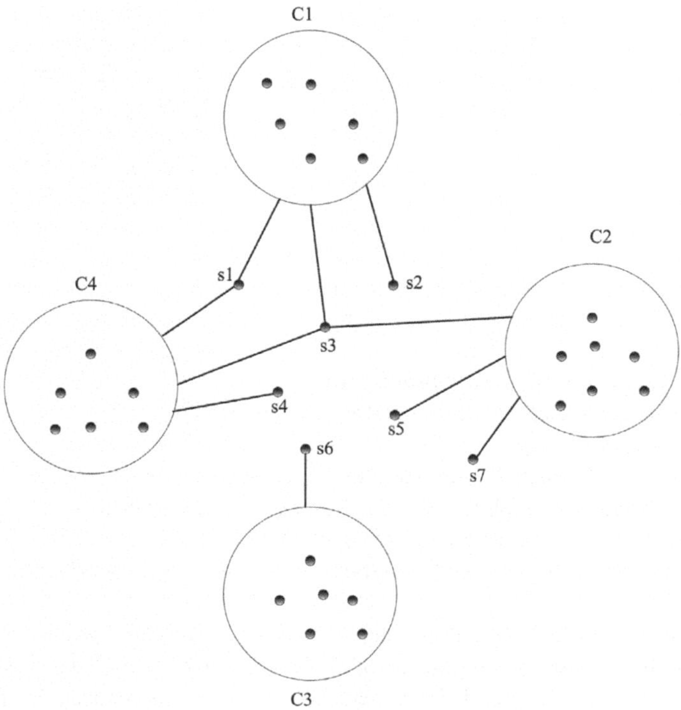

Fig. 14.1 An example of choosing a sensor with maximum connected benefit to C

Minimum spanning cover set algorithm (MSCSA) is described as follows:

MSCSA. Compute a CSC with improved MST.

Input: n sensors, sensing radius of sensors r and communication radius of sensors R

Output: A vertex set C

1 compute all intersections $I = \{t_1, t_2, \ldots, t_m\}$, $m \geqslant 1$

2 for each sensor s_i, compute $G_i = \{t \in I | t \text{ is covered by } s_i\}$, $1 \leqslant i \leqslant n$

3 $C = NULL$

4 while there are intersections uncovered by the sensing area of C do

4.1 add the sensor with the maximum size G_i to C

4.2 remove intersections which have been covered by C from other G_i

5 while the network is not connected do

5.1 run the improved MST algorithm and return T

6 add sensors in T to C

7 return C

In step 5, it checks the connectivity of C. If the graph induced by C is connected, the algorithm ends and the result is C. Otherwise, it runs the improved MST algorithm transformed from MST algorithm [7] to connect all the connected components, and adds the new sensors to C and C is the final result.

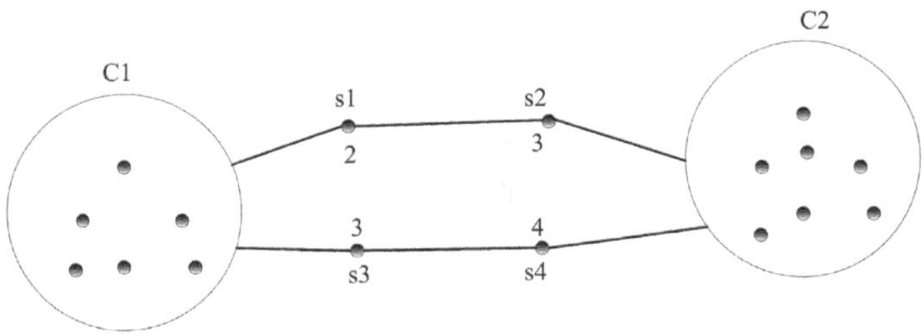

Fig. 14.2 An example of assigning weight for an edge

MSCSA. Compute a CSC with improved MST.

Input: A disconnected graph induced by a vertex set C

Output: A vertex set Q

1 define every connected component of C as a super-vertex

2 create a weighted graph G' in which vertices are super-vertices

3 compute a minimum weight spanning tree of G', T

4 return the set of sensors that lie on the shortest paths corresponding to the edges of T, Q.

In the improved MST algorithm's step 2, it creates a weighted graph G' in which vertices are the connected components of C. A vertex in G' is called a super-vertex. We claim that a sensor s is directly reachable from a super-vertex U if there is an edge in G between a sensor $v \in U$ and s. We compute all the shortest paths between any two super-vertices U and R in G that doesn't contain any sensor which contained in another super-vertex X but not directly reachable from either U or R as the edge between the two super-vertices. In the meantime, we assign each vertex in the shortest paths an attribute called Repeat-times. Repeat-times denotes the times that this vertex is included by the different edges between two super-vertices. The candidate weights of an edge e are computed as below:

$$W_e = \sum_{i \in path} \frac{1}{r_i} \tag{14.1}$$

where W_e is the candidate weight of the edge e, *path* is one of shortest paths between two super-vertices, and r_i denotes the Repeat-times of vertex i. After computing the candidate weights of an edge, we choose the minimum candidate weight as the weight of the edge. In step 3, it computes a minimum weight spanning tree of G', T with Prim algorithm. In step 4, it returns the set of sensors that lie on the shortest paths corresponding to the edges of T.

In Fig. 14.2, $C1$ and $C2$ are two super-vertices. $C1 - s1 - s2 - C2$ and $C1 - s3 - s4 - C2$ are two shortest paths between the set in the example. The value around a sensor is its Repeat-times. We can compute the candidate weights of the two paths: $1/2+1/3 = 5/6$, $1/3+1/4 = 7/12$. We choose $7/12$ as the weight of this edge.

Table 14.1 Simulation terms and notation

Term	Definition	Value
r	Sensing radius	4
n	Number of sensors	100–350
R	Transmission radius	3–9

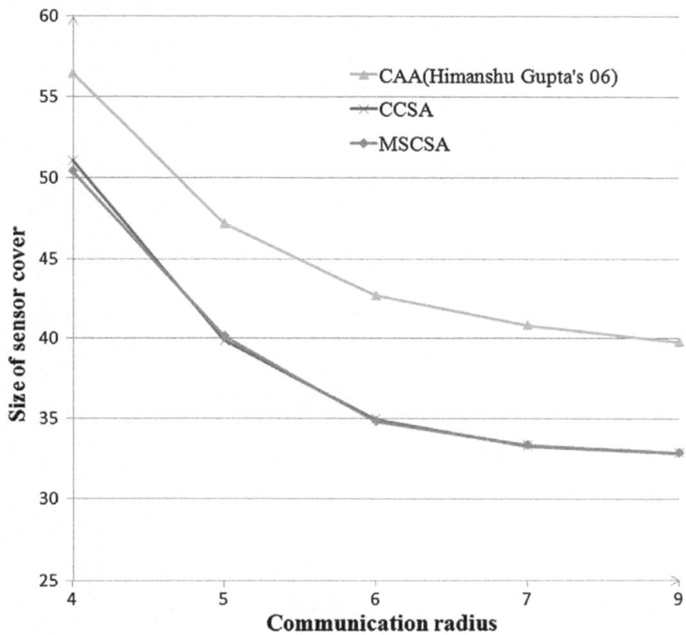

Fig. 14.3 $n = 100$

14.5 Simulation Analysis

In the simulation, we use a 30×30 2D area. Sensors have a sensing radius $r = 4$. We vary the number of sensors (n) and transmission radius (R): n is varied from 100 to 350 and R from 3 to 9. The terms used in the simulation are defined in Table 14.1.

Results show that with lower values of n and R, for instance, $n = 100$, $R = 3$, the network was too sparse that a connected sensor cover could not exist. It is obvious to see that under the condition that $n = 100$ and $R = 3$, the average distance between two sensors is $\sqrt{\frac{30 \times 30}{100}} = 3 = R$, however, it is very difficult to guarantee the connectivity of the network when exploiting random deployment scheme.

When $R < 8$, the MSCSA is better than CCSA, the difference comes from the connected step of the two algorithms. Adopting the MST algorithm, the MSCSA shows its advantages completely. However, when $R \geqslant 8$, the sensors with intersecting coverage range are reachable within one hop in G. In this condition, coverage implies

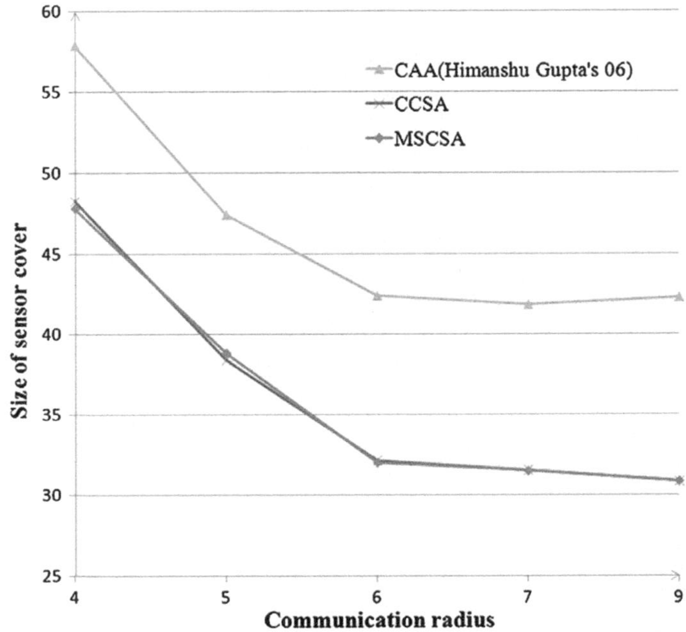

Fig. 14.4 $n = 150$

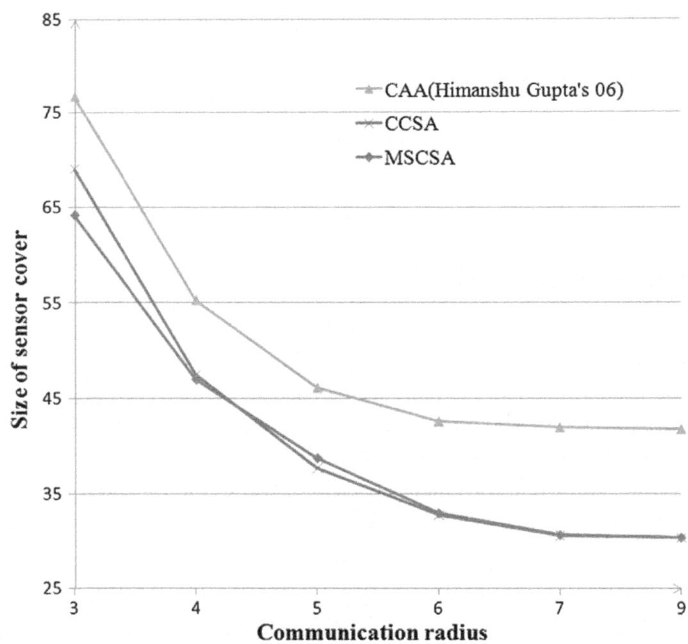

Fig. 14.5 $n = 200$

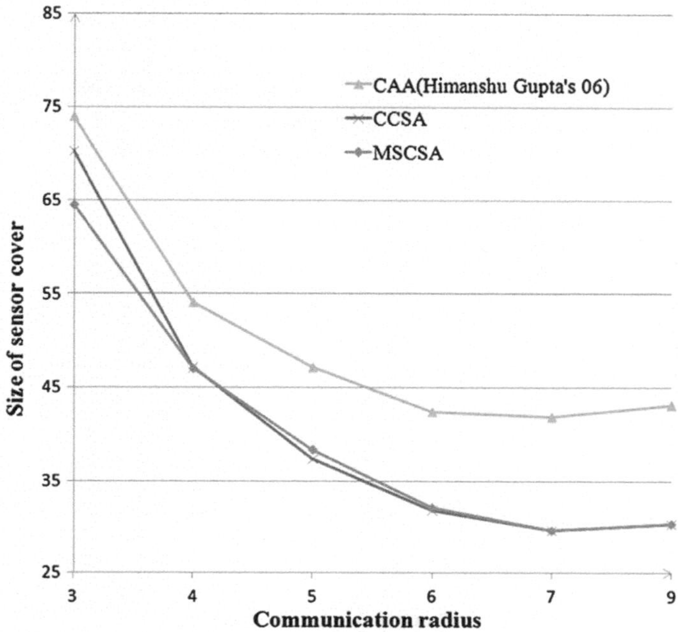

Fig. 14.6 $n = 250$

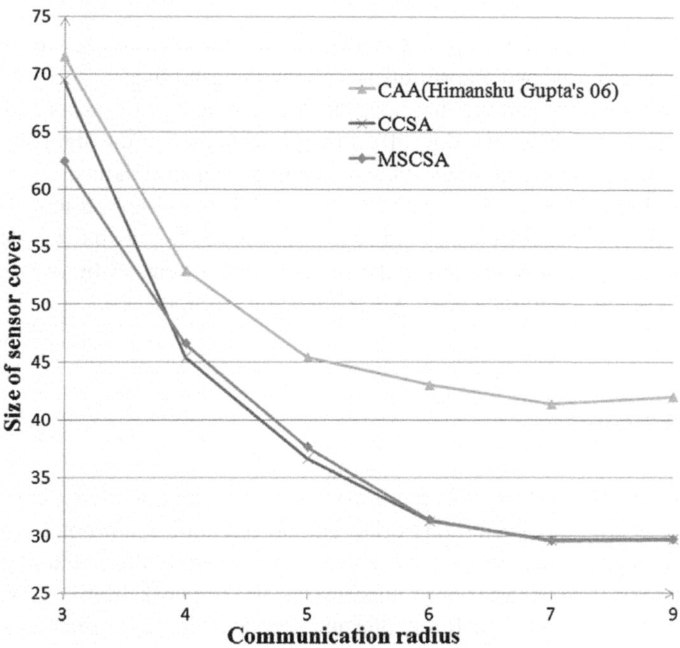

Fig. 14.7 $n = 300$

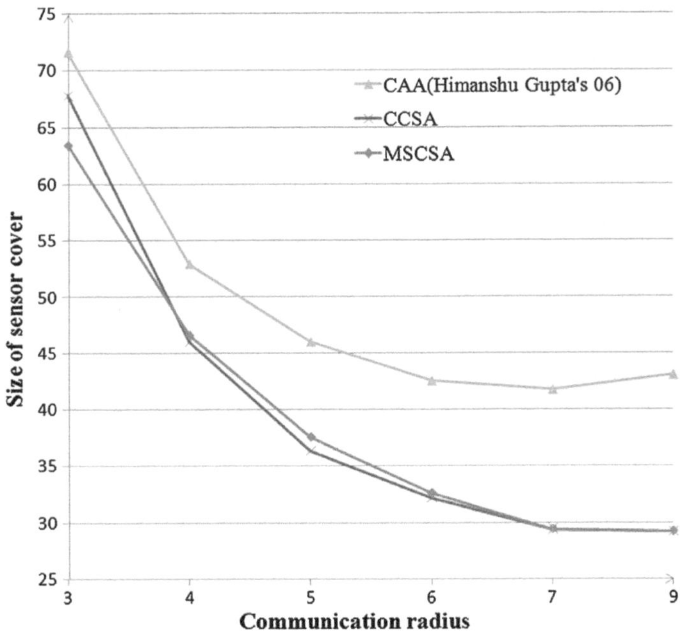

Fig. 14.8 $n = 350$

connectivity [6]. Thus, the proposed algorithms do not need to execute step 5 and the performance of our two algorithms is same with each other.

From simulation we can see that only the spatial density of sensors and the ratio the sensing and transmission radius affect the performance of the algorithms. Thus, it is not necessary to vary the size of the area and the sensing radius.

From the Figs. 14.3, 14.4, 14.5, 14.6, 14.7, 14.8, we observed that the size of connected sensor cover selected by our two algorithms is less than CAA (Himanshu Gumpta' 06). In practical scenarios, the selected active sensors by our algorithms are more energy efficient than the CAA (Himanshu Gumpta' 06).

14.6 Conclusion

In this paper, we study the connected sensor cover problem involving coverage and connectivity in wireless sensor networks. The objective of this problem is to select minimum number of sensors in active mode while keeping the remaining in sleep mode for energy conservation and maximizing the network lifetime. We present two algorithms to produce such a connected sensor cover. Experiments show that the two algorithms are well-performed than the latest algorithm. The further work is to present distributed version of our consideration problems in wireless sensor networks.

Acknowledgments This work was financially supported by National Natural Science Foundation of China with Grant No. 61100191 and 11071271, and Shenzhen Strategic Emerging Industries Program with Grant Nos. ZDSY20120613125016389 and JCYJ20120613151201451, and Natural Scientific Research Innovation Foundation in Harbin Institute of Technology under project HIT.NSFIR.2011128.

References

1. Agre J, Clare L (2000) An integrated architecture for cooperative sensing networks. IEEE Trans Comput 33(5):106–108
2. Bonnet P, Gehrke J, Seshadri P (2001) Toward sensor database systems. Springer-Verlag, Lecture Notes in Computer Science
3. Govindan R, Hellerstein J, Hong W, Madden S, Franklin M, Shenker S (2002) The sensor network as a database. In: Technical Representative, Computer Science Department University, Southern California, Los Angeles
4. Pottie G, Kaiser W (2000) Wireless integrated sensor networks. Commun ACM 43(5):5158
5. Akyildiz IF, Su W, Sankarasubramaniam Y, Cayirci E (2002) Wireless sensor networks: a survey. Comput Netw 38(4):393–422
6. Tian D, Georganas ND (2005) Connectivity maintenance and coverage preservation in wireless sensor networks. Ad Hoc Netw 3(6):744–761
7. Funke S, Kesselman A, Kuhn F, Zvi L, Michael S (2007) Improved approximation algorithms for connected sensor cover. Wireless Netw 13(2):153–164
8. Ajoy K (2006) Datta, Maria Gradinariu, PreethiLinga, Philippe Raipin-Parvedy. Self-* distributed query region covering in sensor networks. J Aerosp Comput Inf Commun 3(8):437–452
9. Javad A-T (2013) An adaptive energy-efficient area coverage algorithm forwireless sensor networks. Ad Hoc Netw 11(6):1655–1666
10. Zhou Z, Das S, Gupta H (2005) Fault tolerant connected sensor cover with variable sensing and transmission ranges. In: Proceedings of the Second Annual IEEE Communications Society Conference on Sensor and Ad Hoc Communications and Networks (SECON), pp 594–604
11. Gupta H, Zhou Z, Quinyi G (2006) Connected sensor cover: self-organization of sensor networks for efficient query execution. IEEE/ACM Trans Networking 14(1):55–67
12. Sai Z, Ming XY, Fei Z, Qi JX, HongWei D, Ying Y (2009) A distributed approximation algorithm for minimal connected cover set problem in sensor networks. WRI international conference on communications and mobile computing, pp 556–562
13. Zhou Z, Das SR, Gupta H (2009) Variable radii connected sensor cover in sensor networks. ACM Trans Sens Netw 5(1):8:1–36
14. Jamali M, Bakhshivand N, Easmaeilpour M, Salami D (2010) An energy-efficient algorithm for connected target coverage problem in wireless sensor networks. In: 3rd IEEE international conference on computer science and information technology, 9–11 July 2010, pp 249–254
15. Shimokawa T, Fujiwara A (2012) Centralized algorithms for the connected target coverage in wireless sensor networks. In: 3rd international conference on networking and computing, 5–7 December 2012, pp 307–310
16. Fethi J (2013) A Lagrangean-based heuristics for the target covering problem in wireless sensor network. Appl Math Model 37(10–11):6780–6785
17. Shen W, Wu Q (2011) Exploring redundancy in sensor deployment to maximize network lifetime and coverage. In: 8th annual IEEE communications society conference on sensors, mesh and Ad Hoc communications and networks, pp 557–565
18. Tahiry R, David S-R (2011) Connectivity preservation and coverage schemes for wireless senor networks. IEEE Trans Autom Control 56(10):2418–2428
19. Luo W, Wang J, Guo J, Chen J (2013) Parameterized complexity of max-lifetime target coverage in wireless sensor networks. Theoret Comput Sci. http://dx.doi.org/10.1016/j.tcs.2013.06.008

20. Du H, Panos PM, Wu W, Wu L (2013) Maximum lifetime connected coverage with two active-phase sensors. J Global Optim 56(2):559–568
21. Wu L, Du H, Wu W, Li D, Lv J, Lee W (2013) Approximations for Minimum Connected Sensor Cover. IEEE INFOCOM
22. Anil Kumar VS, Arya S, Ramesh H (2000) Hardness of set cover with intersection 1. Lecture notes in computer science 1853, Springer-Verlag, pp 624–635

Chapter 15
Partition-Based MAC Protocol for Data Gathering in Wireless Sensor Networks

Bin Cheng, Linlin Ci, Chengping Tian and Taolin Guo

Abstract The limited amount of available orthogonal channels and noise interference are two major issues to constrain performances of multi-channel MAC protocols in wireless sensor networks (WSN). In conjunction with features of data gathering networks, a partitioning-based multi-channel MAC protocol is designed. It partitioned network into several subtrees with lower collision factor, without increasing the total value of interference around the whole network. Simulation results show that the proposed MAC protocol can significantly improve throughput, reduce end-to-end latency, and packet loss, especially in dense networks.

Keywords MAC · Multi-channel · Wireless sensor networks · Network partition

15.1 Introduction

Wireless sensor networks are widely applied into environmental monitoring, intrusion detection and location, medical detection, and so on. Each MAC protocol designed for WSN cannot perform well in every scenario. For example, constant change of environment can lead to many problems, such as frequent sampling and burst transmission. Instead of energy efficiency, in some scenarios, end-to-end delivery latency, throughput or fairness is considered to be more important, that is, special application scenarios (or special performance in one aspect) determine how to design the protocols in wireless sensor networks.

MAC layer often lies in the bottom of protocol stack, and manages to distribute wireless medium resources. So performances of MAC protocols significantly

B. Cheng (✉) · L. Ci · C. Tian
School of Computer Science and Technology, Beijing Institute of Technology, Beijing, China
e-mail: chengbinpaper@gmail.com

T. Guo
Shandong University of science and technology, Tai'an, China

X. Wang et al. (eds.), *Advanced Technologies in Ad Hoc and Sensor Networks*,
Lecture Notes in Electrical Engineering 295, DOI: 10.1007/978-3-642-54174-2_15,

determine the entire network performance. Traditionally, due to its difficulties of energy supplement, energy efficiency often is considered primarily along with the whole design. In order to better accommodating varied application scenarios, a large number of protocols focusing on medium access control, routing, and transmission were proposed.

On basis of the amount of channels being used, existing MAC protocols can be categorized into two main classes: Single-Channel protocols and multiple-channel protocols. For example, S-MAC [1] belongs to Single-Channel protocols, networks performing on them make all nodes communicate on the common channel. Two main negative problems should not be neglected: one is low channel utilization, which is because control packets and real data packets share the common channel; the other is co-channel interference, this problem is more fiercely in dense deployment networks. Additionally, the effect of homogeneous networks is also not neglected, with the widespread application of 2.4 GHz network, the interference caused by unknown homogeneous signals in the background also can degrade network performances.

The wireless transmission module supporting multiple frequencies have been designed, such as CC2420 wireless transmission module equipped on MICAz, Telos, and IRIS sensor nodes, they can provide 16 channels with 5 MHz interval. Recently, some multi-channel MAC protocols were proposed, MMSN [2], TMMAC [3], and MCMAC [4] both realized parallel transmission through making neighbors within two hops perform on different orthogonal channels. Compared to Single-Channel protocols, they achieved better throughput and less interference.

During current researches, most researchers configure their wireless interfaces with these non-overlapped channels only. Owing to the shortage of choices of channels, two nearby nodes may be actually assigned to the same channel, especially under some dense scenarios. For example, the 802.11b standards define 11 channels with only three non-overlapped channels, named 1, 6, and 11. So the shortage of non-overlapped channels become one bottleneck in current multi-channel protocols. Literatures [5, 6] shows that, current multi-channel MAC protocols mostly improve performances under the background of only using non-overlapped channels, although this method can avoid interference between channels and simplifies the complexity of protocol design, the mismatch between the amount of available non-overlapped channels and real demands becomes more irreconcilable, especially in dense and large-scale networks.

Aiming at the two main problems mentioned above, the shortage of channels and the interference, our contributions in that, without degrading the minimum interference of the entire network, we provide an efficient network partition algorithm called NP to divide network into several sub-trees with lower collision factor.

This paper is organized as follows: Sect. 15.2 provides related researches on multi-channel MAC protocols; Sect. 15.3 gives detailed design of the partition-based multi-channel MAC protocol(PM-MAC); Performance evaluations are given in Sect. 15.4; Finally, Sect. 15.5 concludes our work.

15.2 Related Work

Recently, many scholars have done a lot of researches to improve WSN performance, involving MAC protocol, routing protocol, and transmission protocol optimization. In this following, we will focus on analyzing and discussing current researches on multi-channel MAC protocols.

According to whether to adopt one channel or multiple channels, MAC protocols can be divided into two categories: single-channel and multi-channel MAC protocols. In single-channel MAC protocols, for example, S-MAC [1] B-MAC [4] TMAC [5], and so on, control message exchange and real data transmission both perform on one common channel, which largely reduces channel utilization and throughput. To improve throughput, Z-MAC [7] introduced an efficient scheduling scheme in which a specified interval is allocated to each node to avoid conflicts, but the improvement is very limited, owing to only one available channel for use.

With the emergence of wireless transmission module supporting for multiple bands, such as CC2420 wireless communication module used by MICAz and Telos nodes providing 16 independent channels with 5 MHz interval, multi-channel MAC protocols can realize parallel transmission over different channels, which greatly enhances throughput. So, multi-channel MAC protocols is one of current research hotspots in WSN field. Gang Zhou et al. proposed a multi-channel MAC protocol, named MMSN [2], which divides time frame into several time slots based on traditional CSMA protocol, before real data transmission, neighboring nodes compete for channels fairly, then data transmission performs on designated time pots. The number of available channels is a bottleneck. When data transfer begins, each node must frequently switch between its own channel and the channel occupied by its destination to build connection. The frequent switching largely increases end-to-end delivery latency. In addition, MMSN selects one dedicated channel for broadcast transmission, and sets aside a special field for broadcast operations, although this makes broadcast transmission perform well, but low channel utilization also restricts the throughput improvement. On basis of the characteristics of data collection applications, Yafeng Wu et al. proposed TMCP [8], aiming at the tree-based structure applications. It divides the whole tree-based network into several subtrees, and assigns each subtree an orthogonal channel, thus all subtrees can transmit in parallel without interference. Although this kind of coarse-grained channel assignment method can overcome the shortage of available channels, the problem of inner-tree interference cannot be ignored, especially in large-scale dense networks.

Moreover, TMMAC [3] and MCMAC [4] assign different channels to nodes within two hops to avoid potential interferences. However, this node-level assignment mechanism brings more latency owing to frequent channel switching. Aiming at the special application of data gathering, PM-MAC proposes an enhanced MAC protocol for data gathering applications on the basis of TMCP through optimizing the partition procedure and the structure of each sub-tree, reducing the interference of the entire network and the collision within sub-trees.

Fig. 15.1 Basic framework of PM-MAC

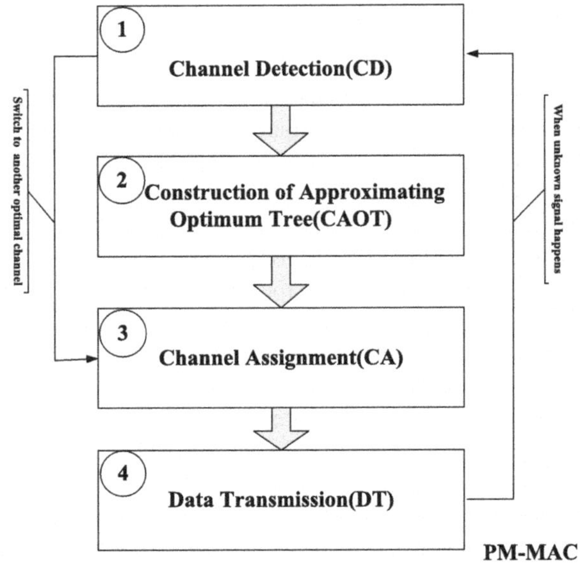

15.3 PM-MAC Protocol Design

15.3.1 Basic Framework

Data gathering as one of the most popular applications usually consisted of a large number of event-driven sensor nodes, when some special events happens in the sensing range of sensor nodes, they would generate data packets in response and transmit data packets to sink along multi-hop route. PM-MAC can be realized by four functional modules: Channel Detection (CD), Construction of Approximating Optimum Tree (CAOT), Channel Assignment (CA), Data Transmission (DT). The basic framework of PM-MAC and the relationship among the four modules are shown in Fig. 15.1.

Channel Detection (CD) During network initialization, two sensor nodes are used to sample the link quality of each channel by transmitting packets to each other, and then select channels with good link qualities.

Construction of Approximating Optimum Tree (CAOT) Divide the whole network into several disjoint subtrees rooted from the base station, meanwhile ensure the sum network interference be minimum. Then optimize subtrees with low collision factor in order to reduce collision at the cross-position, avoiding network congestion.

Channel Assignment (CA) Allocate the available channels to each subtree on basis of graph coloring theory, in order to ensure each subtree work on a relative optimal channel and better avoiding unknown interferences.

Data Transmission (DT) Data packets are forwarded toward the sink by some special strategy, such as aggregation or not.

15.4 Design of PM-MAC

15.4.1 Model and Definition

In this work, we only focus on the data gathering traffic in WSN. Before the design of PM-MAC, we define the interference value of a node in a sub-tree. Reference [9] and [10] both gave us definitions of the interference of one node. For the interference usually happens at the receiving side by senders, so we define the interference value on basis of the number of other nodes potentially disturbed by transmission of this node. In other words, interference is considered to be an issue at the sender instead of at the receiver. Because of the fact that the interference is actually a problem occurring at the receiver, we use a receiver-centric interference definition. The interference value of a node n is the number of other nodes by which the reception at n can be disturbed [11] .

Each node n is denoted as a four-tuple (u_n, s_n, l_n, b_n), where u stands for the interference value; s denotes the collision factor value after partition; l_n represents the layer number where n lies. b_n is used to judge whether the node lies in the overlapped domain between neighboring subtrees, its value is equal to the times it is accessed during Breadth-First search algorithm in the initialization step.

Definition 1 The interference set of a node n is defined as

$$int(n) = |N(n)| \tag{15.1}$$

$$N(n) = \{v|v \in D(n, I_n) \cap (C_v = C_n)\} \tag{15.2}$$

where $D(n, I_n)$ denotes a circle with center at n and radius equal to I_n, and C_v, C_n represent channels used by node v, node n, respectively.

When one node lies in the intersection point of several data streams, the collision rate is proportional to the number of one-hop nodes lying in its former layer. To simplify the design complexity, we denote the number of nodes lying in its former layer and in the same subtree as the degree of collision, named collision factor.

Definition 2 The collision factor of a node n is defined as

$$col(n) = |N(n)| \tag{15.3}$$

$$N(n) = \{v|v \in D(n, I_n) \cap (C_v = C_n) \cap (l_v > l_n)\}. \tag{15.4}$$

Next, we will give the definitions of the sub-tree' interference value and collision factor.

Definition 3 The interference value of each subtree T is equal to the maximum interference value among its member nodes, so we define it as

$$int(T) = \max\{int(n)|n \cap T\}. \tag{15.5}$$

The collision factor of each sub-tree T is equal to the maximum collision factor among its member nodes, so we define it as

$$col(T) = \max\{col(n) | n \cap T\}. \tag{15.6}$$

15.4.2 Network Partition Algorithm

In this Network Partition Algorithm namely with optimal collision factor, namely OCF for short, we assume that the interference sets of every node is already known. For a node n, let C_n denote ns channel, and S_n denote ns node section including all its one-hop neighbor nodes. Like TMCP, OCF algorithm first applies a heuristic Breadth-First search algorithm to compute a fat tree rooted at the base station. There are two important properties of the fat tree. First, nodes keep its height and have multiple parents on the fat tree. Second, the fat tree is actually a shortest path tree, where branches from the base station to each node are paths with the least hop count, because we use BFS strategy to build the tree. The detailed algorithm is as follows.

Algorithm 1 Network Partition Algorithm with Optimal Collision Factor

Input:

The four-tuple (u_n, s_n, l_n, b_n) of each node n;

Output:

The tree-based data gathering network with optimal collision and interference value;

1: If $b_n > 1$, which demonstrates this node stands in the cross-zone of adjacent two subtrees; Calculate the number of two subtrees N_i

2: $N_i = |\{p|(p \in T_i) \cap (l_i \leq l_p)\}|$, then node n join the subtree with less members, then Jump into step 4. If the two subtrees have the same amount, jump into step 3

3: Calculate the whole network interference after node n join perspectively, then n join the subtree with less interference value of the whole network; if the whole network interference is equal, n randomly select one to join; ;

4: If exist more than one parent nodes of n in this subtree, then calculate the network collision factor, and select one node as parent when the network collision factor is smaller after its joining;

5: If there exist any nodes with its $b_n > 1$, jump into step 1

6: **return** E_n;

Theorem 1 The time complexity of the OCF algorithm is $O(d \times k \times n^2)$, where d denotes the radius of the network, n denotes the amount of the entire network, k represents the number of channels.

Proof The time complexity of construing a FAT tree is $O(d \times \delta \times n)$, where δ is the maximum radius of the network. In OCF, Step 4 takes $O(k \times n)$ in the worst case, and the loop beginning at Step 5 may run at most n time. Thus, the procedure within the repeat loop takes $O(k \times n^2)$, and the repeat loop iterates at most d times, because the tree height never exceeds the diameter of the network. So the time complexity is $O(d \times k \times n^2)$ in the worst case.

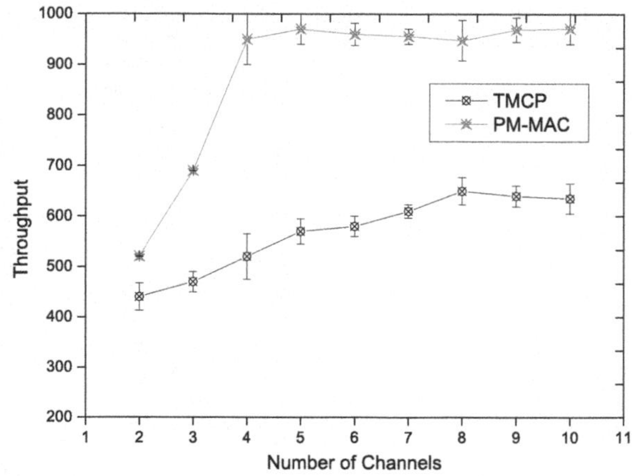

Fig. 15.2 Throughput versus number of channels

15.5 Performance Evaluation

In this section, we evaluate PM-MAC protocol on platform Glomosim. Performance evaluation mainly performs in terms of throughput, packet delivery rate, latency, and energy efficiency, compared with MMSN, TMCP under the condition of varying the number of channels. All protocols run for 1000 times, then calculate average results and set a 95 % confidence interval. All source nodes initiate CBR (continuous bit rate) streams toward the sink and each node generates a packet each 2s (if nodes transmit more frequently, buffer overflow occurs). This scenario can be considered as a periodic data collection. In these simulations, all raw data packets from sources are relayed toward the sink. The number of channels varies from 1 to 10, and 100 nodes are randomly deployed in 150 m × 150 m area. Figure 15.2 shows simulation results in terms of aggregate throughput when the number of channels varies. In this experiment, we calculate the aggregate throughput by counting the number of packet bytes per second at the sink. From the different curves presenting different protocols, we can easily get that PM-MAC achieves higher throughput than TMCP when the number is more than 3. This can verify that PM-MAC benefits from OCF mechanism. When the number of channels is lower than 3, TMCP runs at a higher speed of the increment in throughput than PM-MAC, that is because networking partition makes multiple data streams run in parallel, however, the network partition scheme has its own shortcoming that when the scale of network is fixed, the network reaches saturation even when the number of available channels keeps increasing, that is because interference and conflicts among different subregions degrade performances.

Figure 15.3 shows simulation results in terms of packet delivery rate between PM-MAC and TMCP. As shown in Fig. 15.2, all the trend of four curves are similar to curves of aggregate throughput in Fig. 15.3. In PM-MAC, more channels make more subtrees to perform in parallel when the number of channels increases. The TMCP curve nearly is high than PM-MAC, when the number of channels is less

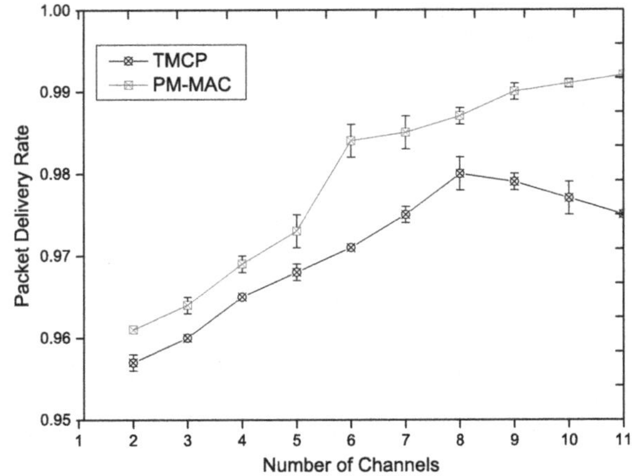

Fig. 15.3 Packet delivery rate versus number of channels

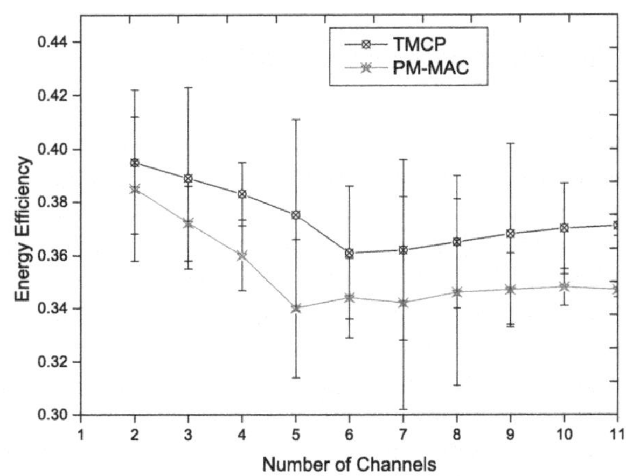

Fig. 15.4 Energy efficiency versus number of channels

than 3, because it can not reach its saturation. When more than 4 channels, TMCP cannot increases continuously, that is because when more the interference inner subtrees degrades its performance in packet delivery rate. PM-MAC largely outperforms TMCP, because its OCF can optimize the structure of each subtrees, in order to reduce congestion.

Figure 15.4 shows the energy efficiency when the number of channels changes. Energy efficiency is a major factor in wireless sensor networks. In this simulation, we evaluate the average energy consumption per byte of packet, which equals to the value of total energy consumed is divided by the total number of successfully delivered packets. As the packet-generated rate increases, both PM-LMAC and TMCP spend more energy than before. PM-MAC largely outperforms TMCP in terms of energy efficiency. Owing to OCF optimizing the structure of subtrees, when the traffic load becomes heavy, its advantages is more significant.

15.6 Conclusions

In oder to improve the performançe of wireless sensor networks, this paper presents an efficient multi-channel MAC protocol named PM-MAC, which proposes a partitioning-based multi-channel MAC protocol was designed. It partitioned network into several subtrees with lower collision factor, without sacrificing the total value of interference around the whole network. Compared to TMCP, simulation results show that the proposed MAC protocol can significantly improve throughput and reduce end-to-end latency and packet loss, especially in dense networks with finite available channels. But the channel assignment mechanism is not high-efficiency, so we will focus on this point in order to further improve network performances.

Acknowledgments Thanks to the support of the National Natural Science Foundation of China (61063042).

References

1. Wei Y, Heidemann J, Estrin D (2002) An energy-efficient MAC protocol for wireless sensor networks. In: Paper presented at 21th Annual Joint Conference of the IEEE Computer and Communications Societies, vol 3. IEEE Press, New York, pp 1567–1576
2. Zhou G, Huang C, Yan T, He T, Stankovic JA, Abdelzaher TF (2006) MMSN: multi-frequency media access control for wireless sensor networks. In: Paper presented at 25th IEEE International Conference on Computer Communications, vol 6. IEEE Press, Barcelona, pp 1–13
3. Zhang JB, Zhou G, Huang CD, Son SH, Stankovic JA (2007) TMMAC: an energy efficient multi-channel MAC protocol for Ad Hoc networks. In: Paper presented at 2007 IEEE International Conference on Communications, IEEE press, pp 3554–3561
4. So W, Walrand J, Mo J et al (2007) McMAC: a parallel rendezvous multi-channel MAC protocol. In: Paper presented at 2007 IEEE Wireless Communications and Networking Conference, IEEE press, pp 334–339
5. Van DT, Langendoen K (2003) An adaptive energy-efficient MAC protocol for wireless sensor networks. In: Paper presented at the 1st international conference on Embedded networked sensor systems, ACM Press, Los Angeles, pp 171–180
6. Keshavarzian A, Lee H, Venkatraman L (2006) Wakeup scheduling in wireless sensor networks. In: Paper presented at the 7th ACM international symposium on Mobile ad hoc networking and computing, ACM Press, Florence, pp 322–333
7. Rhee I, Warrier A, Aia M, Min J, Sichitiu ML (2008) Z-MAC: a Hybrid MAC for wireless sensor networks. IEEE/ACM Trans. Networking 16(3): 511–524
8. Wu Y, Stankovic JA, He T, Lin S (2008) Realistic and efficient multi-channel communications in wireless sensor networks. In: Paper presented at 27th Annual IEEE International Conference on Computer Communications, IEEE Press, Phoenix, pp 1193–1201
9. Zhou G, He T, Stankovic A (2005) RID: radio interference detection in wireless sensor networks. In: Paper presented at the 24th Annual IEEE Joint Conference on Computer and Communications Societies, Vol 2. IEEE Press, Virginia, pp 891–901
10. Padeep K, Nitin H (2005) Vaidya Capacity of multi-channel wireless networks: impact of number of channels and interfaces. In: Paper presented at the 11th annual international conference on Mobile computing and networking, New York, pp 43–57
11. Du S, Saha AK, Johnson DB (2007) RMAC: a routing-enhanced duty-cycle MAC protocol for wireless sensor networks. In: paper presented at 26th Annual IEEE International Conference on Computer Communications, IEEE Press, Alask, pp 1193–1201

Chapter 16
Small World-Based Wireless Sensor Network Power Control Algorithm for Airborne PHM

Wei Zheng and Di Luo

Abstract Aircraft Prognostic and Health Management (PHM) is a system which could diagnose device faults and assess its health status. Due to the complexity of the airborne environment, the aircraft PHM uses Wireless Sensor Network (WSN) technology to collect and transmit data. In this paper, a Power Control Algorithm based on Small World (PCS) is proposed to reduce the network delay. The PCS algorithm adds several shortcuts into the network based on the small world theory and uses genetic algorithm to optimize the shortcuts. Simulation results demonstrate that this method can effectively shorten the average path length and reduce the network delay.

Keywords Aircraft PHM · Wireless sensor network · Small world · Genetic algorithm

16.1 Introduction

Aircraft Prognostic and Health Management (PHM) system takes advantage of sensors to collect all kinds of systemic data. Meanwhile, various intelligent reasoning algorithms such as fuzzy logic, expert system, and neural network are used by aircraft PHM to diagnose the systemic health and predict the systemic problems [1, 2]. Indeed, aircraft PHM is an effective method to enhance the capability of comprehensive support and reduce the aircraft maintenance costs [3].

In the early period, the traditional method of PHM is to monitor various parameters by sensor redundancy technology, and transmit data to tester through the wired network. However, wiring in the limited PHM working space is very difficult and

W. Zheng (✉) · D. Luo
Internet of Things Technology Institute, Nanchang Hang Kong University, Nanchang, Jiangxi, People Republic of China
e-mail: zhengwei_nchu@126.com

X. Wang et al. (eds.), *Advanced Technologies in Ad Hoc and Sensor Networks*, Lecture Notes in Electrical Engineering 295, DOI: 10.1007/978-3-642-54174-2_16,

also has the problems of various monitoring parameters and strong electromagnetic interference [4]. Therefore, we use the Wireless Sensor Network (WSN) technology to collect and transmit data. The WSN is a network comprising a number of sensor nodes and its wireless feature makes it suitable for poor working conditions. Meanwhile, it can reduce the system costs, improve the system reliability, and transmit more real-time [5, 6].

In this paper, we propose a Power Control Algorithm based on Small World (PCS). The main idea of the proposed algorithm includes two parts: (1) choose the power amplifier nodes to be the endpoints of the shortcuts, (2) determine the power magnification of the power amplifier nodes. At the beginning, the PCS algorithm randomly selects several nodes which use long-range power to search their neighbor nodes. Then, the selected nodes which thus are named power amplifier node connect to one of their neighbor nodes to construct shortcuts. Next, these shortcuts are formed as an original individual for genetic algorithm [7] to be iterated and optimized. At last, we can get the determined power magnification through the optimized shortcuts. Simulation results represent that the PCS algorithm can significantly reduce the network delay and greatly meet the demand of real-time transmission in airborne PHM system. In addition, the PCS algorithm has shorter average path length and lower network energy consumption than other algorithms in large-scale wireless sensor networks as well.

16.2 Related Works

Topology control is an effective method to optimize topology, which maintains the detected network area coverage and connectivity as well as balances the network energy efficiency. Meanwhile, it also can infer the network characteristics such as communication interference, network latency, and robustness, etc [8]. Power control is an important way of topology control, it could optimize the network topology by means of changing node transmission power, selecting optimal neighbor set, and removing unnecessary communication links, if the conditions of network coverage and connectivity are satisfied.

In recent years, many power control algorithms have been proposed such as LMN/LMA [9] algorithms based on node degree and XTC [10] algorithm based on information interaction of adjacent nodes. The CLUSTERPOW algorithm based on routing is proposed in [11]. It has all nodes selected one next-hop node in minimum power level, through which they could reach the destination node. Although these methods can optimize the network topology, most of them only take the network energy consumption into consideration. In [12], it proposes a routing algorithm based on cross-layer, which achieves power control on the network and MAC layer. With establishing a number of different power level routings, it makes each node select and group the routings in minimum power level. Nevertheless, it uses different power control strategies to reduce the energy consumption, the hop count will increase and prolong the network delay correspondingly. In [13], it proposes a

WSR routing algorithm which uses weak state along with random directional walks for forwarding packets. When a packet reaches a node that contains a weak state about the destination with higher confidence than that was held by the packet, the walk direction is biased. This algorithm has a balanced energy consumption of the network, but it cannot guarantee the real-time capability. The SCR [14] algorithm introduces the small world theory into WSN, each source node selects its shortcut node, through which the path length to the destination node is greatly reduced, so the SCR has much shorter path length and lower routing overhead. However, it is more effective for the large-scale wireless sensor network.

In [15], Helmy proposes that a network with high efficiency of data querying and other superiorities can be got if the network is based on the small world theory. Therefore, the power control algorithm based on small world is presented in this paper in order to shorten the average path length, reduce the network delay, and cut down the network energy consumption.

16.3 The Power Control Algorithm Based on Small World

In [16], the authors have pointed out the method which is based on the small world theory of having a limited number of nodes with long-range power. In this paper, we make all nodes equipped with two radio frequency modules, one for short-range communications and another with adjustable long-range power [17], which would add shortcuts into the network.

This section describes the proposed algorithm which is based on the small world theory and explains how to choose the optimal power amplifier nodes and how to determine the power magnification of the optimal power amplifier nodes, then fits the small world theory into the algorithm. Moreover, in order to evaluate the proposed algorithm, we make the average path length to be the network performance indicator.

16.3.1 Network Model

Under the assumptions that network is stationary, all sensor nodes are deployed in a two-dimensional grid plane and placed at the lattice points. The network model is shown as follows:

1. Set node u as an example, the location of each node is represented as follow: $\{(u_x, u_y) : u_x \in \{1, 2, ..., n\}, u_y \in \{1, 2, ..., n\}\}$.
2. Euclidean distance is used to represent the distance between two nodes, for example, the Euclidean distance between two nodes u and v is :

$$d(u, v) = \sqrt{(u_x - v_x)^2 + (u_y - v_y)^2} \tag{16.1}$$

3. Neighbor node set: The short-range radius is r and long-range radius is R. Setting power amplifier node u as an example, if $d(u, v) \leq r$, node v is a short-range neighbor node of u and u's short-range neighbor set is G'; if $d(u, v) \leq R$, node z is a long-range neighbor node of u and u's long-range neighbor set is G.
4. The next-hop node set: The ring area $V = G - G'$ is the next-hop node set of each power amplifier node.

16.3.2 Network Performance Indicator

The communication distance is defined as the hop count of the shortest path between two nodes in sensor network; Average Path Length (APL) is defined as the average communication distance of all pairs of nodes, we describe it in formula (16.2):

$$L = \frac{2}{N(N-1)} \sum_{i=1}^{N} \sum_{j=1}^{N} d_{ij} \tag{16.2}$$

where d_{ij} represents the shortest communication distance between node i and j, N is the number of total nodes in the network. In airborne PHM, in order to obtain real-time and accurate information, the network delay should be as short as possible. Therefore, the smaller average path length L is, the better network real-time capability will obtain.

16.3.3 Algorithm Description

The PCS algorithm randomly selects k power amplifier nodes to form a node set as the original solution in genetic algorithm. Then, every selected node chooses one node as its next-hop node in the ring area between short-range and long-range circle. According to M times of power amplifier nodes selection, the PCS algorithm calculates the average path length and keeps comparing to select the optimal individual which makes the average path length shorter. The procedures of PCS algorithm are:

Step 1: The nodes are randomly located in an $m * n$ $(m > n)$ two-dimensional grid plane, the proposed algorithm calculates the distance between all pairs of nodes in the network according to the formula (16.1), then acquires the neighbor set G' of each node with short-range power, and randomly selects $k (k < n)$ nodes as the initial power amplifier nodes.
Step 2: The k randomly selected nodes use long-range power to search their neighbor nodes which formed set G, the next-hop node set of power amplifier node is $V = G - G'$. Next, the PCS algorithm calculates the probability of each optional next-hop node with formula (16.3), and uses roulette to choose the only one.

$$p = d(u, v) / \sum_{w \in V} d(u, w) \tag{16.3}$$

Set power amplifier node u as an example, formula (16.3) represents the ratio of the distance between u and one node in V compare to the sum of distances between u and all nodes in V. Once the next-hop node is selected, the magnification power which makes the selected node u connected to the next-hop node is also ascertained.
Step 3: According to formula (16.2), the PCS algorithm calculates the average path length L of the network.
Step 4: Loop Step1–Step3 S times and compare each average path length L of the topology generated by the different k randomly selected power amplifier nodes, then obtain the final k nodes which make the L shortest.
Step 5: Use genetic algorithm to cross the individual with the given crossover probability and produce a new individual.
Step 6: According to the given mutation probability, the PCS algorithm mutates individual generated in step5 and produces a new individual.
Step 7: Return to step5, the PCS algorithm loops M times and finally gets the optimal k nodes which make the average path length shortest, and also obtains the power magnification of the k power amplifier nodes, respectively.

16.4 Simulation and Results

This section describes the functionality and usability of the PCS algorithm compared with WSR and SCR algorithm, MATLAB is the simulation tool. The PCS algorithm is mainly proposed for the WSN in airborne environment. The general civilian aircraft, such as the A350–800 specification for the fuselage, is about 60.5 m in length and 16.9 m in height. Therefore, we consider the plane is a rectangle and conduct the airborne experiments in an $m * n$ $(m > n)$ two-dimensional plane. The parameters in the PCS algorithm are $r = 10$ and $R = 20$.

As shown in the following figures, the experiments are conducted in a $70 * 20$ plane. In Fig. 16.1, the average path length L of PCS algorithm is declining with the increasing number of network total nodes N. Moreover, it can be seen that the average path length L decreases when the number of power amplifier nodes k increases. This is because when the number of power amplifier nodes k increases, the network density is getting higher and the distances between nodes are getting shorter, then the hop count between two communication nodes is reduced. In spite of the increasing power amplifier nodes, the reduction of the average path length is inapparent for the reason that most nodes can be communicated already from the point $k = 3$ on. Figure 16.2 clearly shows that the proposed algorithm is obviously effective than the WSR and SCR algorithm. The average path length in our algorithm is about 8.6 % lower than that in the SCR algorithm and about 25.4 % lower than that in the WSR algorithm.

The PCS algorithm not only improves the transmission rate in the airborne environment, but also obtains the shorter average path length and lower energy

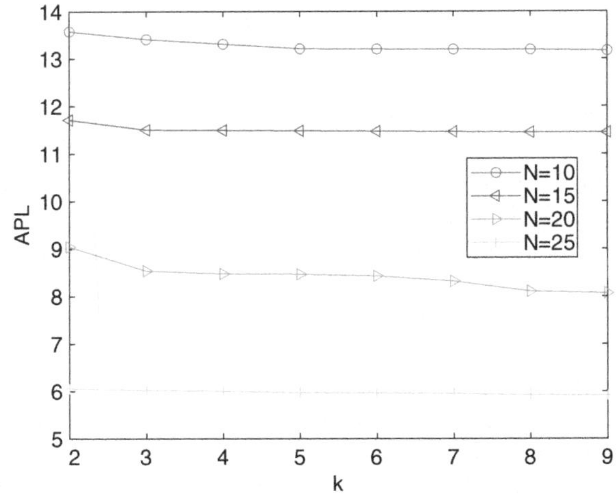

Fig. 16.1 Average path length versus power amplified node numbers

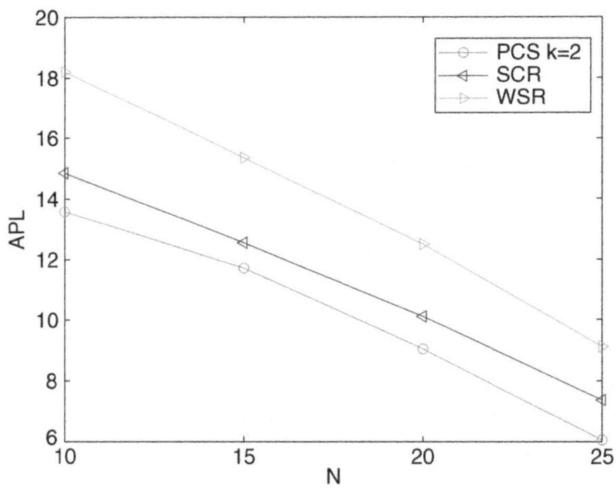

Fig. 16.2 Average path length versus node numbers

consumption in larger sensor network. Similar to the above experiments, Fig. 16.3 describes that in a large-scale sensor network ($N * N$) when the number of total nodes N increases, the average path length gradually become shorter. In Fig. 16.4, it shows that in a $100 * 100$ network, the growth rate of average path length is about 24.4 % lower than the WSR algorithm and about 48.4 % lower than the SCR algorithm. Meanwhile, it also has the shortest average path length. Both the SCR and the PCS algorithm are based on the small world theory, unlike the SCR algorithm, the PCS algorithm uses genetic algorithm to optimize shortcuts which can obtain the optimal node set to shorten the average path length at most. With the increasing number of total nodes, the differences between three algorithms are more remarkable.

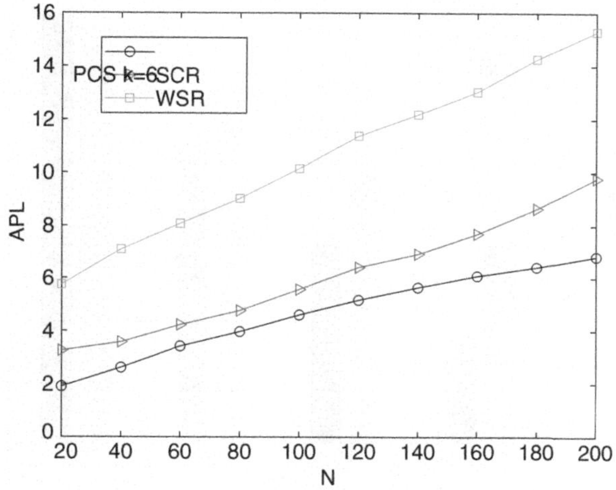

Fig. 16.3 Average path length versus node numbers

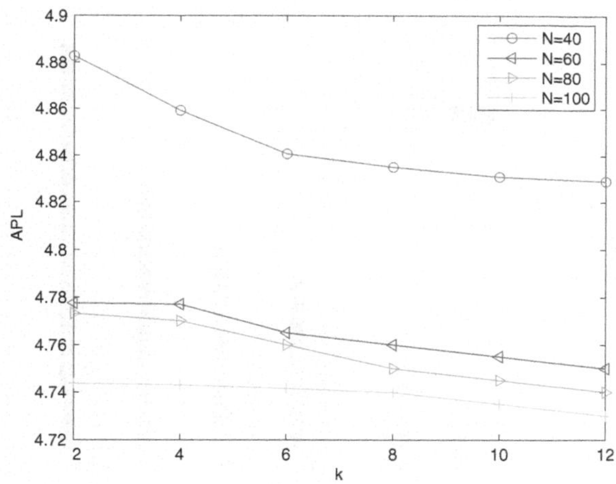

Fig. 16.4 Average path length versus power amplified node numbers

In Fig. 16.5, it represents that the PCS algorithm has larger energy consumption than the SCR algorithm when the number of total nodes is small in the network ($70 * 20$). Then with the number of total nodes increasing, the energy consumption of the PCS algorithm is becoming less than the SCR algorithm. This is because of the increasing number of total nodes, the network density is getting higher and the distance between power amplified node and its shortcut node is relatively getting closer. As a result, the power magnification can dwindle correspondingly. Note that the curve of the WSR algorithm is not given in Fig. 16.5, this is because the energy consumption of the WSR algorithm is rising linearly. In Fig. 16.6, the overheads of the three algorithms are rising with the increasing number of total nodes, especially,

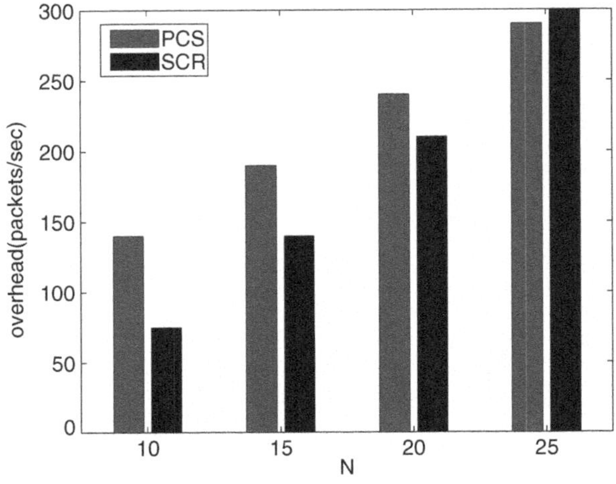

Fig. 16.5 Node numbers versus overheads

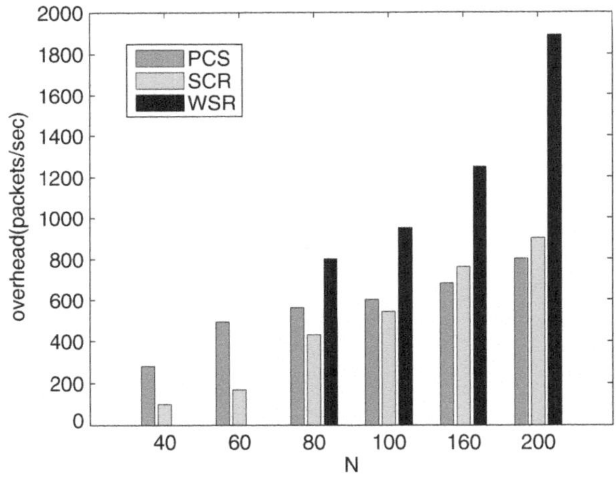

Fig. 16.6 Node numbers versus overheads

the overhead of the WSR algorithm is increasing at most. Whereas, the PCS algorithm has smaller growing rate and its overhead becomes lower than the SCR algorithm gradually.

16.5 Conclusions

In this paper, a power control algorithm based on the small world theory is proposed to shorten the average path length and reduce the network delay in airborne PHM. First, the PCS algorithm applies the small world theory into the network by selecting k

power amplifier nodes with long-range power to construct shortcuts. Second, genetic algorithm is used to optimize the shortcuts, after which the optimal ones can shorten the average path length at most. Finally, the PCS algorithm is clearly verified to reduce the network delay by the experiments conducted in this paper. Furthermore, the algorithm has shorter average path length compared with the WSR and SCR algorithm, and the efficiency of the proposed algorithm is also demonstrated in large-scale sensor networks.

Acknowledgments This work is supported by the National Natural Science Foundation of China under grant NO.61262020; the Aeronautical Science Foundation of China NO.2012ZD56; the Natural Science Foundation of Education Bureau of Jiangxi Province NO.GJJ12460; the Nanchang Hang Kong University doctoral Sustentation Fund NO.EA201120180.

References

1. ZENG S-K, Michael G, Pecht, WU Ji (2005) Status and perspeetives of prognostics and health management technologies. ACTA AERONUTICA ET AST RONAUTICA SINICA 5:25626–25632
2. Zhou Yang, Bo Jing (2011) Application of wireless sensor network on airborne of prognostics and health mangement. J Comput Res Dev S2:338–342
3. Andy H (2001) The joint strike fighter(JSF) prognostics and health management. In: NDIA 4th annual systems engineering conference, pp 2799–2813
4. Harri H, Andre S, Pekka O (2010) Distributed algorithms for lifetime maximization in sensor networks viaMinCMax spanning subgraphs. Wireless Netw 16:875C887
5. Demo J, Steiner A, Friedersdorf F (2010) Development of a wireless miniaturized smart sensor network for aircraft corrosion monitoring. In: Proceedings of IEEE aerospace conference, pp 1–9
6. Cheng SF, Azarian MH, Pecht MG (2010) Sensor system for prognostics and health management. Sensors 10(6):5774–5797
7. Gkoutioudi K, Karatza HD (2012) A simulation study of multi-criteria scheduling in grid based on genetic algorithms. In: Proceedings of IEEE 10th international symposium, IEEE Press, Leganes, pp 317–324
8. Aziz, AA, Sekercioglu YA, Fitzpatrick P (2013) A survey on distributed topology control techniques for extending the lifetime of battery powered wireless sensor networks. IEEE Commun Surv Tutorials 1(15):121–144
9. Kubisch M, Karl H, Wolisz A (2003) Distributed algorithm for transmission power control in wireless sensor networks. In: Proceedings of IEEE WCNC 2003, IEEE Press, New Orleans
10. Wattenhofer R, Zollinger A (2004) XTC: a practical topology control algorithm for ad-hoc networks. In: Proceedings of the international parallel and distributed processing symposium, IEEE Press, New Mexico, pp 216–223
11. Kawadia V, Kumar PR (2003) Power control and clustering in ad-hoc networks. In: Proceedings of the IEEE Conference on Computer Communications, pp 459–469
12. HUANG H-J, HU G-M, YU F-C (2011) Cross-layer power controlled routing in wireless ad hoc networks. Appl Res Comput 28(5):1793–1798
13. Utku GA (2010) Weak state routing for large-scale dynamic networks. IEEE/ACM Trans Networking 54(4):573–588
14. Sheng M, Li J, Li H, Shi Y (2011, June) Small world based cooperative routing protocol for large scale wireless ad hoc networks. In Communications (ICC), 2011 IEEE International Conference on, IEEE, Chicago, pp 1–5

15. Helmy A (2003) Small worlds in wireless networks. IEEE Commun Lett 7(10):490–492
16. Helmy A (2002) Small large-scale wireless networks: mobility-assisted resource discovery. arXiv preprint cs/0207069
17. Cavalcanti D (2004) Exploiting the small-world effect to increase connectivity in wireless ad hoc networks. In: International conference on telecommunications, Fortaleza, p 388–393

Chapter 17
Energy Efficient Multicast Routing with Delay Constraint in Bus-Based DTNs

Lili Ren, Qingshan Wang, Shasha Fu and Yanping Liu

Abstract Delay Tolerant Networks (DTNs) have drawn much attention due to the widely applications deployed in challenging environments. Because of the low node density, high mobility and short radio communication range, a complete path between the source and the destination may not exist. In bus-based DTNs, the buses move following the scheduled trajectories and are equipped with devices which provide two kinds of communication, including short-distance communication and long-distance communication. This paper studies the energy efficient multicast routing problem with delay constraint for each destination. We formulate it as an integer programming. The simulation results show that our optimal model can achieve higher energy saving than the scheme only considering the long-distance communication.

Keywords Multicast · Energy consumption · Integer programming · Bus · DTNs

17.1 Introduction

Delay tolerant networks (DTNs) are emerging as opportunistic networks that experience frequent and long duration partitions due to nodes moving or unpredictable network topology. The vehicular networks, ad hoc networks, military networks and pocket switched networks (PSNs) are some applications of DTNs. Among these types

L. Ren (✉) · Q. Wang · S. Fu · Y. Liu
School of Mathematics, Hefei University of Technology, Hefei 230009, China
e-mail: renlili808826@163.com

Q. Wang
e-mail: qswang@hfut.edu.cn

S. Fu
e-mail: fushasha8161@163.com

Y. Liu
e-mail: enmy0517@163.com

X. Wang et al. (eds.), *Advanced Technologies in Ad Hoc and Sensor Networks*,
Lecture Notes in Electrical Engineering 295, DOI: 10.1007/978-3-642-54174-2_17,

of networks, traditional ad hoc routing protocols which rely on end-to-end connectivity fail to work. Thus a store-carry-forward mechanism is proposed to establish communication in which the mobile nodes are employed to carry the message, then relay and forward to the encountered nodes accidentally with some delay around the networks. In flooding mechanism, when two nodes encounter by random chance, they can copy message from each other. And the message storm can be avoided by limiting the life and quantities of the message in the networks. But this routing mechanism consumes a lot of energy and increases the burden of networks. However, of all the distributed systems, the multicast routing in DTNs has received a certain attention and been widely used. By sharing the common routing paths of the multicast protocols, it costs less energy and guarantees the message to be transmitted with less latency among multiple destinations. Although the existed schemes allow us to minimize the transmission energy and maximize the delivery rate, there is no appropriate method due to the characteristic of DTNs.

In recent years, several papers have studied the research about energy saving and cost in DTNs for multicast. An effective routing algorithm (ERHR) [1] is proposed and its performance is superior to other DTNs routing algorithms on energy constraint. A stochastic optimal approach [2] which considers two decisions at different time scales achieves a good tradeoff between the provable power cost and latency in delay tolerant networks. In term of the energy saving, the message dissemination process [3] is modeled under different mobility velocity by a continuous-time Markov model for discovering the most optimal policy with constant velocity significantly. The K-neighbor broadcast scheme [4] is that the source node waits for K neighbors and then broadcasts the message. This mechanism can reduce the times of transmission and save the energy efficiently. However, the latency is increased. The cooperation of mobile terminals [5] can achieve energy saving by the combination of WiFi and WiMedia in different channel conditions. Because they can take advantage of the good quality of short-range links. Although the remote transmission [6] costs expensive, it can guarantee the connectivity at any time. Mongiovi Misael et al. employ a graph-indexing system to find the set of remote transmission and contact transmission for reducing the multicast cost. They assume the contact transmission does not incur any cost. Obviously, it is idealized.

According to our knowledge, there is no work considering the energy saving of multicast routing by the combination of short-distance communication and long-distance communication. The paper aims at minimizing the energy consumption of multicast routing with delay constraint. Every node has two kinds of communication devices i.e. short-distance communication device and long-distance communication device. The main technology of short-distance communication is WiFi, Bluetooth and etc. And this communication occurs when the nodes locate in the communication range of each other. The long-distance communication often employs the cellular networks to connect such as 3G technology. Obviously, the energy consumption of the two kinds of communication is different. Thus, this paper models the question as an integer programming to find the optimal solution and the main idea is to save the energy by making full use of short-distance communication.

17.2 System Model and Research Problem

In this section, we introduce our network model and the research problem. Here, we model buses as nodes in vehicle network and number them from 1 to n. These nodes $(c_1, c_2, \ldots, c_n)$ move across the scheduled trajectories $(T_1, T_2, \ldots, T_n)$ respectively. The central server is regarded as a special node and numbered 0. At each time instant in the range $(t_0, t_{\max})$, a node corresponds to a certain position in space. Simply, we assume that the communication is so rapid compared to the movement of nodes that the message can be transmitted before the contact ends. When the geographical distance between two nodes is within a certain value d, they can communicate across the Short-Distance Communication (SDC) at a given moment. And d is called the radius of SDC. If node i can communicate with node j at time t, we write for $i \rightarrow_t j$. Moreover, the nodes in the networks can use the Long-Distance Communication (LDC) to get the information from the cellular sever at any time independently. The two kinds of communication consume the specific energy and the LDC consumes more energy. Let E_1 and E_2 be the energy consumption of obtaining one message in LDC and SDC respectively. In order to save energy, we try best to employ SDC as much as possible, rather than LDC. This is the main idea of this paper.

In our model, the message is transmitted to multiple destinations and the sharing message is continuously updated with the time varying. Every destination has a certain delay constraint for the message. Let $I = \{(i, t, \delta)\}$ be the data demand of a message, where i, t, δ represent the destination, the deadline and the latency respectively. That is that destination i can receive the fresher message at the time interval $[t - \delta, t]$. And the instant $t - \delta$ is called the release time that represents the earliest time instant at which destination i receive the message. Here, let the $y_{i,t}$ be the number of LDC, which i represents destination node and t represents the time instant of LDC occuring. The SDC is denoted by a triple (i, j, t), where i, j, t represent the transmitted node, the received node and the time moment of transmission occurring.

Thus, the paper studies the energy efficient multicast routing problem for each destination with delay constraint in bus-based DTNs. And the problem can be described as follows: given the certain trajectories and a data demand I, find the set of LDC and SDC with the minimum energy consumption which satisfies I.

17.3 The Optimized Form of the Problem

In this section, we introduce the energy efficient multicast routing problem with delay constraint and summarize it as IP (Integer Programming) [6, 7].

We define three kinds of Boolean variables. The first variable is of the kind $x_{i,j,t,r}$, where i and j represent nodes, t represents the time instant at which i sends a message to j for satisfying the data need $r = (i_r, t_r, \delta_r) \in I$. This kind of variable models the transmission of message. If node i successfully transmits the message to node j at time t for satisfying the data need r, the variable $x_{i,j,t,r}$ has value 1 otherwise 0. This variable can be substituted for $i \rightarrow_t j$ and $t_r - \delta_r \leq t \leq t_r$. The second variable is of the kind $s_{i,j,t,r}$ that notes the number of SDC between node i and node j at any time t. Because of different kinds of data needs, the number of SDC between node i and node j may be more than one. Thus, we note $n(s_{i,j,t,r})$ as the number of SDC between node i and node j at time t actually. If there is no communication between node i and node j at time t for different kinds of data needs, the variable $n(s_{i,j,t,r})$ has value 0. If the number of SDC between node i and node j at time t for different kinds of data needs is more than one, the variable $n(s_{i,j,t,r})$ has value 1. The reason is that the contact between them only happens once at time t for considering to reduce the energy consumption. The third variable is of the kind $y_{i,t}$ that counts the number of LDC. If there is a transmission between the cellular server and a particular node at a certain time t, the variable $y_{i,t}$ has value 1 otherwise 0. We can formulate the optimized form of our problem as follows:

$$\min\left(\sum_{t\in\tau}\sum_{i=1}^{n} y_{i,t}E_1 + kE_2\right)$$

s.t.

$$\sum_{\substack{t\in\tau \\ t\geq t_r-\delta_r \\ t\leq t_r}} \sum_{\substack{i=0,1,\ldots,n \\ i\rightarrow_t i_r}} x_{i,i_r,t,r} \geq 1 \qquad \forall r=(i_r,t_r,\delta_r)\in I \tag{17.1}$$

$$\sum_{\substack{t'\in\tau \\ t_r-\delta_r\leq t'\leq t}} \sum_{\substack{i=0,1,\ldots,n \\ i\rightarrow_{t'} j_1}} x_{i,j_1,t',r} - x_{j_1,j_2,t,r} \geq 0,$$

$$\forall r=(i_r,t_r,\delta_r)\in I \quad \forall j_1,j_2,t|j_1\rightarrow_t j_2 \tag{17.2}$$

$$s_{i,j,t,r} = \sum_{\substack{t\in\tau \\ t\geq t_r-\delta_r \\ t\leq t_r}} \sum_{\substack{1\leq i,j\leq n \\ i\neq j}} x_{i,j,t,r} \quad \forall r=(i_r,t_r,\delta_r)\in I \tag{17.3}$$

$$k = \sum_{\substack{t\in\tau \\ t\geq t_r-\delta_r \\ t\leq t_r}} \sum_{\substack{1\leq i,j\leq n \\ i\neq j}} n(s_{i,j,t,r}),$$

$$n(s_{i,j,t,r}) = \begin{cases} 0, & s_{i,j,t,r}=0 \\ 1, & s_{i,j,t,r}\geq 1 \end{cases} \quad \forall r=(i_r,t_r,\delta_r)\in I \tag{17.4}$$

$$y_{i,t} \geq x_{0,i,t,r} \quad \forall r\in I \;\; i=1,\ldots,n, \;\; t\in\tau \tag{17.5}$$

$$x_{i,j,t,r}, y_{i,t} \in \{0,1\}$$

Constraint 17.1 indicates that the destination node must receive the message after the release time and before the deadline for each data need r. Constraint 17.2 models the propagation of message. It says that if node j_2 wants to obtain the message from node j_1 at time t for each data need r, then node j_1 must receive the message before t and after the release time $t_r - \delta_r$. Constraint 17.3 represents the number of SDC between node i and node j for different kinds of data needs r at time t. Constraint 17.4 accurately models the number of SDC between node i and node j for different kinds of data needs r at time t. Constraint 17.5 means that if there is a transmission between the central server and node i for each data need r, then the variable of $y_{i,t}$ has value 1 otherwise 0. E_1 and E_2 in the optimized objective represent the energy consumption [10, 11] of the LDC and SDC respectively. The optimization function represents the minimized energy consumption for multicast.

17.4 Simulation

In this section, we compare the performance of the proposed scheme and the scheme which uses the long-distance communication. In urban environment, the movement of buses is often influenced by many factors. Thus, we choose a comprehensive and mobile model which contains more than 1,600 buses in Chicago city for simulation. The duration is 5 h from 7 a.m. to 12 noon on a Monday morning [9]. The trajectory data is obtained from the Chicago Transport Authority (CTA) Bus Tracker API [11]. Due to a lot of variables in our optimized form, we randomly choose 850 buses with two hours trace data to illustrate our scheme in order to simplify the complexity of calculation. The transmission range of short-distance communication is set to 200 m. For each data need, the deadline is randomly generated from the simulation time and the latency is normally distributed ($\mu = 15$ min and $\sigma = 1$). Here, the energy consumption of obtaining a message across the WiFi and 3G interface is 10 and 100 J respectively [5, 8, 10, 11]. The simulator is using C++. Every simulation is run 100 times and the result is its average.

Firstly, we change the number of destination nodes from 250 to 450 with 400 relay nodes unchanged to explore the energy consumption of the schemes. The simulation results are illustrated in Fig. 17.1. Figure 17.1 shows that the energy consumption of the two schemes is increasing with the growth of destination nodes. The reason is that the energy consumption is related to the quantites of destination nodes. Moreover, our scheme can save the energy up to 14.8 % compared with the LDC.

Secondly, we vary the number of relay nodes from 250 to 450 with 300 destination nodes unchanged to explore the energy consumption of the schemes. Figure 17.2 depicts the energy consumption of our scheme and LDC scheme. Figure 17.2 shows that the average energy consumption of our scheme has a 11.67 % decrease by comparing with the LDC scheme. Moreover, we can see that the change of energy consumption is not obvious. The reason is that the number of relay nodes is more enough to find the optimal solution so that it has little impact on our results.

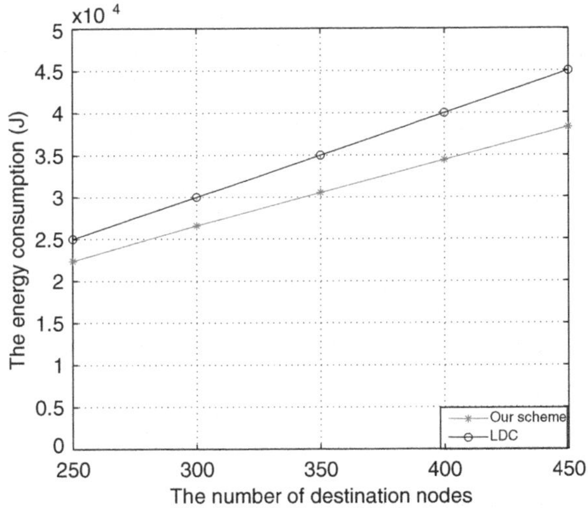

Fig. 17.1 Performance comparison of the schemes under varying the number of destination nodes

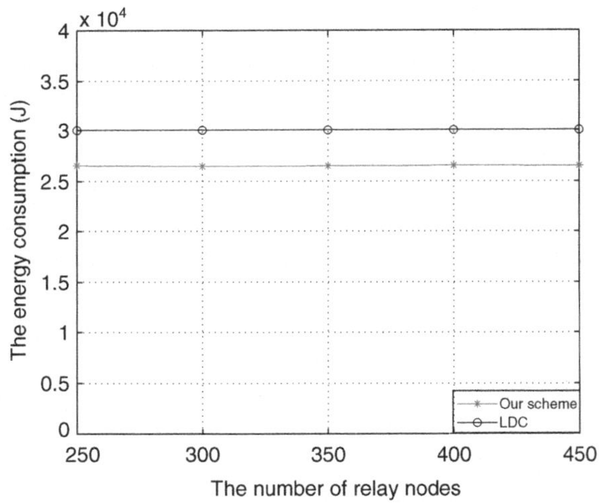

Fig. 17.2 Performance comparison of the schemes under varying the number of relay nodes

Lastly, we vary the simulation time from 1200 to 6000 s. Figure 17.3 presents the energy consumption of the two schemes. Figure 17.3 shows that our scheme can achieve higher energy saving up to 23.7 % improvement by contrast. Moreover, our scheme is gradually close to the LDC scheme with the increasing running time. But the energy consumption of our scheme is still much lower than that of the LDC scheme. The reason is that the buses consume more energy with the longer deadline and the increasing simulation time. As a result, we can observe a new phenomenon that our scheme can save more energy at a shorter deadline.

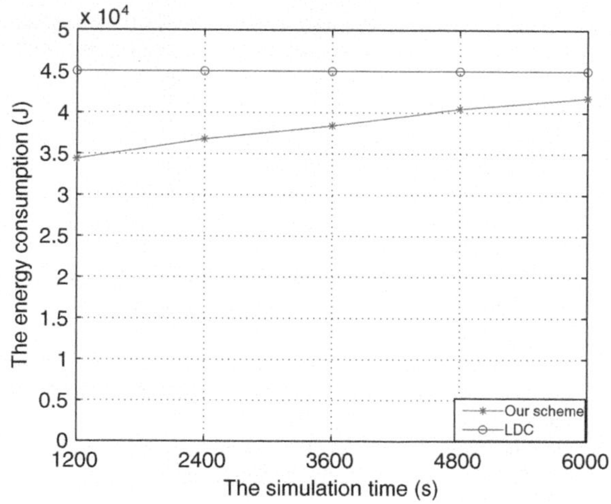

Fig. 17.3 Performance comparison of the schemes by varying the simulation time

17.5 Conclusion

In this paper, we have investigated the energy efficient multicast problem in bus-based DTNs with delay constraint. The main idea is to try to make full use of short-distance communication to get the message for saving the energy consumption. Moreover, we formulate it as an optimization problem to find the optimal solution. The simulations are conducted in the mobile traces of buses in Chicago city, USA. By comparing with the scheme considering the long-distance communication, the simulation results show that our scheme is more energy saving. In our future works, we plan to design an efficient heuristic algorithm to solve our problem and find the relationship between the message propagation and the geographical position to reduce the energy consumption for multicast.

Acknowledgments This work is supported by Anhui Natural Science Foundation, China [1208085MF89, 1308085MF87], SRF for ROCS, SEM [2013JYLH0280], and the Fundamental Research Funds for the Central Universities [2013H GXJ0232, 2013HGXJ0233].

References

1. Wu JJ, Wu RH (2011) A routing algorithm based on energy constraint. In: 3rd international conference on computer research and development, IEEE Press, New York, pp 330–332
2. Yao Y, Huang L, Sharma A, Golubchik L, Michael N (2012) Data centers power reduction: a two time scale approach for delay tolerant workloads. In: Proceedings IEEE INFOCOM, IEEE Press, New York, pp 1431–1439
3. Li Y, Jiang YR, Jin DP, Li S, Zeng LG, Wu DO (2010) Energy-efficient optimal opportunistic forwarding for delay-tolerant networks. In: IEEE transactions on vehicular technology society, vol 59. IEEE Press, New York, pp 4500–4512

4. Goundan A, Coe E, Raghavendra C (2008) Efficient broadcasting in delay tolerant networks. In: IEEE global telecommunications conference, IEEE GLOBECOMIEEE Press, New York, pp 1–5
5. Radwam A, Albano M, Rodriguez J, Verikoukis C (2012) Analysis of energy saving using cooperation use-case: WiFi and WiMedia. In: Future networks and mobilesummit (FutureNetw), IEEE Press, New York, pp 1–10
6. Mongiovi M, Singh AK, Yan XF, Zong B, Psounis K (2012) Efficient multicasting for delay tolerant networks using graph indexing. In: Proceedings IEEE INFOCOM, IEEE Press, New York, pp 1386–1394
7. Cai Y, Zhang LN (2011) Minimum-energy multicast using random linear network coding. In: 7th international conference on mobile ad-hoc and sensor networks, IEEE Press, New York, pp 470–474
8. Doering M, Pögel T, Pöttner WB, Wolf L (2010) A new mobility trace for realistic large-scale simulation of bus-based DTNs. In: Proceedings of 5th ACM workshop on challenged networks, ACM Press, New York, pp 71–73
9. Ristanovic N, Le Boudec J-Y, Chaintreau A, Erramilli V (2011) Energy efficient offloading of 3G networks. In: IEEE 8th international conference on MASS, IEEE Press, New York, pp 202–211
10. Nam M, Choi N, Seok Y, Choi Y (2004) WISE: energy-efficient interface selection on vertical handoff between 3G networks and WLANs. In: 15th IEEE international symposium on PIMRC, IEEE Press, New York, pp 692–698
11. Chicago Transit Authority, Bus Tracker API. http://www.transitchicago.com/developers/bustracker.aspx

Chapter 18
Design and Implementation of an Intelligent Monitoring System Based on ZigBee for the Agricultural Greenhouse

Renfei Wu, Ying Xu, Long Li, Junli Zha and Renfa Li

Abstract This paper describes an implementation of the Internet of things technology to the agriculture production, aiming to design and realize an intelligent monitoring system based on ZigBee a wireless communication technology, in a handmade greenhouse model. In this paper, we also design a control node based on ZigBee, which can control the multiple control equipments in the greenhouse. The system collects the environmental parameters in the greenhouse through a variety of sensors, such as the soil temperature and humidity, the air temperature and humidity, and the light intensity. By applying ZigBee, the collected environmental information is transmitted to a sink node and finally to the monitoring center. The center analyzes the collected parameters and then sends commands to the control node according to the control strategy or sends some warning short messages to the mobile user. According to the received commands, the control node will turn on or off the environmental control equipment in the greenhouse. For example, the control node can turn on or off the exhaust fans to regulate and control the corps' growing environment, which realizes an intelligent agriculture production. The center also provides the remote access mode for users to obtain the real-time environmental parameters and the monitoring videos, which helps the intelligent controlling of various equipments to adjust the corps' growing environment in the greenhouse.

Keywords ZigBee · Internet of things · Intelligent monitoring system

18.1 Introduction

The lack of agricultural resources per capita and the low utilization rate of agricultural resources are the main problems which affect the development of agriculture in China [2]. An effective way to solve this problem is to change the way of

---

R. Wu (✉) · Y. Xu · L. Li · J. Zha · R. Li
School of Information Science and Engineering, Hunan University, Changsha 410082, China
e-mail: wurenfei@hnu.edu.cn

X. Wang et al. (eds.), *Advanced Technologies in Ad Hoc and Sensor Networks*,
Lecture Notes in Electrical Engineering 295, DOI: 10.1007/978-3-642-54174-2_18,

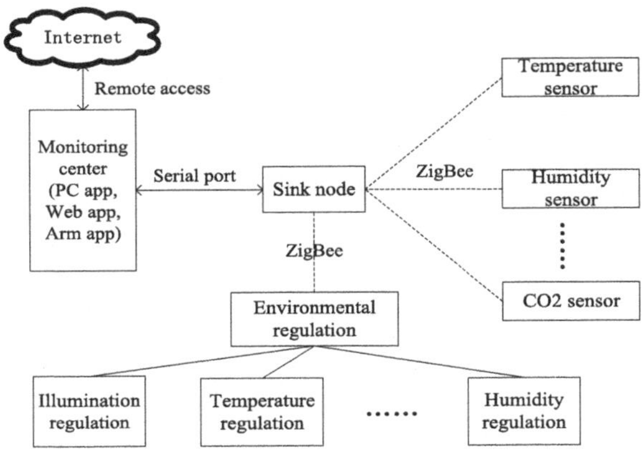

Fig. 18.1 Architecture of the intelligent monitoring system

traditional agricultural production, promote the agricultural technology progress and innovation, and modernize farming equipments. The Internet of things (IoT) technology is an integration of sensing, network communication, and computer science. The application of the IoT technology to the field of agricultural production can effectively improve both the resource utilization rate and the labor productivity. By using sensing technologies, the crop's growth environment information can be obtained, including the illumination, the soil temperature and humidity, and the air temperature and humidity, etc. By using network communication technologies to transmit the sensing information and then using intelligent information processing technologies to process the information, in such a way, both the scientific decision and precision farming standards will be improved (Fig. 18.1).

The ZigBee protocol is based on IEEE802.15.4 standard. It has many advantages, such as a good stability, the low power consumption, and the strong ability of anti-interference and self-organization, etc. [4]. Thus for the IoT and the low-power sensor network, ZigBee has wide application prospects. Greenhouse, as a new crop planting pattern, breakthrough the effects of the traditional agriculture, such as the geography, the natural environment, the climate, and many other factors, is of great significance to the agricultural production. Applying ZigBee in the greenhouse monitoring has become a hot research topic. Zhou et al. designed a ZigBee-based wireless monitoring system used in the greenhouse, focusing on the communication between the coordinator and sensor nodes, and the power consumption of this system. But they did not implement the user graph interface which is very useful for the controlling of the monitoring system [3]. Zhang et al. realized a monitoring hardware system and designed both windows application and Web application for users, but users could only achieve the temperature and humidity information, could not adjust the greenhouse's environment [8]. Di et al. developed application based on Android system which has a good interface, but lack the controlling of the environmental parameters [5]. References [1, 6] studied how to reduce power consumption, considering the system's energy efficiency.

At present, the study of monitoring system based on ZigBee that used in the greenhouse mainly focuses on information collection and power consumption. Current researches are lack of the intelligent controlling and regulation of the environment in the greenhouse. We designed and implemented a monitoring system with both information collection and greenhouse environmental controlling. In addition, to improve the system reliability, a retransmission confirmation mechanism is used to each controlling signal to ensure a safe and reliable control.

The rest of the paper is organized as follows. In Sect. 18.2, we present the brief introduction of the architecture of our designed monitoring system. Section 18.3 gives the details of the hardware and software design. We evaluate our monitoring system in Sect. 18.4. Finally, Sect. 18.5 concludes this paper and brings forward the possible future work.

18.2 System Architecture

The intelligent monitoring system mainly consists of three parts: the information collection & transmission subsystem, the environmental regulation & transmission subsystem, and the monitoring & controlling center. The information collection & transmission subsystem periodically collects the environmental information from sensor nodes in the greenhouse, then transfers these information to the sink node (i.e., the coordinator) using the ZigBee wireless transmission protocol, and finally the sink node sends these information to the monitoring center through the serial port. The environmental regulation & transmission subsystem receives the control commands from the monitoring center, and then transfers these commands to turn on or off the controlling equipment such as the exhaust fans to regulate the greenhouse's environment. To make the system more flexible, we design the monitoring center with three client software, namely the PC (personal computer) client, the Web client, and the ARM-based client. Due to the common use of PC, designing the PC client can effectively utilize this resource. Through the Web client, users can conveniently access the monitoring center by Web pages on the Internet. The monitoring center can also be accessed through Web pages by smartphone users. For the high performance and low price, ARM chips become common chips used for embedded systems. We thus implement the monitoring center of three clients to provide more selections for end users. The functions of the monitoring center are as follows: First, it stores the collected information to a database in real-time and provides users with a good graph interface; Second, it sends control commands according to collected information and the controlling strategy; Finally, if there are abnormal environmental situations in the greenhouse, the monitoring center can send short messages to users' mobile to remind users.

18.3 System Design

18.3.1 Hardware Design

Both the information collection & transmission subsystem and the environmental regulation & transport subsystem use CC2530 as the system's transmission chips. CC2530 chips integrate advanced RF (Radio Frequency) transceiver, rich on-chip peripherals, and powerful DMA (Direct Memory Access) controller [5]. Because of its rich on-chip peripherals, for example, USB, UART, and SPI communication interfaces and the A/D (Analog to Digital) converter interface, it provides convenience for the development of various applications based on ZigBee. Powerful DMA controller can ensure the real-time response and improve the CPU utilization. We use wireless dragon company's CC2530 sensor nodes [7]. The sensor node is composed of two parts : the RF module and the sensing module, as shown in Fig. 18.2. Sensors collect the environmental information and convert it into analog signals. The RF module converts analog signals into digital signals through A/D interface and sends them out.

The environmental control subsystem design is also based on CC2530 RF module, which receives the control commands from the monitoring center and then controls the equipments via the GPIO (General Purpose Input/Output) ports, as show in Fig. 18.3. There are several control equipments connected by the CC2530 RF module via the GPIO port, for example, the filament lamp which is a lighting equipment, the exhaust fan which is used to regulate CO_2 concentration and emission of combustible gas, the sun shading board, the electronic water valve, and the alarm buzzer, etc. The artificial lighting equipment can adjust the brightness in the greenhouse. When the AC voltage reaches the zero point, CC2530 RF module input port (P20) will appear a jump. CC2530 RF module can make the thyristor (BTA06) conduct through the output port P01 and then turn on the lamp. By adjusting the time delay between the zero point and the thyristor conduction, CC2530 RF module can control the brightness of the light. The smaller the time delay, the brighter the light. The speed of the exhaust fan can be adjusted by Pulse Width Modulation(PWM). We use a stepper motor to turn on or off the sun shading board.

We choose wireless dragon company's RF2-V210 development board platform as the ARM board. The core processor of RF2-V210 platform integrates Cortex-A8 core chip with 1GHz, the graphics engine, video accelerators, and rich multimedia peripherals. The sink node based on CC2530 RF module, is used for sending commands and receiving collected information through the ZigBee protocol. TC35 GSM Modem is used for short messages alert and control (Fig. 18.4).

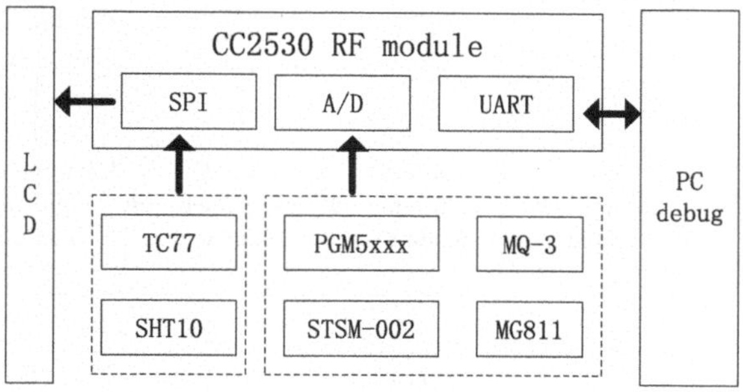

Fig. 18.2 Hardware design of sensor nodes

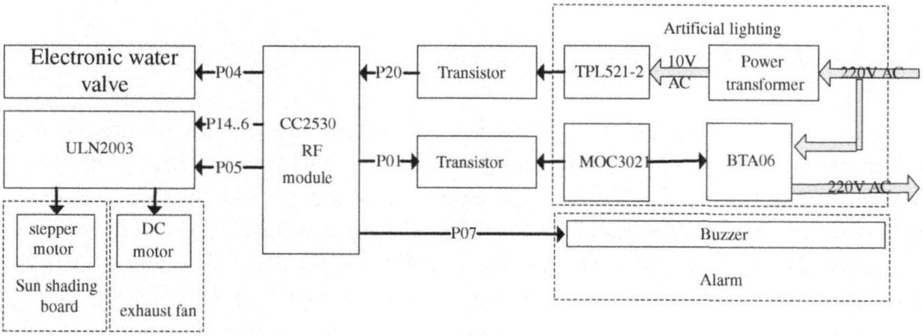

Fig. 18.3 Hardware design of our designed control node

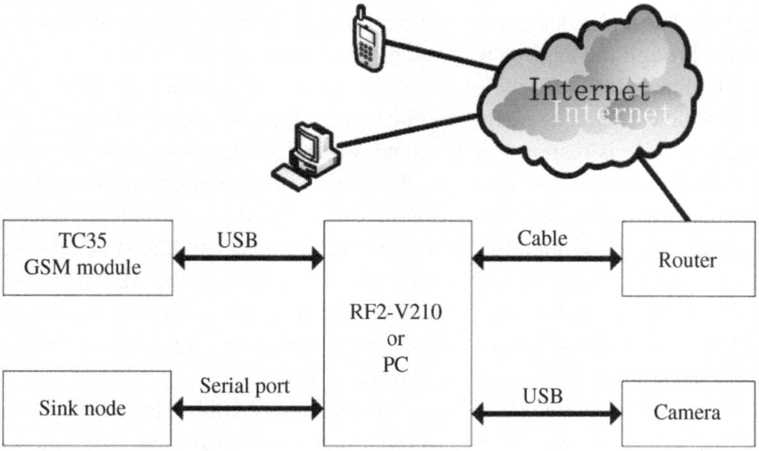

Fig. 18.4 Hardware design of the monitoring center

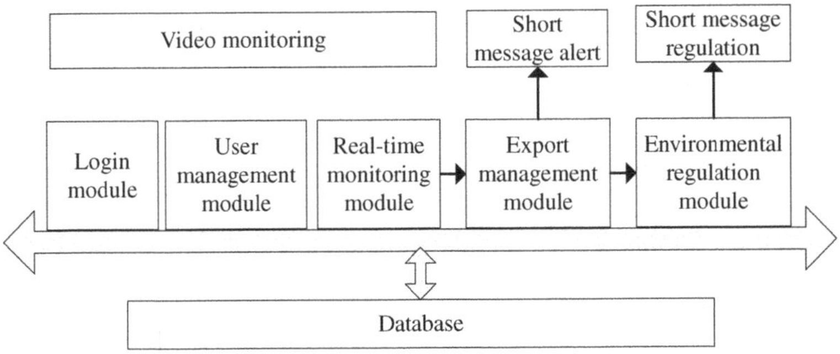

Fig. 18.5 Software design of the monitoring center

18.3.2 Software Design

The information collection & transmission subsystem and the environmental regulation & transport subsystem are based on ZigBee sensor nodes, using Operating System Abstraction Layer (OSAL) to manage resources. OSAL is able to achieve multitask scheduling. The main tasks of OSAL are user tasks and Media Access Control (MAC) system tasks. The tasks have different priorities. The priority of system task of all nodes is the same and the priority of user task is different. Sensor nodes can receive configuration commands from the monitoring center and periodically reads and transfers the environment information from sensors to the monitoring center. The control node receives commends from the monitoring center, then calls corresponding control functions, and returns the execution results to the monitoring center at last. The sink node manages the routing table and forwards information and commands through the ZigBee technology and serial ports.

The functional design of the monitoring center is shown in Fig. 18.5. The login module and the user management module control the user's access rights. Real-time monitoring module is responsible for handling and storing collected environmental information in database for agricultural research, and provides users with a good interface. The environmental regulation module is used for automatically or manually sending control commands. The expert management module stores the suitable growth conditions of various crops. After a certain crop is selected, the expert management module will set thresholds to adjust the greenhouse environment to achieve the suitable growth conditions of the selected crop; Users can observe crops in the greenhouse through the real-time monitoring video.

The PC client is implemented by using Visual Studio 2008. The software interface of the PC client is shown in Fig. 18.6a which displays the curve of the current soil moisture. We develop the Web software interface by Java, which is shown in Fig. 18.6b. The ARM monitoring center is designed based on WinCE 6.0 system, using Microsoft Foundation Class (MFC) for the software development. The ARM software interface with the current greenhouse monitoring video screen is shown in Fig. 18.6c.

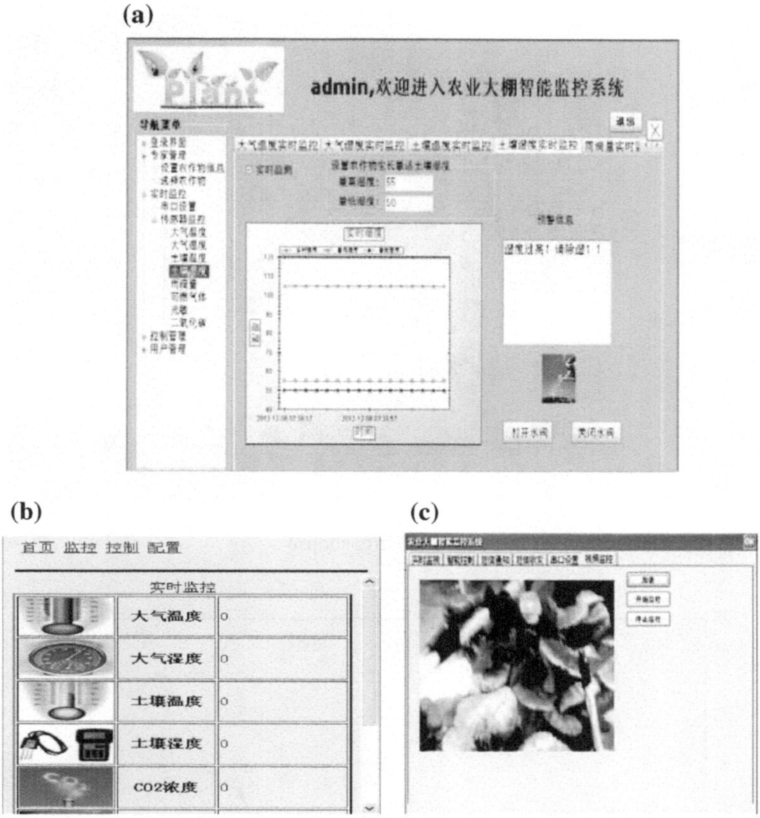

Fig. 18.6 Implementation of the monitoring center software. **a** The PC client, **b** The web client, **c** The ARM client

18.4 System Evaluation

In order to test the intelligent monitoring system, we built a test model of the agricultural greenhouse, as shown in Fig. 18.7. The model consists of seven nodes, including five sensor nodes, one control node, and one sink node.

As an example, we test the illumination regulation of our monitoring system. The sampling period is 5 min. We put the model in a dark room with a light, which is the simulation of the sun. The suitable illumination intensity interval is set between 10 and 15. When the room light is turned off, the illumination intensity is below 10. Then the system automatically turns on the artificial lighting equipment through the light sensor feedback information and dynamically adjusts the brightness of the light. After the room light is turned on, the illumination intensity exceeds 15. The sun shading board will rotate and keep out the light. When a combustible gas appears,

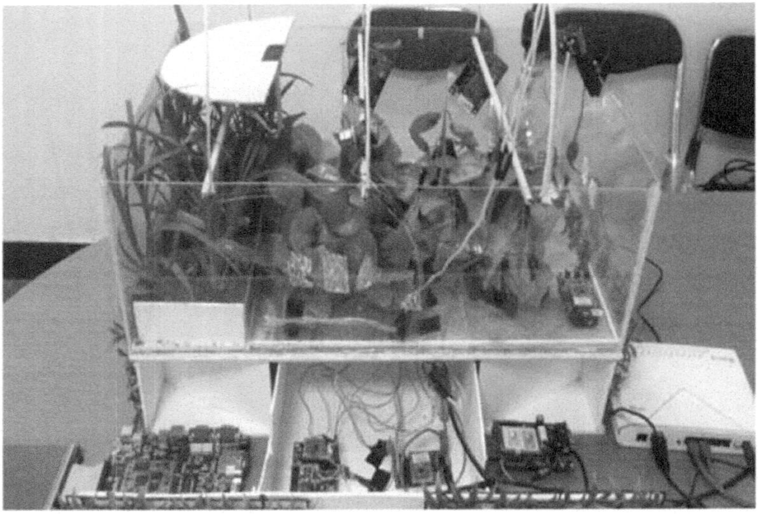

Fig. 18.7 System test model of the agricultural greenhouse

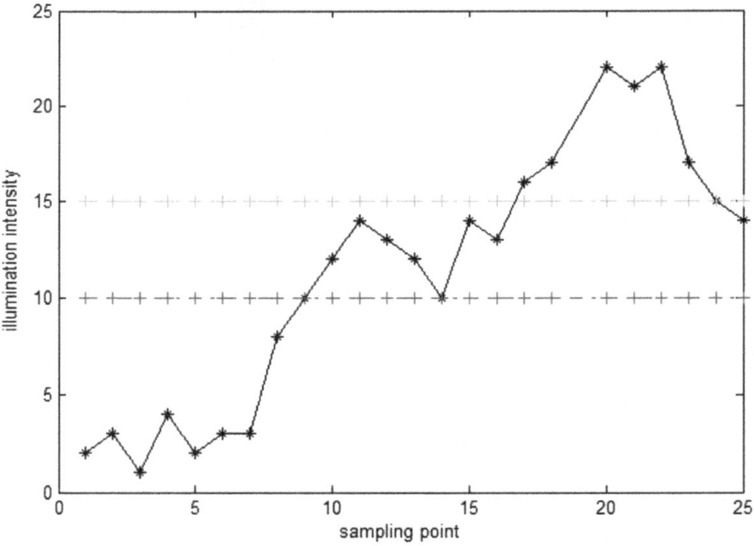

Fig. 18.8 Illumination regulation

the exhaust fan will be turned on until the combustible gas concentration return to be normal (Fig. 18.8).

18.5 Conclusion

This paper designed and implemented an intelligent monitoring system based on ZigBee used in agricultural greenhouse. The system provides multiple client software, including the PC client, the Web client and the ARM client. In addition, users can use short messages to get the crop's growing environmental information and adjust the environment through a control node designed by ourselves. We designed a handmade test model of agricultural greenhouse and tested the system. The results show that the system can monitor the environmental parameters in the greenhouse and regulate the environment simultaneously.

The functions of our system could be further improved. For example, we could add RFID tags to track agricultural products from growing to marketing in the future. We also could use some intelligent learning algorithms to optimize the controlling performance. More regulation equipments may be added, such as the heating equipment, the air humidifier, and the CO_2 or Oxygen generator, and so on.

Acknowledgments This work was Supported by the National Natural Science Foundation of China under Grant No.61173036 and No.61202289.

References

1. Gong S, Zhang C, Ma L, Fang J, Wang S (2010) Design and implementation of a low-power ZigBee wireless temperature humidity sensor network. Computer and computing technologies in agriculture, Nan Chang, 615–622
2. Li Z (2011) Discussions on the internet of things applications in agriculture and key technical requirements. China Institute of Communications, Zheng Zhou, pp 631–634
3. Liu Y (2007) Wireless argricultural information sensor network based on the ZigBee prototype (Master thesis). University of Science and Technology of China
4. Liu J, Jiang C (2007) Design of wireless embedded temperature sampling instrument based on ZigBee technology. J Southwest Univ Sci Technol, 76–79
5. Qu D, Yang J, Chen W (2012) Design of intelligent mobile monitor system based on ZigBee and Android. Microcontrollers embedded systems, pp 10–12
6. Su W, Shao J, Yu C, Wang X, Yang Y (2012) Low power design for greenhouse monitoring system based on ZigBee. Computer Measurement and Control, pp 1812–1815
7. Wireless Dragon Company. RF-CC2530 Experiment Box. http://ec.eepw.com.cn/mfmember/showproduct/userid/34376/id/622370
8. Zhang X (2010) The design and implementation of an agricultural greenhouse control system based on ZigBee (Master thesis). Beijing University of Posts Telecommunications

Chapter 19
Does Such a Large-Scale Wireless Sensor Network Work?

Yanliang Jin, Lina Xu, Zhishu Bai, Xuqin Zhou, Wei Ma and Muxin Wang

Abstract In order to further explore the mechanism about large scale wireless sensor networks, a novel model with deterministic design is proposed. As the network grows, the whole system eventually evolves into a large-scale one. The theoretical analysis and simulation show that the tail of its accumulative degree distribution of nodes along off-diagonal line observes power law property, the average shortest path length linear and the clustering coefficient monotonically increasing. Although building such kind of wireless sensor network might be costly, its overall performance is desirable, and can draw us much more enthusiasm to keep it as upper bound of statistical analysis.

Keywords Large-scale wireless sensor network · Complex networks · Scale-free

19.1 Introduction

Recent advances in wireless communication and MEMS-based sensor technology have enabled the development of relatively a large number of inexpensive and low-power wireless sensor nodes. Nodes sense the environment information and send their reports toward a processing center that is called "sink" or base station (BS), where the data will be made available to the end user. Such networks have a wide range of potential applications, from military surveillance to habitat monitoring [1, 2]. Large-scale is one of the main tendencies for future wireless sensor networks, with hundreds even thousands of nodes jointly cooperating to complete a particular task, and an essential question has attracted the attention of many researchers: Does a wireless sensor network which contains hundreds or even thousands of nodes really work just as it was expected? There have been several large-scale sensor network

Y. Jin (✉) · L. Xu · Z. Bai · X. Zhou · W. Ma · M. Wang
Key Laboratory of Specific Optical Fiber and Light Access of Ministry of Education, Shanghai University, Shanghai 200072, China
e-mail: jinyanliang@staff.shu.edu.cn

X. Wang et al. (eds.), *Advanced Technologies in Ad Hoc and Sensor Networks*, Lecture Notes in Electrical Engineering 295, DOI: 10.1007/978-3-642-54174-2_19,

deployments during the past years, including Vigil-Net [3], Motelab [4], SensorScope [5], Trio [6] and GreenOrbs [7]. The above-mentioned problem, however, is still not resolved [8]. It is of great necessity to gain a general view of such complex networks in order to know better about their macroscopical feature and dynamic nature.

Complex network theory is very effective to study the problem of the large-scale network. Large-scale complex networks contain a wide range of systems existing in nature and society, such as the metabolic network [9, 10], the World Wide Web (WWW) [11], scientific collaboration network, social network and so on, in which nodes are defined as the elements of the system and their links represent the interactions among them. Be it stochastic or deterministic, those are all very complex and total different topological structures, but in the form of degree distribution they have certain similarities. In the theory of complex networks, the degree is defined as the number of a node's links, the degree distribution denotes the probability $P(k)$, which is the ratio of nodes' number whose degree are k to the total numbers of the nodes in the network. In scale-free network, it follows. One of the most classical scale-free networks is first being introduced by Barabasi A and Albert R as the BA model [12, 13]. Networks in the real world usually are not stochastic, only a handful of nodes have many links, but most of them with rather few connections, these with such degree distribution are called scale-free networks. Scale-free's inherent features of robustness and efficiency have been a desirable concern ever since its emergence, and various methods from different research fields have invoked to draw knowledgeable design guidelines.

Such consideration greatly inspires us to carry on designing network from the topological and degree distribution respects to further know network's inherent feature since from the existing literature there is not much theoretic analysis so far. Based on the complex network theory, this paper presents a large-scale cellular network generation model. The theoretical analysis proves that the scaling exponent of its accumulative degree distribution varies from 1 to 2, the average shortest path length is linear and the clustering coefficient is monotonically increasing. The results carried out in this paper would provide the guidelines for network deployment and protocol design in the future applications.

For the remainder of this paper, we organize it as follows. Section 19.2 is cellular topological structure model presentation. In Sect. 19.3, we analytically calculate the accumulative degree distribution, the average shortest path length and the clustering coefficient according to previous definitions. In Sect. 19.4, we test and verify them via simulation and finally conclusion is drawn for this paper in Sect. 19.5.

19.2 Model Presentation

The cellular topological structure model we construct is based on the common virtual hexagonal topology [14, 15], as is shown in Fig. 19.1. Assuming there are N wireless sensor nodes being distributed in a selected domain who self-organize to generate a cellular network and choose the BS (Base Station) as the centralized control. We

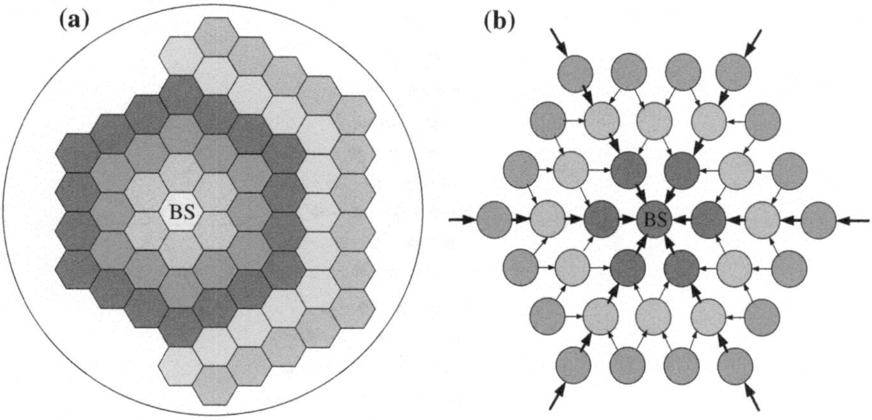

Fig. 19.1 Cellular topological structure of WSNs

use a series of virtual hexagons to cover such field, and every hexagon is called a cell. In order to transmit data as soon as possible, undoubtedly we define that data is being relayed by these nodes from outer layer to the inner using minimal hops, so the data will not be transmitted among nodes in the same lay (cells of the same color in Fig. 19.1a).

For the sake of analyzing and distinguishing easily, we draw another graph as Fig. 19.1b, in which the whole field is being compartmentalized by three overstriking lines, namely the diagonal lines. They each have a 60° angle of intersection and their intersection point is right at the BS. Since data is transmitted among nodes of different layers and can only being relayed by neighbors within one hop, we define that these nodes have links and draw a line between them, while nodes in the same layer do not have such links with each other.

The construction of this cellular topological model follows an iterative form. Right at the initiative, there is only one cell as the BS. The network's scale is growing by adding one layer of cells, always choosing BS as the center. The time steps can be sketched and showed in Fig. 19.2 as follows.

Step 0: There is only one cell at the center of a particular domain, which is the BS.

Step 1: We start building the cellular network from a single BS and six others presented around it, each with 60° angle of intersection, then connect them with the BS.

Step 2: We add one layer of cells at the outer sphere, that is, we add another 12 cells by the similar arrangement, from diagonal lines to off-diagonal ones in order of 60° angles. Then add links to the cells at inner layer with a distance of one hop.

Step 3: Follow step 2 and add one more layer of cells and respective links as the previous iteration.

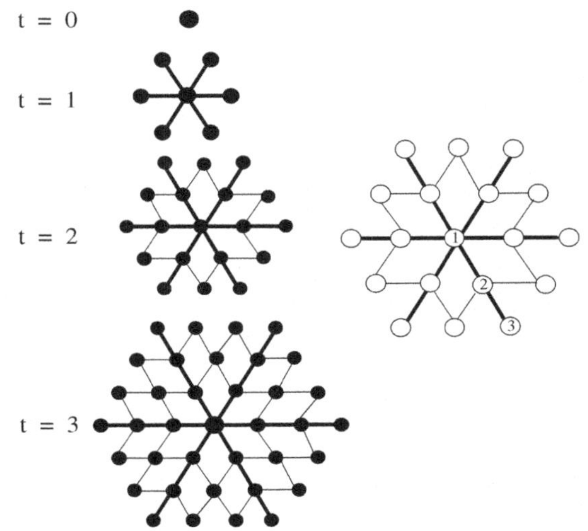

Fig. 19.2 Growth of WSN model

19.3 Theoretical Analysis

With such growth mode, we can see that after several time steps, there would emerge four kinds of cells, respectively the BS, we call them the root node, the bottom node, nodes along the diagonal lines and nodes along the off-diagonal lines.

19.3.1 Degree Distribution

We define that the accumulative degree of a certain node is the total number of the paths from the nodes of outer layers to itself. For example, at step $t = 2$, there are 24 paths to the sink (node 1), and 3 paths to node 2, and no path to node 3 (reference to Fig. 19.2), so their accumulative degrees are respectively as $k = 24, 3, 0$. We enumerate the first five steps in the following Table 19.1. In this paper, n denotes the number of a particular kind of nodes and k denotes the corresponding accumulative degree.

As the virtual cellular wireless sensor network continues to grow, it finally self-organizes into a large-scale one, we calculate that at time step t, there will be $(3t^2 + 3t + 1)$ nodes and $(2t - 1)$ kinds of degrees in all, we list them as follows:
(a) The root node:

$$n = 1, \quad k = 6(2^{t+1} - t - 2) \tag{19.1}$$

(b) The bottom nodes:

$$n = 6t, \quad k = 0 \tag{19.2}$$

Table 19.1 Degrees and numbers for the first five time steps

t	k n								
$t=1$	$k=6$	$k=0$							
	$n=1$	$n=6$							
$t=2$	$k=24$	$k=3$	$k=0$						
	$n=1$	$n=6$	$n=12$						
$t=3$	$k=66$	$k=10$	$k=3$	$k=1$	$k=0$				
	$n=1$	$n=6$	$n=6$	$n=6$	$n=8$				
$t=4$	$k=156$	$k=25$	$k=10$	$k=6$	$k=3$	$k=2$	$k=0$		
	$n=1$	$n=6$	$n=6$	$n=6$	$n=6$	$n=12$	$n=24$		
$t=5$	$k=342$	$k=56$	$k=25$	$k=14$	$k=10$	$k=6$	$k=3$	$k=2$	$k=0$
	$n=1$	$n=6$	$n=6$	$n=6$	$n=6$	$n=12$	$n=6$	$n=18$	$n=30$

Table 19.2 Nodes' k and n along the off-diagonal lines

k	2	6	$\cdots$	$2^{t-3}-2$	$2^{t-2}-2$	$2^{t-1}-2$
n	0	$6(t-3)$	$\cdots$	18	12	6

(c) Nodes along the diagonal lines ($t \geq 2$):

Each of them has a fixed $n = 6$, and there are $(t-1)$ kinds of degrees which ranges in the following order:

$$k = 3, 10, \cdots, 2^{t-1} - t - 1, 2^{t} - t - 2, 2^{t+1} - t - 3 \tag{19.3}$$

(d) Nodes along the off-diagonal lines ($t \geq 3$):

There are $(t-2)$ kinds of degrees but each of them contains different numbers of n. We list them below Table 19.2.

Under these considerations, we have calculated the exact accumulative degree distribution of the whole network ausing statistical definition of accumulative degree distribution, which denotes

$$P(k) = \frac{\text{the number of cells whose accumulative degree is } k}{\text{the total number of cells}} \tag{19.4}$$

19.3.2 Average Shortest Path Length

In WSNs, all the destination of data transmisson is the sink node. In this paper, we denote average hops from the node to sink node as the whole network's average shortest path length. From the above calculation, we can get it:

$$\overline{L} = \frac{1}{N_t} \sum_{i=1}^{N_t} d_{iBS} = \frac{2t+1}{3} \tag{19.5}$$

19.3.3 Clustering Coefficient

Clustering coefficient is an indicator being used to evaluate nodes' gathering state in a network, it denotes the degree to which a certain node's neighbors themselves have neighboring relations. In this paper, we set:

$$C_i = \frac{E_i}{k_i(k_i - 1)/2} \tag{19.6}$$

in which E_i is the number of out-layer nodes' links reaching the sink node via node i. k_i means the actual out-layer node number of node i, while $k_i(k_i - 1)/2$ means the maximum links available of node i's out-layer nodes themselves to have neighboring relations. Then for the whole network, its clustering coefficient is set as the average value of all the nodes' clustering coefficients.

$$C = \frac{1}{N_t} \sum_{i}^{N_t} C_i = \frac{1}{N_t} \sum_{i}^{N_t} \frac{2E_i}{k_i(k_i - 1)} \tag{19.7}$$

19.4 Experimental Results

Simulation via MATLAB shows in Fig. 19.3, we put variable k and its probabilities $p(k)$ in double logarithmic coordinates. With different time steps t, namely t = 6, 100, 500, 1000, we get four curves made of various of discrete data points, which clearly show that the tail of the accumulative degree distribution is power damping, and their scaling exponents are respectively as γ = 1.9095, 1.0137, 1.0026, 1.0013.

For other time steps t, using complementary distribution, we gain that the scaling exponent γ varies from 1 to 2 without exception. According to certain research, there are compact relationships among scale-free networks' degree distribution, its topological structures and spreading dynamic behaviors [16].

When the network's γ is within the scope of 1 to 2, it denotes that such kind of network can not only virtually designed, but can also exist in the real world. Since it contains too much hubs, if building it, the cost will be rather expensive to make it maintain a good performance. We also can see that those nodes, along the diagonal lines, have fixed numbers for a certain k, as their n are constantly fixed at n = 6 and their accumulative degrees varies greatly.

The Fig. 19.4 shows the average shortest path length of the network is linear and proportional to the time step t.

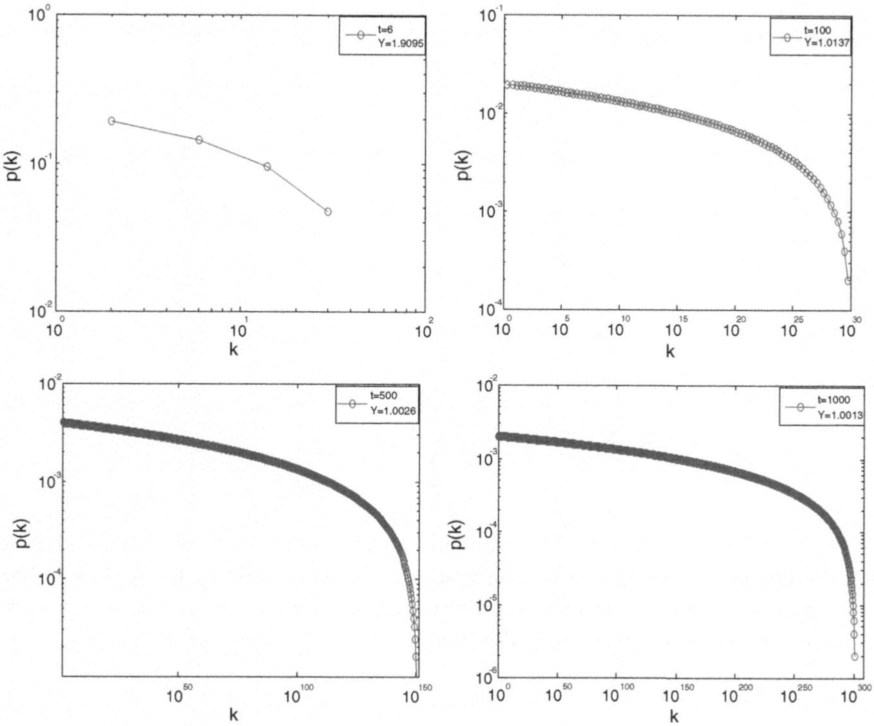

Fig. 19.3 Probabilities $P(k)$ versus k at different time steps

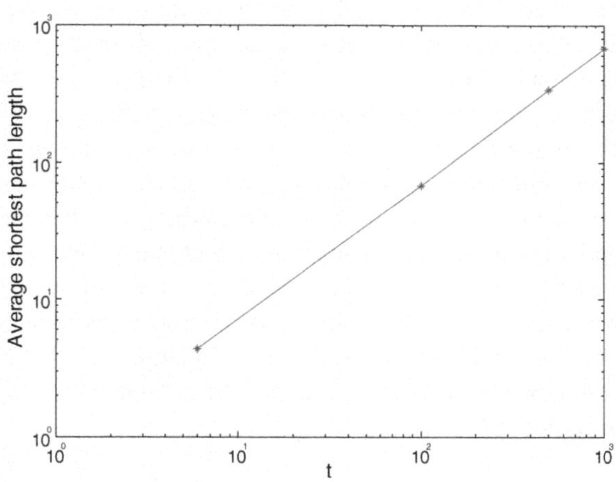

Fig. 19.4 Average shortest path length versus time steps

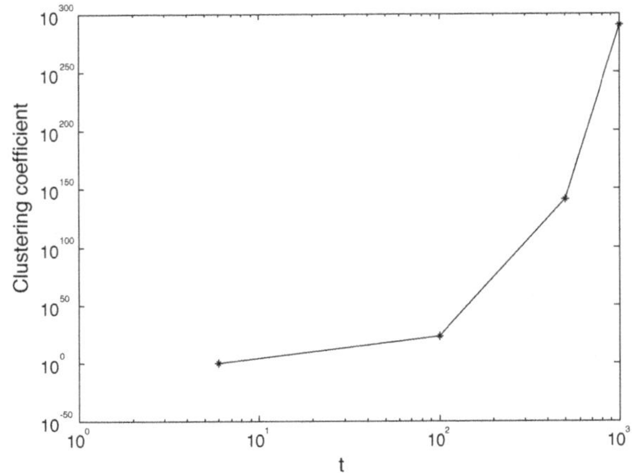

Fig. 19.5 Clustering coefficient versus time steps

The clustering coefficient of the networt is given in Fig. 19.5. We can see clearly that the clustering coefficient is monotonically increasing when the node number is increasing. And with the increase in the size of the network, the entire network gradually increased tendency to gather into a group.

19.5 Conclusions

The model we proposed generates a network whose nodes' accumulative degree distribution along the off-diagonal lines observes power law property. It has scale-free feature to some extent, following $P(\mathrm{k}) \propto k^{-\gamma}$, with degree exponent γ between 1 and 2. The network's average shortest path length and clustering coefficient have also being present. Such network does exist, but can be costly to build and different from traditional network structures to keep it functioning well. Close observe and calculation hint that all the results in this paper have actually reaching the theoretic upper bound, which means if deployed in real environment it will definitely perform even worse. However, this is just an initial step we take in our research, for future work we will focus on more tough issues, including the formation of hexagonal topology and statistics deployment in large-scale WSNs. These will surely be more complex, but can provide the guidelines for network deployment and protocol design in future applications.

Acknowledgments The authors wish to express our sincere gratitude to Professor Dinghua Shi for his kind help and support during the whole research.

References

1. Pottie G, Kaiser W (2000) Wireless integrated network sensors. Commun ACM 43(5):51–58
2. Jin YL, Lin HJ, Zhang ZM (2008) Estimating the reliability and lifetime of wireless sensor network. In: Proceedings of the 4th IEEE international conference on wireless communications, networking and mobile computing, 2008 (WiCOM2008)
3. He T, Vicaire P, Yan T, Cao Q, Zhou G, Gu L, Luo L, Stoleru R, Stankovic JA, Abdelzaher TF (2009) Achieving long-term surveil-lance in VigilNet. TOSN 5(I):1–39
4. Werner-Allen G, Swieskowski P, Welsh M (2005) MoteLab: a wireless sensor network testbed. In: Proceedings of ACM/IEEE IPSN, 2005
5. Barrenetxea G, Ingelrest F, Schaefer G, Vetterli M, Couach O, Parlange M (2008) SensorScope: out-of-the-box environmental monitor-ing. In: Proceedings of ACMI IEEE IPSN, 2008
6. Dutta P, Hui L, Jeong L, Kim S, Sharp C, Taneja J, Tolle G, Whitehouse K, Culler D (2006) Trio: enabling sustainable and scal-able outdoor wireless sensor network deployments. In: Proceedings of ACM/IEEE IPSN, 2006
7. Liu Y, Zhou G, Zhao J, Dai G, Li X, Gu M, Ma H, Mo L, He Y, Wang J, Li M, Liu K, Dong W, Xi W (2010) Long-term large-scale sensing in the forest: recent advances and future directions of GreenOrbs. Front Comput Sci China (Special Issue on Cognitive Sense of China) 4(3):334–338
8. Liu Y, He Y, Li M, Wang J, Liu K, Mo L, Dong W, Yang Z, Xi M, Zhao J, Li XY (2011) Does wireless sensor network scale? a measurement study on GreenOrbs. Paper presented at INFOCOM 2011
9. Erdos P, Renyi A (1960) On the evolution of random graphs. Publ Math Inst Hung Acad Sci 5:17–61
10. Jeong H, Tombor B, Albert R, Oltvai Z, Barabsi A (2000) The large-scale organization of metabolic networks. Nature (Lond) 407:651–654
11. Albert R, Jeong H, Barabasi A (1999) Diameter of the world-wide web. Nature 401:130–131
12. Barabasi A-L, Albert R (1999) Emergence of scaling in random network. Science 286:509–512
13. Wang HA, Dai GZ (2009) 'Complex networks' scale-free features, scale-free phenomenon and control. Science Press, Beijing, pp 54–65
14. Rappaport TS (2005) Wireless communications principle and practice. Prentice Hall, New Jersey
15. Prabh, KS, Deshmukh C, Sachan S (2009, September) A distributed algorithm for hexagonal topology formation in wireless sensor networks. In Emerging technologies and factory automation, 2009. ETFA 2009, IEEE Conference on, IEEE, pp 1–7
16. Dinghua S (2011) Theory of network degree distributions. Higher Education Press, Beijing, pp 34–5

Chapter 20
A Peer Selection Algorithm Based on Tolerance and Behavior Capacity in P2P Streaming Media System

Ruichun Tang, Junwei Hou and Shuangle Guo

Ruichun Tang—CWSN 2013

Abstract Considering the long start-up delay and load unbalance, based on the relation model between user behavior and peer selection, a Peer Selection algorithm based on Tolerance and Behavior Capacity (PSTBC) in P2P streaming media system is proposed. First, tolerance of requested streaming media data is computed and sorted. Second, the behavior capacity of the peer is computed according to the user online time. Then request peer selects the peer which has the best behavior capacity as provider. Simulation results demonstrate the validity of the designed algorithm.

Keywords Peer selection · P2P · PSTBC · Behavior capacity · Tolerance

20.1 Introduction

With the constant improvement in the Internet infrastructure, streaming media research and application based on P2P has been greatly developed. There are tremendous advantages in P2P technology such as higher resource utilization, speed, and lower cost. So the P2P technology, which is in the network file sharing, distributed computing, collaborative systems, and electronic commerce, has a great application prospect.

R. Tang (✉) · J. Hou
College of Information Ocean University of China, Qingdao, China
e-mail: tangruichun@ouc.edu.cn

S. Guo
College of Computer Science Hengshui University, Hengshui, China
e-mail: guoshuangle@ouc.edu.cn

X. Wang et al. (eds.), *Advanced Technologies in Ad Hoc and Sensor Networks*, Lecture Notes in Electrical Engineering 295, DOI: 10.1007/978-3-642-54174-2_20,

Currently, there are already some peer selection strategies in P2P streaming media system. For example, Reference [1] has proposed the random strategy, for vacant streaming media data blocks, the peer randomly selects a peer from the neighbor peers which hold the streaming media data blocks and then requests the peer. The strategy is simple. Reference [2] has advanced the round-robin strategy, which allocates all requests for streaming media data blocks to a peer according to the proportion; if one peer has the streaming media data blocks, then it gets from it, otherwise, from peers which have the maximum available bandwidth. The strategy considers the bandwidth and achieves a certain load balance. Reference [3] has given the Rarest-first strategy, which requests streaming media data blocks that have less potential providers. If a streaming media data block has many potential providers it selects the one that has the maximum bandwidth and enough available time.

The above strategies do not consider user behavior. Reference [4] proposed the algorithm which selected providers considering the bandwidth problems only. Reference [5] proposed the algorithm which selected providers that have more available bandwidth and time. According to research and analysis on user behavior, the user behavior is predictable. Above all, a peer selection algorithm based on tolerance and behavior capacity is proposed. First, the requested streaming media data blocks will be sorted according to tolerance, and then according to user online time, service of provider and bandwidth, calculate the behavior capacity, and select the peer that has the requested streaming media data blocks and has the biggest behavior capacity as the provider. The simulation result shows that the proposed method maintains system stability and improves the system efficiency.

The rest of this paper is organized as follows. Section 20.2 describes the related work. The primary method based on tolerance and behavior capacity P2P streaming media scheduling strategy is detailed in the third section. The fourth section evaluates experimental result. The final section draws the conclusions.

20.2 Related Work

20.2.1 Research Environment

As is shown in Fig. 20.1, P2P streaming media system consists of one or more tracking servers and a number of peers. Each peer has a globally unique label ID. If two peers communicate with each other directly, they are called neighbor peer. Each peer needs to know the ID of neighbor peer and streaming media data information, which is stored in an adjacent table.

20.2.2 Problem Statement

User behaviors include several factors such as popularity, requesting rate, and user online time. The investigation finds that not all users exit after the program ends; user online time is similar to the lognormal distribution. According to this rule, it can

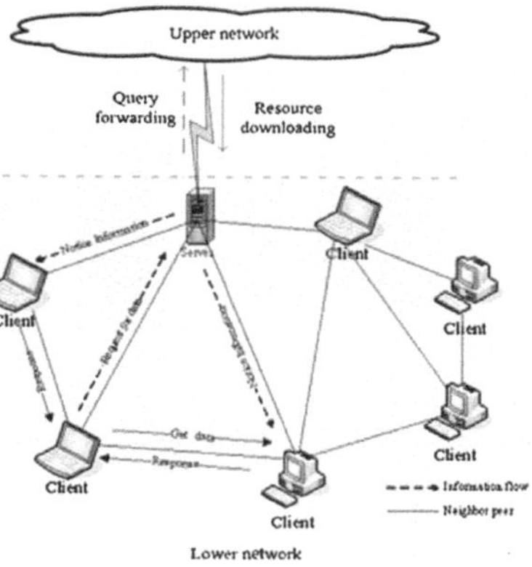

Fig. 20.1 P2P streaming media system environment

be seen that logging users have lower probability of being online, but the probability slowly becomes larger and reaches a certain threshold, then the probability will quickly reduce. In the algorithm request peer should select the peer whose online time is similar to threshold as the provider.

20.3 Peer Selection Algorithm Based on Tolerance and Behavior Capacity in P2P Streaming Media System

20.3.1 Tolerance

When streaming media data transmits between peers, it is divided into equal size streaming media data blocks. Each block is specified a unique sequence number which is used to represent the play order. Each peer has a sliding window [6] to know the situation of playing streaming media; sliding window is used to buffer the streaming media data as shown in Fig. 20.2. Sliding speed of sliding window is the same as the normal playback rate of media content. In the window, the streaming media data information available uses a two-dimensional matrix WindowBuffer (abbreviated as WB). If WB[i, j] = 1, then peer i has a streaming media data cache block j; or not.

To improve the efficiency of streaming media system, the real-time requirement must be taken into account. Therefore, the design of algorithm considers two aspects. One is that streaming media data with fewer providers is likely to cause bottlenecks in the system; another is when the streaming media data is playing, each streaming media data block has a certain order and deadline. To reduce the start-up delay, streaming media data block which has relatively small sequence number should be requested first. In summary, a tolerance for each streaming media data block is given. Tolerance for the streaming media data block *j* and peer *i*:

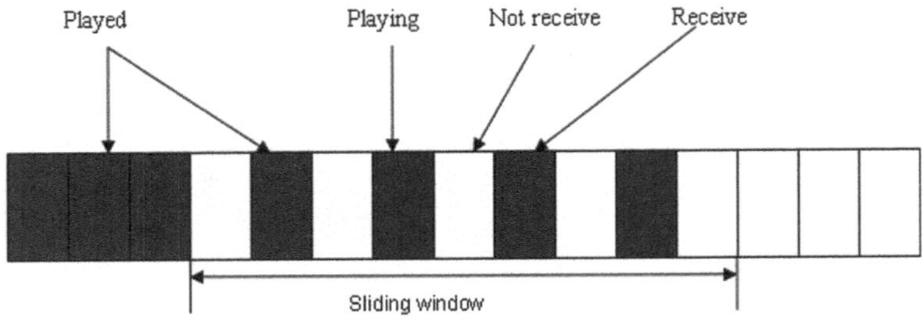

Fig. 20.2 Schematic diagram of sliding window

$$T_i^j = \alpha \times \frac{j - \text{CurrentId}}{WB} + (1 - \alpha) \times \frac{M_j}{K} \tag{20.1}$$

where *CurrentId* is the current playing block number. *WB* is the size of sliding window. *K* is defined as the total number of neighbor peers. M_j is the number of neighbor peers that have streaming media data block j .

Remark 1 $\alpha = 1$. When the peer joins in the system and requests a streaming media data block, in order to reduce the start-up delay, the urgency of the streaming media data block is a main factor affecting tolerance. Then the tolerance only considers the urgency of the streaming media data block. It shows that the smaller the number sequence of streaming media data blocks, the smaller the tolerance, and the higher the priority to request. Therefore, the request peer waits for short time to start-up.

Remark 2 $\alpha = 0$. After the streaming media data of the sliding window is buffered, the peer gets into a state of relative stability. The urgency of the streaming media data block is not the first consideration factor, then modify $\alpha = 0$. Tolerance only considers the rarity of streaming media data blocks. If the streaming media data blocks in the system are scare resources, then the smaller the value of tolerance, the higher the priority to request. This makes the streaming media data be more shared in the system and makes system performance better.

20.3.2 Behavior Capacity Value

Service of provider is defined as the number of peers that the provider services at a time. It affects the service peer calculation of behavior capacity value. If a peer is providing services for some peers, it has relatively small service capacity. If another peer requests service, it has relatively small possibility to become service peer. Therefore, the probability to be a service peer is proportional to the remaining service capacity; in other words, inversely proportional to the number of peers it served.

In order to reach relative fairness, the algorithm guarantees that all peers have equal probability to become a service peer. So, for a streaming media data block i, assuming N_i peers cache it, its service probability P_i is defined as:

$$P_i = \frac{1}{N_i} \tag{20.2}$$

N_i changes as the peers joining in or leaving the system. If N_i increase, P_i decrease, which means that the probability that the peer becomes provider gets small as the peers increase. The value of peer behavior capacity can be dynamically adjusted.

After request sequence of streaming media data blocks is sorted, according to the sequence the peers request to the tracking server and send the streaming media data block id. Tracking server will transmit the information to other peers. The peers which have the requested streaming media data block transmit information to the request peer. Then the request peer selects an optimal peer as a provider. For measuring the peer service capabilities, define the behavior capacity value of the peer i:

$$C_i^k = \begin{cases} \omega P_i + (1-\omega)(\alpha L_i + \dfrac{\beta}{S_i+1} + \gamma \sum\limits_{l=1}^{n} \lambda_l B_i[n-l]),\ L_i < L \\ \omega P_i + (1-\omega)(\dfrac{\alpha}{L_i} + \dfrac{\beta}{S_i+1} + \gamma \sum\limits_{l=1}^{n} \lambda_l B_i[n-l]),\ L_i > L \end{cases} \tag{20.3}$$

where k is streaming media data block sequence number. P_i is the probability of the peer i which can be selected as the provider at first. S_i is defined as the number of peers that peer i services. L_i is online time of peer i. L is threshold. $B_i[n-l]$ is bandwidth of the peer i in the n-l cycle. And $\sum_{l=1}^{n} \lambda_l = 1, \lambda_l > \lambda_2 > \lambda_3 > \ldots > \lambda_n$, $\alpha + \beta + \gamma = 1$. Through $\sum_{l=1}^{n} \lambda_l B_i[n-l]$, the current available bandwidth can be estimated.

20.3.3 Algorithm

The specific algorithm is described as follows:

```
   1) According to the formula of tolerance, choose streaming
media data blocks set.
   Requestdata_Set: sequence of streaming media data, at first
it is null.
   While(k<MAXNUM)
   {According to formula (1), calculates tolerance of streaming
media data block k.
   Find position by tolerance and save it.
   For(each neighbor peer that has streaming media data block k)
   {Get behavior capacity values of neighbor peer i, Capacity[i].
   Send Capacity[i] to requester.
   }
   2) While(Requestdata_Set is not null)
```

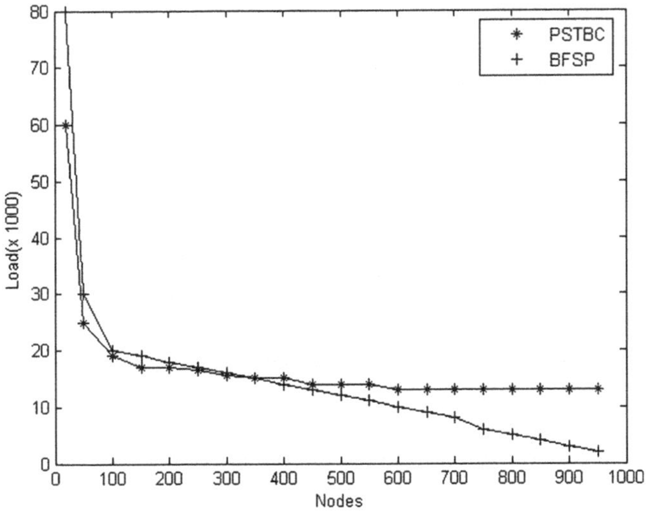

Fig. 20.3 The load of peer

```
   {Select a streaming media data block that has minimum tolerance
from Requestdata_Set to request
   T[k]<T[m],m=0,1,2,L ,k!=m
   For streaming media data block k, select the peer i that have
maximum behavior capacity values
   Capacity[i]>Capacity[m],m=0,1,2,L,i!=m Peer i join in request_Queue,
then delete k from Requestdata_Set.
   Output request_Queue.
```

20.4 Performance Evaluation

To evaluate the algorithm PSTBC, use Microsoft Visual C++ 6.0 to simulate it. The number of peers is 1,000, and the peers are placed randomly. Get the following data from simulation: the average start-up delay of the peer and the peer participation rate. Comparison with Bandwidth First Strategyon Plethora in Plet hora network.

Figure 20.3 shows the comparison result of load of the algorithm BFSP and the algorithm PSTBC in the same environment. In this figure the value of y-coordinate denotes the total number of information of peer routing forwarding in 300 s. It can be seen from Fig. 20.3 that the load distribution of algorithm BFSP is very uneven and load of some peers is low. In contrast, load of these peers is almost the same in algorithm PSTBC, and from 300, remains at a constant.

Figure 20.4 shows comparison result that BFSP algorithm and algorithm PSTBC change in start_up delay in the same condition. The value of y-coordinate is the average peer start_up delay, it denotes time from hit to playing. As can be seen from Fig. 20.4, the request rate changes, algorithm PSTBC and algorithm BFSP has no obvious changes, but start_up delay of algorithm BFSP is smaller than that of algorithm PSTBC 2 s.

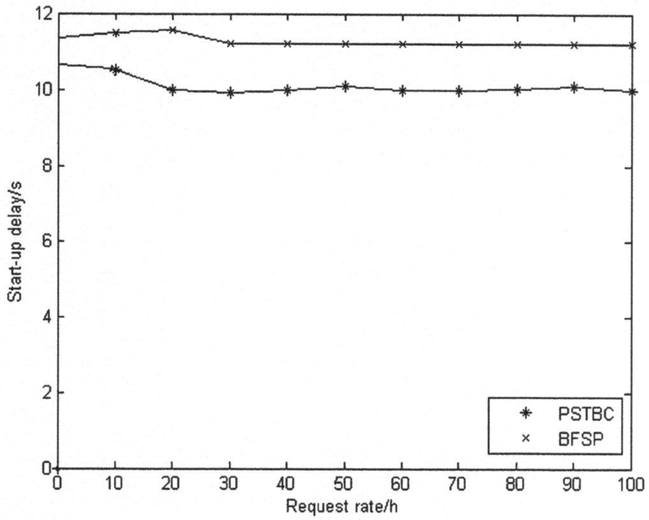

Fig. 20.4 Start_up delay variation as request rate changing

20.5 Conclusion

This paper has researched on the long start-up delay and load unbalance in the P2P streaming media system, and raised based on tolerance and behavior capacity, a peer selection algorithm called PSTBC. First, according to the urgency and rarity of streaming media data, tolerance is computed. Then according to predicting user online time, PSTBC calculates behavior capacity of peer and then selects the best peer, reducing the system start-up delay. Simulation results show the effectiveness of the algorithm.

Acknowledgments This work is supported by the National Science and Technology Support Program Project under Grant No.2012BAH12F00;the National Science and Technology Support Program Project under Grant No.2011BAH16B03.

References

1. Bononi L, Conti M, Gregori E (2000) Design and performance evaluation of an asymptotically optimal backoff algorithm for IEEE 802. 11 WirelessLANs. In: Proceedings of the 33rd Hawaii international conference on system sciences, vol 8. IEEE ComputerSociety, Washington DC, p 8025
2. Wang G, Shu Y, Fan Y (2003) An efficient resource reservation mechanism in AdHoc networks. Canadian Conference on E-lectrical and ComputerEngineering, l IEEE Press, Montrea
3. Terzis A, Srivastava M, Zhang L (1999) A simple QoS signaling protocol formobile hosts in the integrated services Internet. In: Proceedings of IEEE INFOCOM99. IEEE Press, NY, pp 1011–1018

4. Ferreira AR, Jagannathan S, Grama A. (2006) Locality in st ructured peer-to-peer networks. J Parallel Distrib Comput 66 (2) : 257–273
5. Stefan Zoels, Zoran Despotovic, Wolfgang Kellerer (2008) On hierarchical DHT systemsan analytical approach for optimal designs. Comput Commun 31(3):576–590
6. Zheng X, Zheng Q, LI Q (2010) Comprehensive factors-based scheduling algorithm for P2P media streaming. Comput Syst Appl 19(1): 99–103

Chapter 21
Research of Deployment Strategy of Barrier Coverage for Limited Speed Mobile Target

Yunhong Lu, Zhongwen Guo, Qing Liu and Like Qiu

Abstract We propose a deployment strategy to detect the mobile targets based on disk sensing model of barrier coverage and the adjustment of the distance between adjacent nodes. After analyzing the cost and the power consumption of sensor nodes, we propose a strategy by considering the critical conditions of the node radius, sensing time, and the speed of mobile target. An efficient scheduling policy between adjacent nodes is used to improve the maximum speed of detected mobile targets. The simulation results show that the proposed strategy can effectively improve the lifetime of the network.

Keywords Wireless sensor networks · Disk sensing model · Barrier coverage

21.1 Introduction

Wireless sensor network (WSN) is composed of a large number of sensor nodes, which have limited storage capacity and energy. Energy consumption is a major problem in sensor networks. In order to improve the lifetime of the network, many scheduling policies related to power conservation have been proposed.

Coverage is a fundamental issue in wireless sensor network research. Barrier coverage is a special kind of coverage [1]. By constructing belt barriers or torus barriers around protected region, every movement crossing the protected region can be detected [1].

Y. Lu (✉) · Z. Guo · L. Qiu
Department of Computer Science, Ocean University of China, Qingdao 266100, China
e-mail: luyunhong@gmail.com

Z. Guo
e-mail: guozhw@ouc.edu.cn

Q. Liu
Ocean University of China Library, Qingdao 266100, China

X. Wang et al. (eds.), *Advanced Technologies in Ad Hoc and Sensor Networks*,
Lecture Notes in Electrical Engineering 295, DOI: 10.1007/978-3-642-54174-2_21,

Disk sensing model and probabilistic sensing model are common models used to study barrier coverage. Strong barrier coverage and weak barrier coverage are defined based on disk sensing model and an optimal deterministic deployment strategy for k-barrier coverage is proposed by Kumar [2]. As regards random deployment with Poisson distribution, a critical condition for weak k-barrier coverage and a centralized construction algorithm are also presented in [2]. In [3], Liu proposes a critical condition for strong k-barrier coverage and a distributed construction algorithm by means of Divide and Conquer strategy. In [4], Chen proposes a distributed algorithm to judge whether there is local barrier coverage among node deployment region. In [5], Saipulla studies the influence of variance on the barrier coverage when sensors are deployed along a line following a normal distribution. Moreover, the influence of distance between adjacent lines and node offset on the barrier coverage when sensors are deployed along multiple lines is also analyzed. Based on disk sensing model, the relationship of the network detection ability with the node location known or unknown, the shape, and the node density of deployment region is shown in [6]. The node density threshold to detect target is derived based on probabilistic sensing model. Based on disk sensing model and restricted node movement, Saipulla proposes a two-stage node mobile strategy, which is first vertical and then horizontal [7]. The energy consumption between nodes is balanced by minimizing the largest mobile distance of any node, and the effect of the ratio of the mobile node and movement distance on the barrier coverage is analyzed.

Based on disk sensing model, Barr concludes that there is no strong barrier coverage for underwater sensor networks when static nodes are deployed according to Poisson distribution with limited density [8]. Moreover, when the power consumption of any node is minimized, he puts forward a three-stage centralized algorithm for three-dimensional strong barrier coverage with the help of mobile nodes. Based on probabilistic sensing model, a virtual node is built by exploiting mutual cooperation and data fusion of neighbor nodes, converting barrier coverage into line segment coverage [9]. The lower bound estimation for projection length of the virtual nodes is derived, and a centralized algorithm for constructing node sets of barrier coverage is designed. In [10], Xing shows that probabilistic sensing model can effectively improve the network coverage performance than disk sensing model when deployed nodes are with Poisson distribution. It also points out that the disk sensing model is only available for the case of high SNR, while the probabilistic sensing model is suitable for lower SNR by means of data fusion.

Wang [11] shows that probabilistic sensing model with the help of data fusion among adjacent nodes requires fewer nodes for monitoring than disk sensing model. The upper bound for regional coverage and lower bound for energy consumption according to the physical node number of virtual node is presented, and a centralized cooperative detection algorithm with virtual node for regional coverage is proposed.

The above works describe the model of node energy attenuation based on disk sensing model or probabilistic sensing model. When there is an intersection point between the barrier coverage formed by physical nodes or virtual nodes and the trajectory of mobile target, the target is determined to be detected. However, the speed of mobile target is not considered in the above models. In this paper, a deployment

Fig. 21.1 Strong barrier coverage

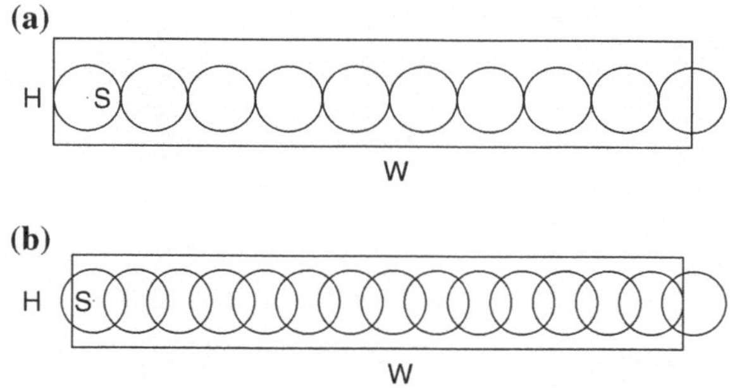

strategy based on disk sensing model is proposed to detect the mobile target with a speed less than critical speed. The lifetime of the network can be improved by adjusting the distance between adjacent nodes.

21.2 System Model

21.2.1 Disk Sensing Model

We assume that the sensor nodes are homogeneous and the sensing region of the nodes is a circular region. The sensing radius of each node is denoted as *r*, and the center is denoted as *S*. For any point *p* (x, y) on the plane, its sensing probability is given by:

$$P_r = \begin{cases} 1 & d(S, P) \le r \\ 0 & d(S, P) > r \end{cases} \tag{21.1}$$

where d(*S*, *P*) is the Euclidean distance from node *S* to point *P*.

21.3 Barrier Deployment

21.3.1 Deployment Strategy

We assume the deployment region of sensor nodes is a belt region, and just concerns the regular rectangular region, $W \times H$, as shown in Fig. 21.1. In [2], Kumar proposed a deterministic deployment strategy for k-barrier coverage, in which *k* rows of nodes were deployed parallel to *W*. The number of sensors needed in this region is $k \times \lceil W/2r \rceil$. The case of $k = 1$ is shown in Fig. 21.1a. We call it deployment strategy one, denoted as DS1.

Fig. 21.2 Intersection area of adjacent nodes

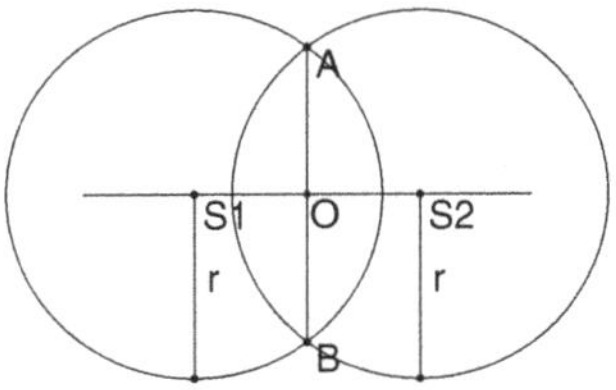

Deployment strategy two depicted in Fig. 21.1b, is denoted as DS2. There are intersected coverage areas between adjacent nodes, thus the required node number of DS2 is more than that of DS1. In DS1, the fastest and least likely to be detected way to pass through the rectangular area is along the path parallel to H and through the intersection point of the adjacent circles. While in DS2, the fastest and least likely to be detected way to pass through the rectangular area is along the intersecting chord of adjacent circles. In order to detect the mobile target, the DS1 requires continuous monitoring, while the sensor nodes in DS2 just need to detect the target during the time period through the intersecting chord. As such, the nodes in DS2 can switch between sensing state and sleep state to save energy.

21.3.2 Lifetime

The energy consumption per unit of time for the sensor nodes is η in sensing state, the total working time is T, and the energy for detection can be expressed as $E_\eta = \eta \times T$. We assume the cost of each sensor node is C, so the cost of barrier coverage in DS1 is $C_1 = C \times \lceil W/2r \rceil$. If $W \gg r$, then $C_1 \cong C \times W/2r$. We define the lifetime per unit cost of the sensor node as T/C, then the lifetime per unit cost of barrier coverage network is given by: $T_1 = T/C = 2rT/CW$.

Figure 21.2 shows the intersection area of adjacent nodes in DS2. We assume the distance between the two adjacent nodes is $2d$, where $0 < d \leq r$, the speed of mobile target is v, where $0 < v \leq v_{\max}$. The distance between point A and point B is $d\,(A, B) = 2 \times \sqrt{r^2 - d^2}$. The number of nodes is $\lceil W/(r + d) \rceil$, and the cost of barrier coverage is $C_2 = C \times \lceil W/(r + d) \rceil$. If $W \gg r$, then $C_2 = C \times W/(r + d)$. When a mobile target pass through AB with the speed $v_{\max}$, the require time is $T_{\min} = 2 \times \sqrt{r^2 - d^2}/v_{\max}$. We assume the sensing time is ΔT, where $0 < \Delta T < T_{\min}$, then the total detection time of each node is $T \times T_{\min}/\Delta T$. The lifetime per unit cost of barrier coverage network is given by:

$$T_2 = T \times \frac{T_{\min}}{\Delta T}/C_2 = \frac{2T\sqrt{r^2 - d^2}(r + d)}{\Delta T\, v_{\max} C W} \tag{21.2}$$

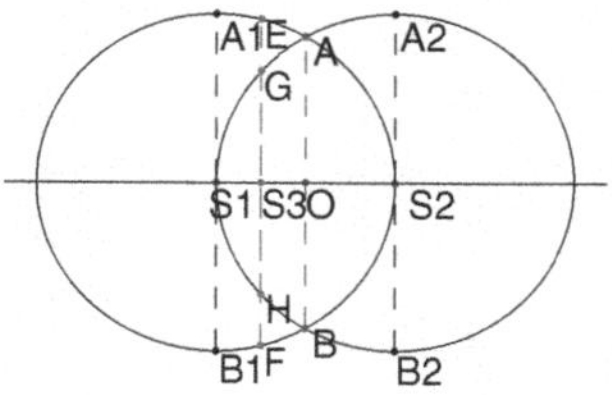

Fig. 21.3 Adjacent nodes when DS2 is optimal

The maximum value of T_2 is $\frac{3\sqrt{3}Tr^2}{2\Delta T V_{\max} CW}$ with $d = r/2$. When $T_2 > T_1$, that is $\frac{3\sqrt{3}Tr^2}{2\Delta T V_{\max} CW} > \frac{2rT}{CW}$, deployment strategy DS2 is better than DS1. The condition can be further simplified as

$$\frac{3\sqrt{3}}{4} \frac{r}{\Delta T v_{\max}} > 1 \tag{21.3}$$

The above condition indicates that DS2 is better than DS1 when the sensing radius is big, the speed of mobile target and the sensing time is small.

21.3.3 Node Scheduling Scheme

Due to the symmetry of the region in DS2, we just need to study the case crossing the region from one of the point during the arc B_1FB, as shown in Fig. 21.3. When the target passes through the region with the speed $v_{\max}$ along AB, the require time is $T_{\min}$. When the target passes through the region along A_1B_1 with the require time of $T_{\min}$, the require speed is $v_{A_1B_1} = \frac{2}{\sqrt{3}} v_{\max}$. When the target passes through the region from any point on the arc B_1FB with time $T_{\min}$, the require speed is greater than or equal to $v_{\max}$. Thus the maximum speed that can be detected is $v_{\max}$.

The deployment strategy DS3 is similar to DS2. In DS3, the sensing start time of the two adjacent nodes has a difference of $T_{\min}/2$, and the maximum speed that can be detected is $\frac{2}{\sqrt{3}} v_{\max}$. Due to doubling the frequency of monitoring, when the mobile target pass through the region along AB, the maximum speed of the target is $2v_{\max}$. In the following case, the mobile target pass through the region along EF, as shown in Fig. 21.3. We assume $d(S_1, S_3) = d$, where $0 < d < r/2$, then $d(S_3, F) = \sqrt{r^2 - d^2}$, $d(S_3, H) = \sqrt{2rd - d^2}$, $d(G, F) = \sqrt{2rd - d^2} + \sqrt{r^2 - d^2}$, and $r < d(G, F) < \sqrt{3}r$. When a mobile target passes through the region from F along EF with the speed $\frac{2}{\sqrt{3}} v_{\max}$, the travel distance is r at time $T_{\min}/2$, which is less than $d(G, F)$. Thus, the mobile target can be detected by one of the two nodes. The maximum travel speed along EF is greater than that along the A_1B_1.

To sum up, the mobile targets can be detected when they pass through the region from one of the points on the arc B_1FB along the line parallel to AB with speed less than $\frac{2}{\sqrt{3}} v_{\max}$. The maximum speed of mobile target is $\frac{2}{\sqrt{3}} v_{\max}$ when the time difference of adjacent nodes is set to $T_{\min}/2$.

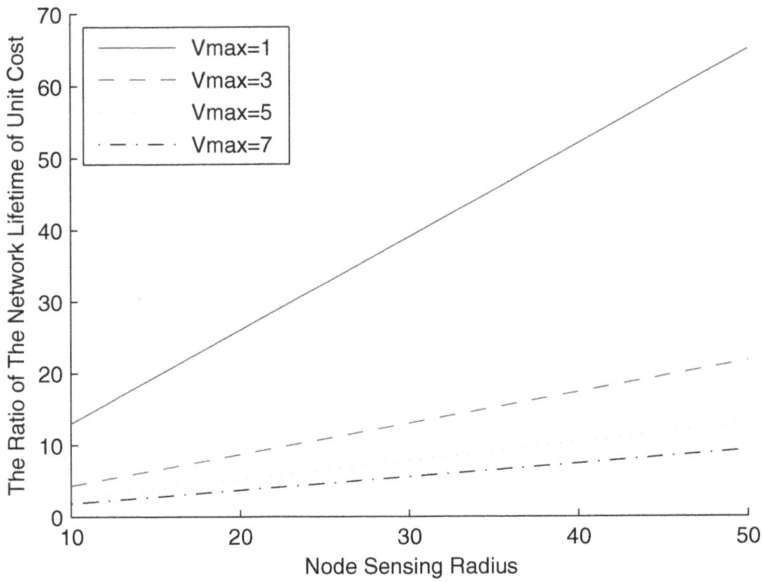

Fig. 21.4 The network lifetime, node sensing radius, and critical speed of mobile target

21.4 Performance Evaluation

We implement three deployment strategies DS1, DS2, and DS3 based on matlab. The relation of barrier coverage network lifetime T per unit cost to node sensing radius r and critical speed $v_{\max}$ of mobile target is analyzed. We assume the deployment region of nodes as 20000×200, the cost per node as 1, the detection time ΔT as 1, the energy consumption per unit of time as 0.01, and the total working time as 10,000.

Figure 21.4 compares the deployment strategy DS2 and DS1. The ratio of the network lifetime per unit cost increases with the increment in sensing radius, while decreasing with the increment of critical speed of mobile target. When the sensing radius is large, the required travel time of the mobile target is long. Thus the sleeping time of node is long, which prolongs the lifetime of the network. When the travel speed is large, the required travel time of the mobile target is short. So the sleeping time of node is short, which reduces the lifetime of the network.

Figure 21.5 compares the deployment strategy DS2 and DS3. The critical speed of the mobile target increases with the increment of sensing radius, while decreasing with the increment of the detection time ΔT. Due to the scheduling policy between adjacent nodes, the maximum critical speed of strategy DS3 is greater than that of strategy DS2.

Figure 21.6 shows the relation of the network lifetime to node sensing radius and the speed of mobile target with DS1, DS2, and DS3 deployment strategies. The node

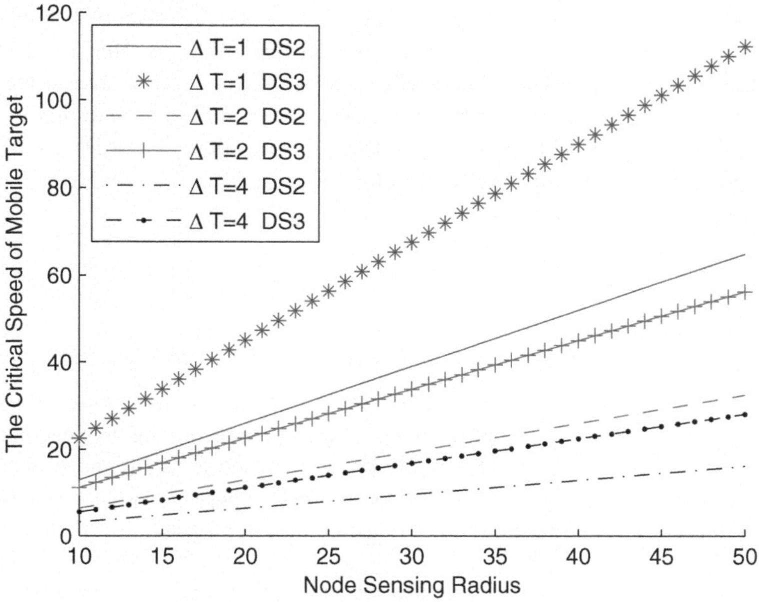

Fig. 21.5 The critical speed of mobile target of DS2 and DS3

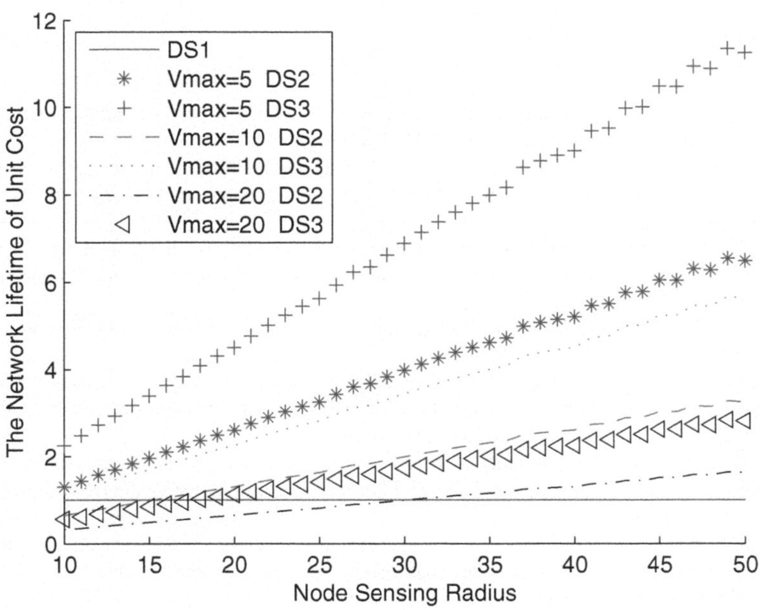

Fig. 21.6 Normalized lifetime of different strategies

detected time ΔT is set to 2. As shown in Fig. 21.6, when the speed of mobile target is set to 20 and the node sensing radius is greater than 30.8, the lifetime of DS2 is larger than that of DS1. When the speed of mobile target is set to 20 and the node sensing radius is greater than 17.8, the lifetime of DS3 is larger than that of DS1. This implies strategy DS2 requires larger sensing radius than that of DS3 to achieve better performance when the speed of mobile target is fast. Thus DS3 is suitable to detect fast-speed mobile targets.

21.5 Conclusion

In this paper, we study the barrier coverage of belt region based on disk sensing model. A deterministic node deployment strategy by adjusting the distance between adjacent nodes is proposed to effectively detect the mobile target and prolong the lifetime of the network. An efficient scheduling policy between adjacent nodes can further improve the maximum speed of detected mobile targets.

Acknowledgments This research was supported by National Natural Science Foundation of China under grant No. 61170258.

References

1. Gage DW (1992) Command control for many-robot systems. Naval command control and ocean survelillance center RDT and E DIV. San Diego, CA
2. Kumar S, Lai TH, Arora A et al (2005) Barrier Coverage with Wireless Sensors. In: Proceedings of the 11th ACM annual international conference on mobile computing and networking, pp 284–298
3. Liu B, Dousse O, Wang J, et al (2008) Strong barrier coverage of wireless sensor networks. In: Proceedings of the 9th ACM international symposium on mobile ad-hoc networking and computing, pp 411–420
4. Chen A, Kumar S, Lai TH (2010) Local barrier coverage in wireless sensor networks. IEEE Trans Mob Comput 9(4):491–504
5. Saipulla A, Westphal C, Liu B, Wang J et al (2009) Barrier coverage of line-based deployed wireless sensor networks. In: Proceedings of IEEE INFOCOM, Institute of Electrical and Electronics Engineers Inc., Rio de Janeiro, Brazil
6. Liu B, Towsley D (2004) A study of the coverage of large-scale sensor networks. In: The 1st IEEE international conference on mobile ad-hoc and sensor systems(MASS), Fort Lauderdale, Florida, USA
7. Saipulla A, Liu B, Xing G, et al (2010) Barrier coverage with sensors of limited mobility. In: Proceedings of the 11th ACM international symposium on mobile ad-hoc networking and computing, pp 201–210
8. Barr SJ, Wang J, Liu B et al (2011) An efficient method for constructing underwater sensor barriers. J Commun 6(5):370–383
9. Yang G, Qiao D (2009) Barrier information coverage with wireless sensors. In: IEEE INFOCOM, pp 918–926

10. Xing G, Tan R, Liu B, et al (2009) Data fusion improves the coverage of wireless sensor networks. In: Proceedings of the 15th ACM annual international conference on mobile computing and networking, pp 157–168
11. Wang W, Srinivasan V, Chua KC, et al (2007) Energy-efficient coverage for target detection in wireless sensor networks. In: Proceedings of the 6th ACM international conference on information processing in sensor networks, pp 313–322

Chapter 22
ZTDMA: A Multi-zones and Multi-objectives Channel Allocation Protocol Based on TOA Real-Time Geolocation System

Cheng Xu, Jie He, Fei Liu, Mowei Lv and Qin Wang

Abstract In order to meet the needs of real-time and dynamic joining of nodes in wireless real-time geolocation system, fully improve the channel utilization rate, propose a specific ZTDMA protocol that divide the geolocation system coverage area into several independent subdomain which can work in parallel. ZTDMA can shorten the average localization period of nodes effectively, and detect the joining and exiting of target nodes in time, dynamically allocate time slot for them, ensure real-time of the system. Then simulate the ZTDMA protocol using OMNet++ simulation platform, in order to prove the superiority of real-time characteristic.

Keywords Multi-zones · Multi-objective · Real-time geolocation system · Channel utilization

22.1 Introduction

Real-Time Location System (RTLS) is a kind of short-distance wireless communication system that provides related services based on the geolocation results through accessing to the information of target nodes in the monitoring area, and it is one of the

C. Xu · J. He (✉) · F. Liu · M. Lv · Q. Wang
School of Computer and Communication Engineering, University of Science and Technology Beijing, Beijing 100083, China
e-mail: hejie1983@gmail.com

C. Xu
e-mail: 377164753@qq.com

F. Liu
e-mail: happy326111@sina.cn

M. Lv
e-mail: lvmowei@gmail.com

Q. Wang
e-mail: wangqin@ies.ustb.edu.cn

X. Wang et al. (eds.), *Advanced Technologies in Ad Hoc and Sensor Networks*, Lecture Notes in Electrical Engineering 295, DOI: 10.1007/978-3-642-54174-2_22,

most promising applications of WSN (Wireless Sensor Network) [1, 2]. RTLS can provide both the identity information of the target and accurate location information, and real-time event detection with the use of data mining, pattern recognition, and other technology. It has extremely important value of application in the scenarios that due to the limitation of environmental factors could not have real-time access to the target location information manually such as the scenarios of underground work, prison management, and fire disaster. RTLS not only has the advantages of low cost, high accuracy, detailed data, and with easy and fast operation, has already been widely used [3].

In practical applications, number of targets is often more than one, namely multiple objectives. Multi-objectives localization process can be divided into target detecting and node task allocation, network data integration, and predicting target state. However, what networking and channel allocation method you take in multi-objective location network with more than one stations will directly affect the efficiency and precision of geolocation system.

Among the existing real-time geolocation systems, Time Division Multiple Access (TDMA) has been more widely adopted. It divides cyclical time into frames, then further into time slots assigned to the target nodes in the location area. Each target node ranges with every base station (BS) node respectively within their assigned time slot and then uploads ranging information to location server to complete the follow-up positioning services. TDMA protocol has the characteristics of simple implementation, each target node working in its own time slot, less work conflicts between nodes, suitable for application in the scenarios in which the number of target nodes is less, and location area is smaller. However, TDMA has the following two main shortcomings: geolocation cycle's fast increase along with the increase of nodes number worsens the system's real-time performance; ordinary TDMA protocol assigns time slot to each node, when some nodes do not work properly or withdraw from the system due to other reasons, it cannot change the original time slot allocation, and in this specific time slot no work is done which leads to the waste of time slot.

Moreover, TDMA cannot implement dynamic joining of new nodes, if there are new nodes to join the network; it must reset the slot number and assign the time slot which brings trouble to the network management and maintenance.

Real-time is the primary and the most important performance indicator in wireless real-time geolocation system, which appears especially important in emergency scenarios such as firefighters positioning and field operation personnel positioning [4–6]. For multi-objective real-time geolocation system, how to effectively handle node dynamic accessing is also one of the important problems which need to be solved. How to effectively solve real-time and node dynamic accessing at the same time, is the key to evaluate multi-objectives real-time geolocation system performance. Meanwhile, for larger positioning scenario, the communication radius of base stations tends not to overlap and noninterference of RF radius can be done. If these stations can work in parallel, it will greatly shorten the cycle time and improve the channel utilization. But above all of these requirements are beyond ordinary

TDMA protocol's abilities, which greatly restrict the geolocation system's overall performance.

To address the problems, this paper presents a Multi-zones and Multi-objectives Channel Allocation Protocol Based on TOA Real-Time Geolocation System ZTDMA, which divides the system coverage area into several subareas that can work in parallel and has solved the problems in TDMA such as that positioning cycle's fast increasement along with the increase of nodes number worsens the system real-time performance, and the recycling of nodes' time slot is not in time that leads to the waste of channel resource. ZTDMA can ensure the targets' geolocation accuracy and improve the geolocation system's overall performance.

In Sect. 22.2, a summary of TOA real-time geolocation system and its working process are displayed, and it is the background and basis of ZTDMA protocol. In Sect. 22.3, analyze the working process and the main defects of TDMA protocol, and based on this, propose the ZTDMA protocol, give the implementation in two aspects: network topology and time slot allocation concrete. In Sect. 22.4, simulate the ZTDMA protocol and analyze its superiority; Sect. 22.5 is the conclusion of this paper.

22.2 TOA Real-Time Location System

Time of Arrival (TOA) is a method used to estimate distance by measuring the time of a radio signal transmission. According to whether needed to synchronize between the sender and receiver, it can be divided into synchronous TOA and asynchronous TOA. The synchronous TOA method needs to measure the signal transmission time accurately, it needs to execute high-precision time synchronization between the sender and receiver, which is difficult to achieve, the relevant international standards [7] and actual TOA chip design [8] usually use the synchronous TOA. Its implementation principle is as follows: shown in Fig. 22.1, Node A and Node B are two nodes for TOA ranging. T_1 is the moment of Node A sending Ranging Data to Node B; T_2 is the moment of Node B receiving the Ranging Data; T_3 is the moment of Node A sending ACK to Node B; T_4 it the moment of Node A receiving ACK; Node A and Node B measure t_{round} and t_{reply} time length through the local clock, respectively, and calculate the signal transmission time t_p between the two devices. Computation formula of the distance between the devices is as follows:

$$\hat{d} = t_p * C = \frac{t_{\text{round}} - t_{\text{reply}}}{2} * C = \frac{(T_4 - T_1) - (T_3 - T_2)}{2} * C \qquad (22.1)$$

The general structure of TOA indoor geolocation systems is shown in Fig. 22.2a, including the reference nodes whose position are fixed and known, the target nodes bounding to certain targets, server running location algorithm and software. In a wireless indoor geolocation system, the reference node and the target node are short-range wireless communication equipment which can measure the wireless signal

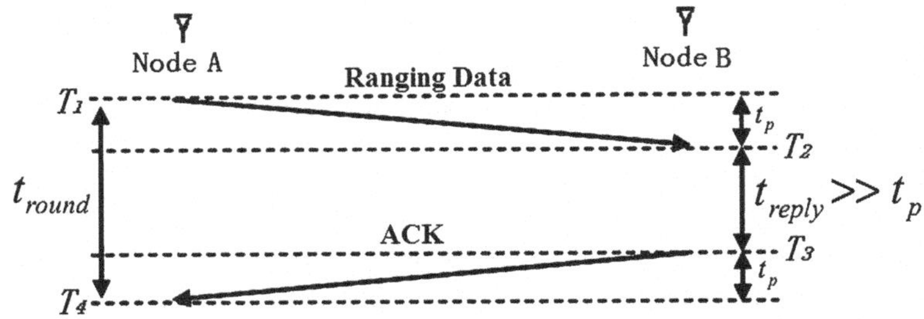

Fig. 22.1 Principle of asynchronous TOA

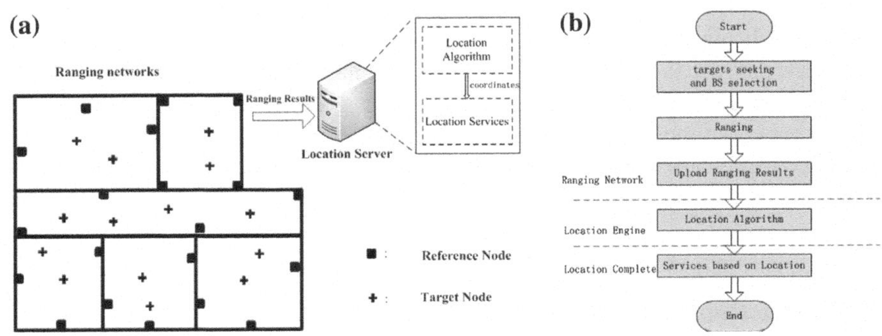

Fig. 22.2 Architecture and procedure of TOA indoor geolocation system. **a** Architecture. **b** Procedure

characteristic parameters associated with distance. Location server can be either a server with powerful computing capabilities, a personal computer, can also be a PDA, smartphones, tablet computers, and other portable computing devices.

The workflow of wireless indoor geolocation system is shown in Fig. 22.2b. The target node and several reference nodes do wireless signal transmission and measure time of signal transmission, the measurement results are then uploaded to the location server by reference nodes network; Server runs location algorithm, works out the coordinate of the target node according to the coordinates of reference nodes, measure characteristic parameters and the relationship between certain parameters and the distance.

The ZTDMA protocol proposed in the paper mainly aims at the ranging network layer in Fig. 22.2b, optimizes the performance in the sessions of targets seeking and BS selection, from basic level manages the network mode and time slot allocation policy of ranging network, and makes as many nodes work in parallel as possible, shortens the location cycle, and improves the channel utilization real time.

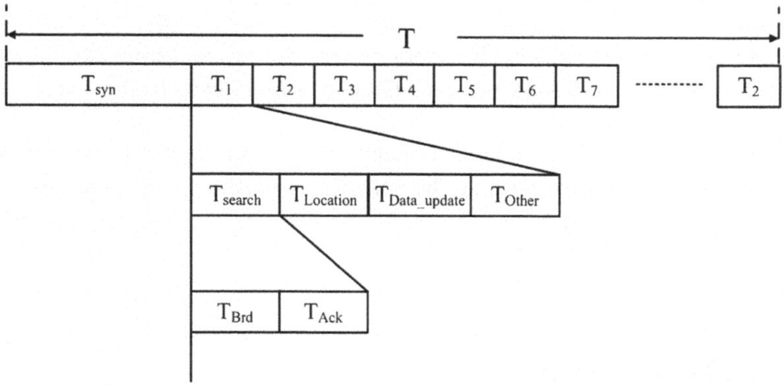

Fig. 22.3 Framework of ordinary TDMA protocol

22.3 ZTDMA Design

22.3.1 Analysis of Ordinary TDMA

TDMA divides cyclical time into frames, then further into several time slots. Under the conditions of timing and synchronization, each BS can, respectively, receive the signal from TAG in its own time slot without confusion. At the same time, the signals from BS to multiple TAGs are arranged in the scheduled time slot in sequence, and each TAG can distinguish and receive the signal sent to it among others as long as they receive within a specified time slot.

If MAC protocol based on TDMA is adopted in wireless geolocation systems, it can be designed as follows: divide the general period T into N time slots, and complete the corresponding work in each time slot. The geolocation system protocol's framework is given in Fig. 22.3, and concretely analyzed as follows:

- T_{syn} slot: complete the entire network's clock synchronization. Sink node starts synchronization at the beginning of cycle initiation, BSs forward synchronization information in predefined time slot.
- T_n slot: target nodes' time slot. Target node n completes location, data uploading and transmission. It can be divided into smaller time slots:
  - T_{search} slot: Target node initiates broadcast searching for base station;
  - T_{Location} slot: Target node ranges with selected BSs in proper sequence;
  - $T_{\text{Data\_update}}$ slot: Upload the ranging results to location engine through BS network;
  - T_{Other} slot: Transmit other relevant data.

However, in the actual wireless geolocation system, TDMA has the following two main shortcomings: positioning cycle's fast increase along with the increase of nodes number worsens the system real-time performance; ordinary TDMA protocol assigns time slot to each node, but when some nodes don't work properly or withdraw from the system due to other reasons, it cannot change the original time slot allocation, and in these specific time slots no work is done, which leads to the waste of time slot.

Otherwise, ordinary TDMA protocol cannot implement dynamic joining of new nodes, if there are new nodes to join the network, it must reset the slot number and assign the time slot which brings trouble to the network management and maintenance.

22.3.2 ZTDMA Design

Amouris [9] draws lessons from the spectrum reuse method of mobile communication network and sets forward a way of Space-time Division Multiple Access (STDMA) aimed at Ad Hoc network data transmission. Its basic idea is to divide the coverage area into subsystems, and it is noninterfering between the subareas that work in parallel in the collection, and it helps improve the channel utilization.

In this paper, on the basis of TDMA, we improve the STDMA time slot allocation method and put forward a new kind of channel allocation protocol: ZTDMA (Zones-Time Division Multiple Access), which effectively solves the TDMA's shortcomings, at the same time improves the channel utilization.

The juche idea of ZTDMA is to divide a large localization area into small subareas, then divide each subarea into parallel working groups, resulting in the shortening of location cycle. The key of channel utilization's improvement is the increase of parallel working groups' work efficiency, so that how to place the BS enables more area to work in parallel has become a key problem.

22.3.2.1 ZTDMA's Network Topology Construction

(1) The Location of Base Stations

Our geolocation system has two main kinds of nodes: BSs and mobile TAGs. The location of BS is directly related to the performance of ZTDMA. BS includes Manager nodes, subarea management nodes, and location BS nodes.

The Manager node is responsible for the allocation of time slots of the entire location area, it connects to each subarea management node via wired links so where exactly to put the manager node is out of question.

Subarea management nodes communicate with location BS nodes by radio. We suppose that the coverage of wireless node is a node centered round, the radius of this circle is the maximum communication distance of this node. Therefore, subarea management node ought to be put in the center of this circle in order to cover the whole subarea as shown in Fig. 22.4a.

The placement of location BS nodes is one of the key problems in the design of ZTDMA, which directly affects how to work in parallel between subareas. In this

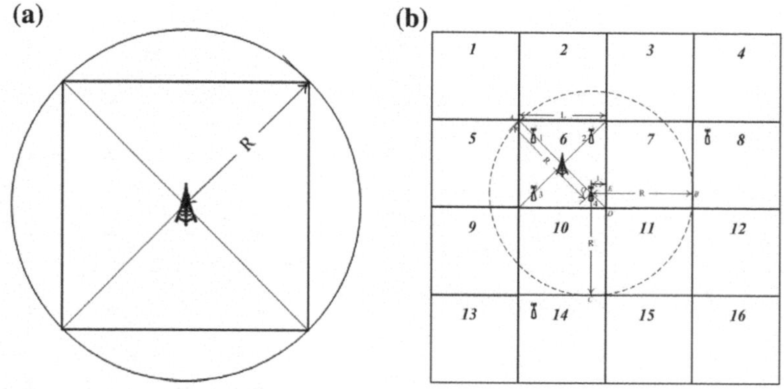

Fig. 22.4 The position of nodes' **a** Sub regional management node, **b** base station node

process, three problems should be considered: first cover its entire small location subarea; second try to minimize the interference between other areas; third be away from the center of subareas as much as possible, otherwise it will affect the location accuracy. In view of the above-mentioned facts, calculation results show that if the small location area is a square with length L, the coverage radius of location BS node is about $1.2L$, so the location BS node should be placed in the point $0.2L$ from edge of the small area.

(2) The division of parallel regions

Through the analysis of previous section, the wireless signal in a subarea can affect the communication quality of its surrounding areas, so we can divide the subarea into four parallel working groups. As shown in Fig. 22.4b, we represent the jth small location area in the ith subarea by the symbol of Z_{ij}, the small location areas with the same label in each subarea contribute to a parallel working group that can do locating work at the same time (The mesh regions shown in Fig. 22.5a make up a parallel working group).

In each locating cycle, it starts from the nodes in areas with symbol 1 in all subareas of Z_{i1}, after all target nodes in small areas with symbol 1 finish their locating work, it comes to Z_{i2}, Z_{i3} and Z_{i4}.

In conclusion, the network topology is defined as follows (as shown in Fig. 22.5b): at least four location BS nodes are deployed in each small location area, one of which is determined to be the management node. There is a Sink node in the center of the whole network and all manager nodes communicate with the Sink via 900 MHz mesh network.

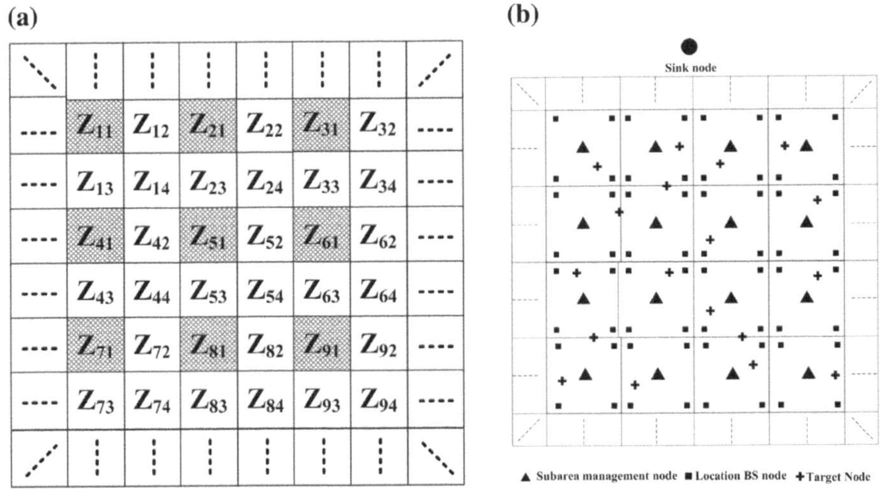

Fig. 22.5 Topology architecture. **a** Division of parallel orientation area, **b** network topology

22.3.2.2 ZTDMA's Time Slot Assignment

We represent the jth small location area in the ith subarea by the symbol of Z_{ij}, $G_i = Z_{1i}, Z_{2i}, \ldots, Z_{ni}$ stands for a collection of parallel location areas, namely a parallel working group. There can be no interference between subareas Z_{ij} in G_i and it is possible to do data transmission at the same time, so all subareas in the same working group share a same time slot.

ZTDMA's time slot allocation method is shown as Fig. 22.6, T_{ZTDMA} is a refresh cycle. t_{C} stands for control time slot in which mainly complete the following work:

i. At the beginning of each control time slot, the Sink node sends entire network synchronization information to the management node of each subarea;
ii. Each management node of location subarea broadcasts for target nodes within its range after receiving the synchronization information sent by Sink;
iii. The target node shall reply confirmation including its ID information by the method of CSMA/CA after receiving the broadcast sent by the management node of certain subarea which covers the target node itself.
iv. The management node in each area continues receiving confirmation replied by target nodes and stops after T_{Ack}, then uploads the number of target nodes within his reach to Sink;
v. Sink works out the time slot allocation table of each parallel area after receiving all subareas' targets number report, and then spreads the table to every management node.
vi. The management node calculates time slot allocation table of the target nodes within its range and broadcast it to all accessible target nodes after it receives

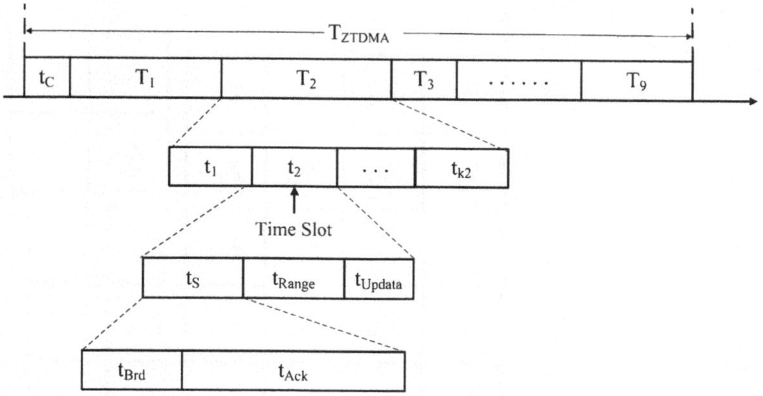

Fig. 22.6 Channel time slot allocation methods of ZTDMA

the time slot table sent by Sink; At the end of the control time slot t_C, continue to the step vii.

vii. All the target nodes within certain localization area complete locating work in proper sequence within its assigned time slot.

All above steps together make up the whole refresh cycle T_{ZTDMA}, and complete all target nodes' location process.

In Fig. 22.6, T_i stands for the time assigned to G_i, including k_i time slots while k_i is the maximum of target numbers among G_i, namely

$$k_i = \max\{N_{i1}, N_{i2}, ..., N_{i9}\} \tag{22.2}$$

The relationship between the location cycle and the target node number in ZTDMA is shown as follows:

$$T_{\text{ZTDMA}} = t_C + (k_1 + k_2 + ... + k_m) * t_{\text{Location}} \tag{22.3}$$

22.4 Simulation and Analysis

This paper simulates and evaluates ZTDMA protocol's average geolocation cycle under various TAG numbers in typical application scenarios by OMNet++. And by contrasting the performance of ZTDMA and TDMA, it proves ZTDMA's superiority of real-time characteristic in TOA based real-time geolocation system.

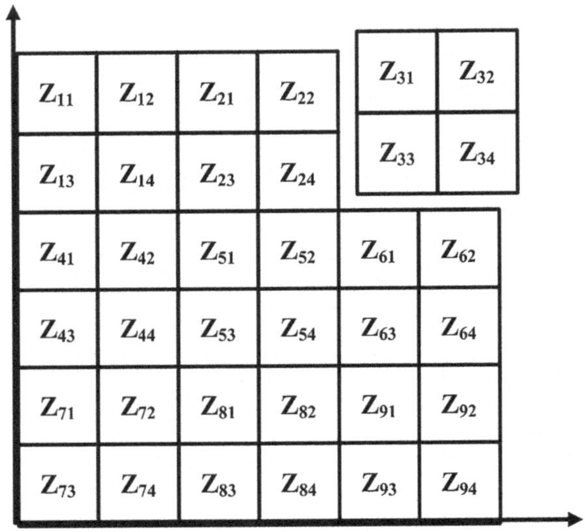

Fig. 22.7 Parallel area division with space frame length 4

22.4.1 Simulation Scenario

Space frame of simulation scenario is divided as shown in Fig. 22.7, with frame length 4. The factors that may influence ZTDMA's performance mainly include the number of target nodes and the number of subareas. This paper, respectively, evaluates these two factors' influence on the average system geolocation cycle through simulation. In simulation experiments, geolocation system model parameters come from StarLOC geolocation system [10], a single node's localization time is 40 ms, the data rate is 1 Mbps.

22.4.2 Results Analysis

The influencing factors of geolocation cycle: We verify the distribution of target nodes' influence on ZTDMA geolocation cycle, assuming the simulation conditions as follows: 16 location subareas, 100 target nodes.

This paper uses for reference the dispersion concept in non-ranging positioning algorithm in literature [11], we define dispersion in ranging positioning S_D (Scatteredness Degree) as the reciprocal of average square centroid distance value between all TAGs whose location are unknown and BS nodes, namely

$$S_D = \frac{1}{\frac{1}{n}\sum_{i=1}^{n} i^2 ||x_i - \frac{1}{m}\sum_{k=1}^{m} a_k||^2} \tag{22.4}$$

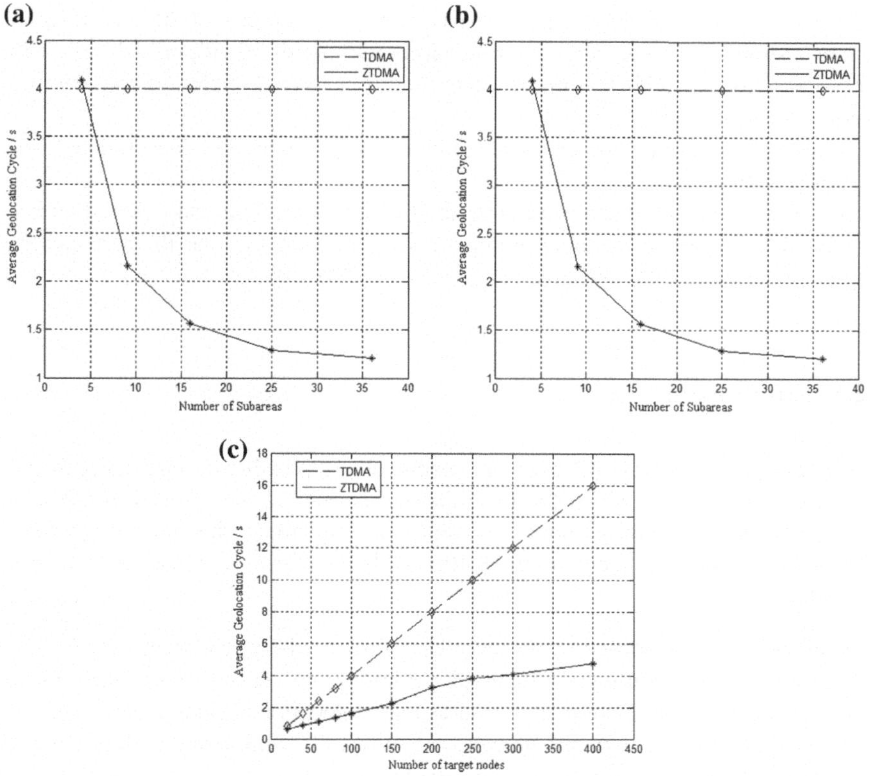

Fig. 22.8 Simulation results. **a** Scatteredness degree versus geolocation cycle. **b** Numbers of subareas versus average geolocation cycle. **c** Number of target nodes versus average geolocation cycle

As shown in Fig. 22.8a, the more dispersed the target nodes distribute, the shorter time it needs to finish locating work; otherwise the more concentrated the target nodes distribute, the longer time it needs to finish locating work.

The factors affecting the average geolocation cycle: In order to verify the influence of location subareas' separations on average geolocation cycle of ZTDMA, we assume the simulation conditions as follows: the length of fixed target area is 1,200, the width is 1,200, the number of geolocation target nodes is 100, and divide the certain area into 4, 9, 16, 25, and 36 subareas, respectively, to do simulation experiments each of which runs 10 cycles to calculate the average location cycle. Results are shown in Fig. 22.8b.

Performance Compared with TDMA: According to Sect. 22.3, analysis of ZTDMA performance shows that the geolocation cycle time of ordinary TDMA increases linearly along with the numbers of target nodes, while to obtain the relationship between ZTDMA protocol's geolocation cycle and the numbers of target nodes simulation experiments need to be done. Assume simulation conditions as follows: the length of fixed target area is 1,200, the width is 1,200, 16 location subareas simulate, respectively, with the target number of 20, 40, 60, 80, 100, 150, 200, 250, 300, and 400 and under each condition run 10 cycles to calculate the average location cycle.

Modify geolocation parameters do simulation experiment and then come up with the curve as Fig. 22.8c.

It is observed that the geolocation cycle of ZTDMA is significantly less than normal TDMA. The greater number of target nodes, the more obvious the performance improvement of ZTDMA relative to the normal TDMA protocol.

22.5 Conclusion

This paper proposes ZTDMA protocol specifically applicable to wireless real-time geolocation system, in which the location area is divided into groups of subareas that can work in parallel without confusion and implement channel multiplexing, shorten the location cycle and improve effectively the real-time. It has the following characteristics:

- Applicable to network with constantly changing topology structure;
- Shortens the location cycle effectively, improves the speed of updating target nodes' location information, and ensures the geolocation system's real-time;
- When applied to the dynamic network, identifies the joining and withdrawing of target node and dynamically allocates time slot for it.

References

1. He J, Wang Q, Zhang Q, Yu Y, Liu, B (2013) Modelling ranging error of indoor TOA-based position system based on receive signal strength. Int J Sens Netw 13(3):162–172
2. He J, Yu Y, Wang Q (2013) RSS assisted TOA-based indoor geolocation. Int J Wireless Inf Netw 20(2):157–165
3. He J, Yu Y, Wang Q RSS Assisted TOA-based indoor geolocation. Int J Wireless Inf Networks
4. Li C, Wang G (2007) IEEE802.15.4 MAC protocol research in wireless sensor network. Comput Sci 01:34
5. Pahlavan k, Levesque A (2005) Wireless information networks, 2nd edn. Wiley, New York
6. Guillemette MG, Fontaine I, Caron C (2008) Hybrid RFID-GPS real-time location system for human resources: development, impacts and perspectives. Proceedings of the 41st Hawaii international conference on system sciences, USA, 1530–1605
7. IEEE 802.15 Group, IEEE 802.15.4a Standard, 2007
8. Dardari D, Conti A, Ferner U, Giorgetti A et al (2009) Ranging with ultrawide bandwidth signals in multipath environments. Proc IEEE 97(2):404–426
9. Amouris KG (2001) Space-time division multiple access (STDMA) and coordinated, power-aware MACA for mobile ad hoc networks. IEEE Symposium on ad hoc networks (SAWN2001), Texas

10. He J, WANG Q, YU M et al (2009) StarLOCa multi-objective real-time geolocation system based on TWR. CWSN'2009 3rd China conference on wireless sensor networks. 314–317, 333
11. Shi Q, He C (2008) A SDP approach for range-free localization in wireless sensor networks. IEEE technology management council IEEE communications society. Proceedings of the symposium on wireless communication and systems of ICC 2008. Department of Electronic Engineering Shanghai Jiao Tong University, Shanghai, P.R.China, p 5

Chapter 23
MLOC: A Multiple Service Fusion Self-Organizing Geolocation System

Cheng Xu, Jie He, Fei Liu, Mowei Lv and Qin Wang

Abstract Self-organizing geolocation system does not depend on infrastructure, can build coordinate system automatically and calculate reference point coordinates position, has very high application value in emergency situations. Geolocation service alone often cannot meet all application requirements, while multiple service fusion can improve the practicability of the system. In this paper, we design a multiple service fusion self-organizing geolocation system–MLOC, in addition to the geolocation service, it can also provide voice communication, short message communication and elevation measurement services. This paper deeply analyzes the challenges for multiple service fusion mechanism and self-organization accurate geolocation mechanism; design an unpreemptible hierarchical scheduling mechanism and TDMA/CSMA time slot allocation strategy to solve the problem of multiple services fusion; propose a self-organizing accurate geolocation mechanism based on maximum likelihood centroid localization algorithm; build up a set of prototype MLOC system and verify the MLOC solution through the actual test.

Keywords Real-time geolocation system · Accurate geolocation · Self-organizing geolocation system · Multiple service fusion

C. Xu · J. He (✉) · F. Liu · M. Lv · Q. Wang
School of Computer and Communication Engineering, University of Science and Technology Beijing, Beijing 100083, China
e-mail: hejie1983@gmail.com

C. Xu
e-mail: 377164753@qq.com

F. Liu
e-mail: happy326111@sina.cn

M. Lv
e-mail: lvmowei@gmail.com

Q. Wang
e-mail: wangqin@ies.ustb.edu.cn

X. Wang et al. (eds.), *Advanced Technologies in Ad Hoc and Sensor Networks*, Lecture Notes in Electrical Engineering 295, DOI: 10.1007/978-3-642-54174-2_23,

23.1 Introduction

In recent years, with high accuracy ranging technology maturing, the real-time accurate geolocation system has been being brought into effect step by step, and its positioning methods and modes of service have drawn more and more attention [1, 2]. Self-organizing geolocation system doesn't depend on infrastructure, builds its own coordinate system automatically, and it's suitable for emergency deployment scenarios. Due to the nature of positioning with fast organizing, the self-organizing positioning system has great application value in fields such as military commanding, fire controlling. For example, positioning system for firefighters [3] can not only provide for on-site command with firefighters' location information, improve the efficiency of commanding, but also can ensure that firefighters in distress can be rescued in time. In addition, the single location service often cannot meet the requirements of many applications. It can greatly improve the practicality of the geolocation system with the fusion of voice communications, sensing and other services.

Multiple service fusion self-organization geolocation system has the following two main challenges: The key of multiple service fusion mechanism is how to ensure the quality of service; The key of precise positioning mechanism is how to automatically build accurate topology, how to overcome the influence of base station coordinate error and the ranging error on localization accuracy.

This paper proposed a multiple service fusion self-organizing geolocation system: MLOC. In terms of multiple service fusion, we design multiple service fusion mechanism from aspects of program scheduling and network communication, realize the mergence of positioning, voice and short message communication, elevation measurement, etc; In terms of self-organization accurate positioning, propose a kind of precise positioning mechanism based on maximum likelihood centroid localization algorithm, including fast neighbor discovery protocol, topology construction algorithm and the target localization algorithm.

The rest of this paper is structured as follows: Sect. 23.2 describes the system framework of MLOC; Sect. 23.3 shows the design of MLOC hardware nodes; Sect. 23.4 introduces the multiple service fusion mechanism; Sect. 23.5 intrdouces self-organization accurate geolocation mechanism; Sect. 23.6 covers prototype system and test validation; Sect. 23.7 comes to the conclusions.

23.2 Related Works

At present, most of the research on self-organization positioning is set backgrounds on sensor networks [4, 5].

The existing anchor-free localization algorithm includes AFL [6], KPS [7], and ABS [8]. AFL algorithm is anchor-free, fully distributed and it firstly gets a no-folding layout using a heuristic principle, then balance and correct the positioning error based on the mass-spring model optimization algorithm minimizing the

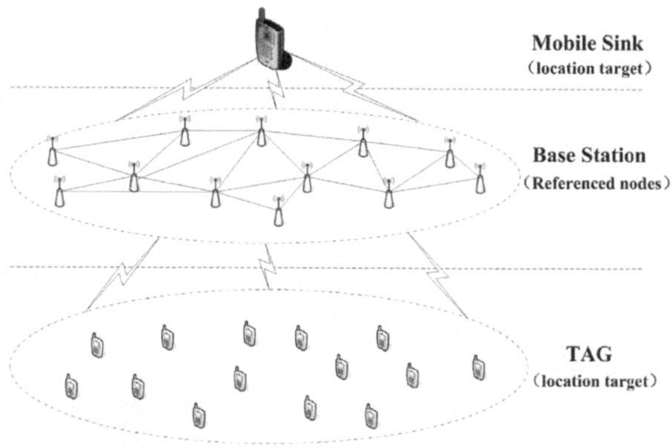

Fig. 23.1 System architecture of MLOC

energy function of corresponding position. KPS algorithm needs a prior assumed node configuration model, each node observes the number of member nodes of its adjacent group then realizes self-location according to the prior assumption. ABS algorithm calculates one node's coordinates at a time in sequence by using the communication links between nodes, continuously corrects redundant calculation and reduce positioning error. But all above three algorithms still have problems, such as that AFL doesn't present a clear method of calculating one hop distance, KPS is on the premise of grouping probability model, in ABS 4 anchor's coordinates need to be known or preestablished. All these algorithms are designed for RSSI geolocation and not suitable for precise geolocation.

In terms of multiple service fusion, realizing positioning function in sensor network is the most common way [4, 5], but it has low precision and real-time requirements as an additional function. Besides, literature [4] says that it had already realized the fusion in zigbee, but it didn't give an actual solution.

All above research's common feature is that self-organizing geolocation is a kind of auxiliary means, with low precision and real-time requirements and not suitable for scenarios needing high accurate positioning and tracking. Given the reason above this paper presents a self-organization geolocation system aimed at high precise positioning which at the same time provides services such as voice and short message communication, etc.

23.3 System Architecture

The system architecture of MLOC is shown as Fig. 23.1. MLOC consists of target nodes (TAGs), base stations(BSs) and mobile sink node. TAGs are target nodes with PDA that can topology according to the ranging results, calculate and display their own positions. BSs as referenced nodes at the same time build up a data transmission network. Mobile sink node is a special kind of target node which cannot only realize

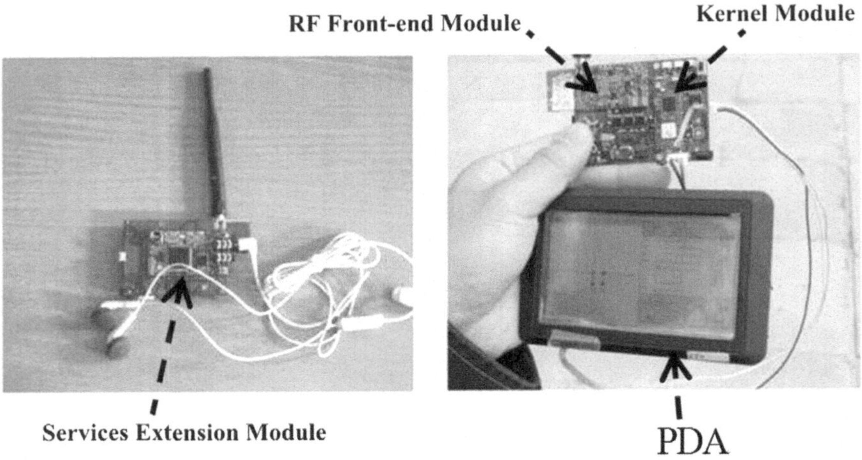

Fig. 23.2 Prototype of nodes

self-geolocation but also can receive data packets of all nodes' coordinates, monitor all TAGs and is usually equipped with conrtollers.

MLOC working process can be divided into the following two stages:

- ***Topology construction:*** BS ranges with each other, and then broadcast the packet of topology with all distance information to the entire network; according to the topology packets, nodes with ability of positioning calculation establish the relative coordinate system.
- ***Target positioning:*** The system supports both positioning and other services. TAG ranges with BSs and calculates its own coordinates, then uploads the result to Sink via the BS network.

23.4 Hardware of Nodes

MLOC node consists of the location core module, the RF front-end module, services extension module. Nodes connect with PDA via a serial port and contribute to the TAG and mobile Sink node.

The location core module provides accurate distance measurement function and mainly consists of RF chip NanoLOC [9] based on IEEE 802.15.4A standard and the ATmega644V-10AU microprocessor. NanoLOC has the function of TOA ranging and supports both TWR and SDS-TWR [10] ranging methods (Fig. 23.2).

The RF front-end module provides the long distance communication ability, can expand the coverage of the system and includes power amplifier AP1046, low noise amplifier AN5905 and RF switch HMC574MS8. This module has 30 db sending power gain and 3 db of receiving sensitivity gain, the maximum transmission distance outdoor is up to 1,000 m.

Services extension module is used to implement various service functions, mainly including AMBE1000 voice codec chip, altitude meter HOP3.

PDA is used to perform positioning calculation, display topology and location information, send and receive messages, with processor ARM10, running WinCE6.0 operating system.

23.5 Self-Organizing Precise Geolocation Mechanism

23.5.1 Main Challenges

Location errors of self-organizing geolocation system mainly come from the following two aspects:

- ***Ranging Error:*** Include time error caused by hardware timer measuring the time of signal arrival and transmission error caused by multipath and non-line-of-sight (NLOS). NLOS has the biggest influence on precision.
- ***BS Coordinates Error:*** As BS coordinates of the self-organizing positioning are automatically calculated the by algorithm that may leads to errors, it may affect the precision of location.

The main purpose of self-organizing geolocation is reducing the influence on positioning accuracy caused by the above two kinds of error.

23.5.2 Slot-Divide Round Based on CSMA Neighbor Discover Protocol

Slot-Divide Round Based on CSMA Neighbor Discover Protocol (SDRBC) is used for searching Anchor without consideration for TAG.

SDRBC uses CSMA protocol for collision avoidance and divides time into time slots which is divided into four sub-slots. Clock synchronization occurs at Slot0 and it is the initial stage of the protocol. Figure 23.3 shows a protocol timing diagram. Each slot is organized as follows:

Slot1: After receiving clock synchronization, Anchor nodes select random delay $t_i(t_i \in [0, \omega - \delta]$, δ is transmission delay of node) in Slot1 (Length of ω)and then broadcast Hello Message to inform other nodes in CSMA way.

Slot2: Anchor nodes select random delay, then broadcast their neighbor list (NT) in CSMA way. NT packet format is shown in Fig. 23.3b. Anchor receives neighbor node NT and detects whether it's in its own NT. If not, add it into its own NT and in slot3 broadcast its own NT again. Check whether the received NT contains itself, and if not, then in slot4 broadcast its own NT.

Slot3: Anchor node based on the previous slot randomly sends its new NT and also selects random t_i to broadcast in CSMA way. The Node receives a new NT, if

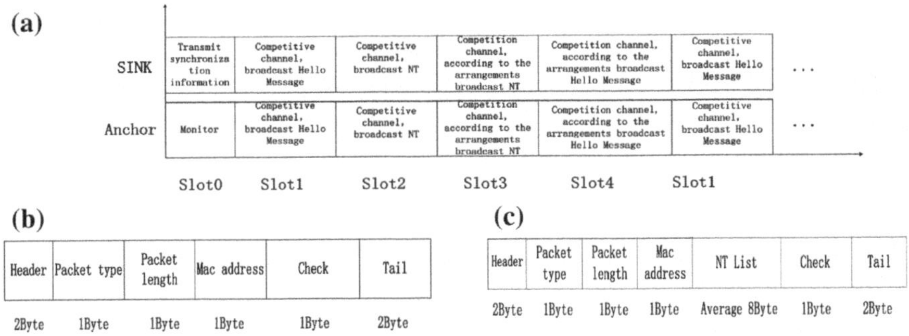

Fig. 23.3 Topology accuracy test. **a** Protocol timing diagram, **b** Packet format of hello message, **c** Data packet format including NT

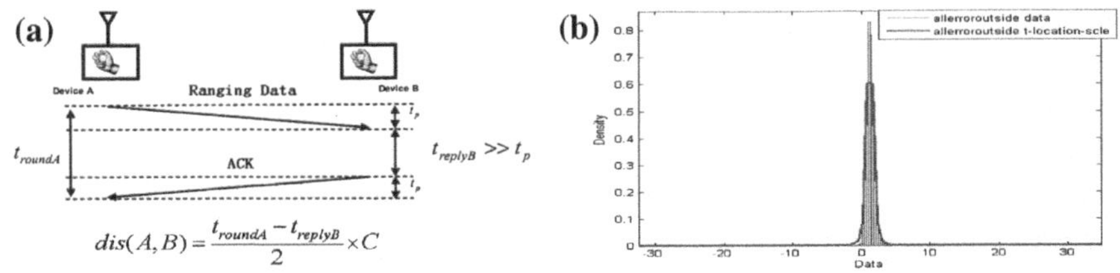

Fig. 23.4 Modeling ranging error. **a** Principle of TWR, **b** Probability density distribution

finding itself in the NT list of its neighbor node then accordance with rules to cancel the broadcast in slot3. If not, it continues broadcasting on demand.

Slot4: If the node needs to broadcast NT at this slot, it randomly selects t_i to broadcasts its Hello Message in CSMA way.

In SRDBC protocol, time is divided into units of slots and each time slot is further divided into four smaller slots to reduce unnecessary transmission in each small slot and achieve the purpose of reducing the node discovery time.

23.5.3 Modeling Ranging Error Based on Receive Signal Strength

This paper uses the TWR TOA ranging algorithm [10] to complete the distance ranging between nodes, and its principle is shown as Fig. 23.4a [11].

Amounts of distance experiments with NanoLOC have been done, and the results through the Matlab Dfittool fitting are shown in Fig. 23.4b.

It turned out that under the condition of line of sight (LOS), TWR ranging error probability density is approximately in line with the T distribution whose average is 1.2 and freedom degree of 9.

LOS range error 95 % occurred in (0, 2)m, and 5 % in (2, +∞)m, namely

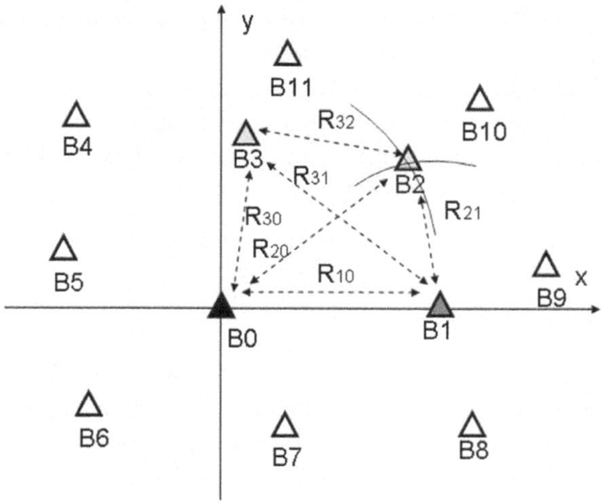

Fig. 23.5 Topology construction of BS

$$\begin{cases} P(d_n > r_n) = 0 \\ P(r_n - 2 \le d_n \le r_n) = 0.95 \\ P(d_n < r_n - 2) = 0.05 \end{cases} \tag{23.1}$$

d_n stands for the actual distance between TAG and Anchor, r_n stands for the measurement distance between TAG and Anchor.

23.5.4 Method of Topology Construction

During the stage of topology construction, BS ranges with each other for many times and the results are used for establishing the reference coordinate system and the calculation of BS's coordinates. MLOC BSs are deployed artificially, thus it can guarantee the line-of-sight between BSs, so the ranging error meets LOS-TWR model, the mean distance measurement error is 1.2 m.

Assume that B_i stands for the ith BS, r_{ij} stands for the average measurement distance between the ith and the jth BS, R_{ij} is fixed distance value, namely $R_{ij} = r_{ij} - 1.2$. Topology construction of BS consists of the following steps:

Step1: Base Station Classification According to all the ranging results, if all BS collection is $\{B_0, B_1, B_2, ..., B_m\}$, choose any two BS from the collection B_0, B_1, B_2, ..., B_n in which between any two there is a certain ranging result. B_0 and B_1 are reference BS of the coordinate system; B_2, B_3, ..., B_n are Level 1 reference location BS. The other BSs are Level 2 reference location BS.

Step2: The Establishment of the Coordinate System As shown in Fig. 23.5, take the BS B_0 as the origin of coordinates; take the line of B_0 and B_1 as X axis; take the line coming through B_0 and perpendicular to the line of the $B_0 - B_1$ as Y axis, positive direction is shown in Fig. 23.5.

Step3: Calculation of Base Station Coordinates Reference BS is shown in Fig. 23.5, $B_0(0, 0)$ and $B_1(R_1, 0)$;

(1) Level 1 reference location BS

B_2 coordinates: Respectively, draw circles centered on B_0, B_1 with radius of R_{20} and R_{21}, compute intersection points and take the intersection on the side of *Y* axis as the coordinate of $B_2(x_2, y_2)$, namely

$$\begin{cases} x_2^2 + y_2^2 = R_{20}^2 & y_2 > 0 \\ (x_2 - R_{10})^2 + y_2^2 = R_{21}^2 \end{cases} \tag{23.2}$$

$B_3, B_4, \ldots, B_n$ coordinates: take B_3 for example, respectively, draw circles centered on B_0, B_1 with radius of R_{30} and R_{31}, compute intersection points (x'_3, y'_3), (x''_3, y''_3). Calculate the distance between them and B_2 and respectively, take that's nearer to R_{32} as the coordinate of B_3 (x_3, y_3), namely

$$|R_{32} - d'_{32}| < |R_{32} - d'_{32}| \tag{23.3}$$

Through a lot of outdoor experiments, the average coordinate error distribution of the measured coordinates refer to the B_0, B_1 and Level 1 reference location BS is 0, 0.23, 0.35 m; the maximum error is 0, 0.66, 1.5 m.

(2) Level 2 reference location BS

Based on the coordinate error characteristics, through the improved maximum likelihood centroid localization algorithm, calculate the coordinates of Level 2 reference location BS.

Through a lot of outdoor experiments, coordinate error distribution characteristics of the measured coordinates Level 2 reference location BS: average error of station coordinates is about 1.15 m, maximum error is 1.79 m.

23.5.5 Improved Maximum Likelihood LOS-TWR Centroid Localization Algorithm

This algorithm has two main steps:

Step1: Region Segmentation In an anchor-based geolocation system, region segmentation with the maximum likelihood LOS-TWR centroid localization algorithm based only on distance measurement error model, regardless of the base station coordinate error. In this paper, the improvement is taking the coordinate error of BS as a parameter of region segmentation, that makes it apply to self-organizing geolocation system. Details as follows:

The calculated coordinate error of BS is defined as the distance between calculated coordinates and the actual coordinates. Assume that in self-organizing geolocation system, the error of base station B_n is l_n, $l_n \in [0, \Delta l_n]$. As shown in Fig. 23.6a, the location area is divided into three small ones:

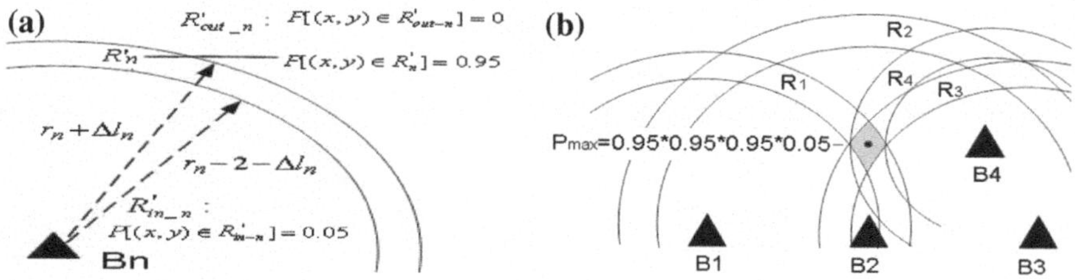

Fig. 23.6 Improved algorithm. **a** Region segmentation with BS coordinate error, **b** The maximum likelihood centroid

- $R'_{\text{out}-n}$: the ring region with the center B_n, radius of inner circle $r_n - 2 - \Delta l_n$ and radius of outter circle $r_n + \Delta l_n$
- $R'_{\text{in}-n}$: the region inside of the inner circle of R'_n
- $R'_{\text{out}-n}$: the region outside of the outer circle of R'_n

The probability of target being in each region is shown as follows:

$$\begin{cases} P[(x, y) \in R'_n] = p(r_n - 2 - \Delta l_n \le d_n \le r_n + \Delta l_n) = 0.95 \\ P[(x, y) \in R'_{\text{in}-n}] = 0.05 \\ P[(x, y) \in R'_{\text{out}-n}] = p(d_n > r_n) = 0 \end{cases} \tag{23.4}$$

Step2: The maximum likelihood centroid If m BSs take part in the location, then the whole area consists of regions R_1, R_2, ..., R_m, $R_{\text{in}-1}$, $R_{\text{in}-2}$, ..., $R_{\text{in}-m}$ and $R_{\text{out}-1}$, $R_{\text{out}-2}$, ..., $R_{\text{out}-m}$. Any subarea can be described as the intersection of the above three areas. As shown in Fig. 23.6b, work out the probability value of each area, and the centroid of area with the maximum probability is taken as the location result.

23.6 Multiple Service Fusion Mechanism

MLOC provides positioning, voice communication, short message communication, and elevation measurement services, of which elevation measurement is periodic, voice and short message communication is paroxysmal. The voice and short message are used for commanding and dispatching, with higher priority than positioning and elevation measurement.

There are two main challenges in fusion of periodic and paroxysmal services:

Node Level: The challenge in node level lies in how to ensure task real-time. Services have different response time to tasks, and execution sequences are also different needing reasonable task scheduling mechanism to ensure the real-time.

Network Level: The challenge in network level lies in the fusion of periodic and paroxysmal data transmission. By reasonable time slot allocation strategy, meeting the different needs of a variety of services for data transmission, is the key to ensure the quality of services in network level.

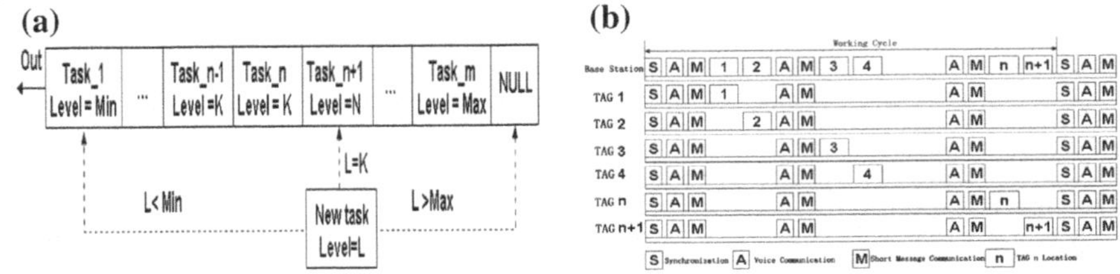

Fig. 23.7 Test and verification. **a** Task scheduling with priority, **b** TDMA time slot allocation

(1). Multiple Service Fusion in Node Level

Tasks in MLOC are sequenced by importance as follows: voice communication, short message communication, positioning, elevation measurement. Task priority level can be graded into five level by its associated service, scheduled as 1-voice, 2-short messages, 3-positioning, 4-elevation, 5-other.

In this paper, we design a unpreemptive priority scheduling mechanism to ensure all kinds of real-time services. The basic idea is that each task has a certain priority attribute and the executing task cannot be preempted.

Because the task cannot be preempted, if the currently executing task costs too long time, it will affect higher-priority task's response time delay. By granularity partition limiting each task's execution time can effectively reduce the influence, guarantee the task response time delay.

Figure 23.7a shows waiting task queue with a new task inserted, Level standing for priority, the smaller Level value equals the higher priority. PS, $N > K$.

(2). Multiple Service Fusion in Network Level

Network level of service integration is mainly manifested in the time slot allocation of MAC layer. In this paper, we design a TDMA/CSMA hybrid MAC protocol to realize the multiple service fusion: eliminate mutual interference between the various services to guarantee service quality by TDMA; CSMA is used by moving targets when searching for BS within range, and it fits the dynamical network topology changes caused by movement.

Detailed time slot allocation of TDMA is shown as Fig. 23.7b. At the beginning of the whole cycle, base station broadcast synchronization packet to the entire network. All TAGs work out their own positioning time slot according to their own node number. In order to ensure continuity and real-time of voice and short message, time slots of voice and short message are distributed at fixed intervals throughout the work cycle.

23.7 Test and Verification

23.7.1 Prototype System and Experiment Scenario

According to the above design, this paper constructed a prototype MLOC system, in which the positioning slot of each TAG is 50 ms. Time slots of voice and short message communication are both 10 ms.

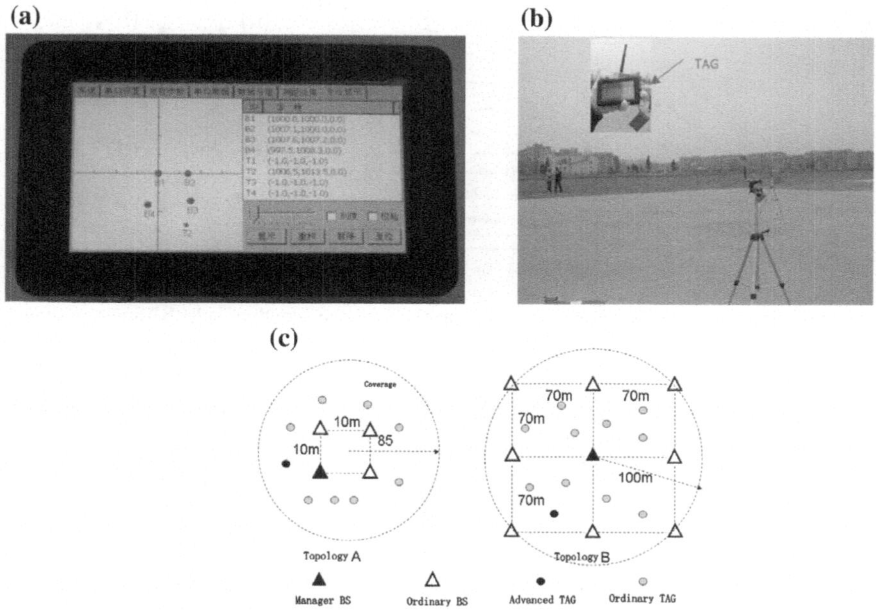

Fig. 23.8 Test and verification. **a** Software interface, **b** Experiments scene, **c** Topology of testing system

Test experiment of MLOC is done in the Olympic Park, squared 300 m. Each BS is placed on a tripod, 1.5 m high. TAGs are bounded to testers, 1.5 m high above the ground. The sending power of nodes is set 0 dbm with effective transmission distance of 100 m.

The topology of testing system is shown as Fig. 23.8c, A is center deployed and B is deployed on edges. Comparison test has been done.

23.7.2 Positioning Accuracy Test

(1). Topology Accuracy Test

Coordinate error of BS is defined as Euclidean distance between calculating coordinates and real coordinates. Assume that calculating coordinates of B_i is (x_i, y_i) and real coordinates (x_i, y_i), then coordinate error of B_i is:

$$E_i = \sqrt{(x_i - x_i')^2 + (y_i - y_i')^2} \tag{23.5}$$

Topology accuracy is mainly reflected in the accuracy of BS coordinates. In this paper, we use the average error of BS coordinates to represent the precision of

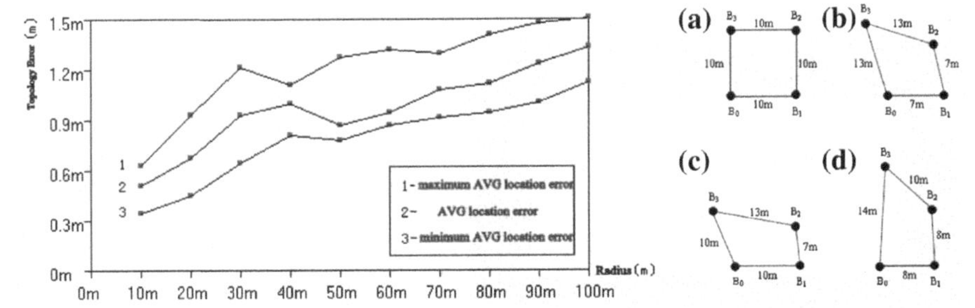

Fig. 23.9 Topology accuracy test. **a** Topology precision versus length of side, **b** Topology shape

topology construction. If there are n BSs in the system then the precision of topology construction can be calculated like this:

$$E_T = \frac{\sum_{i=0}^{n} E_i}{n} \tag{23.6}$$

- *The precision of Topology A and B*
  Repeat the construction process 100 times, and then statistic BS coordinate error, get the topology error results. Minimum error: A-0.35 m, B-0.78 m; Average error: A-0.51 m, B-0.87 m; Maximum error: A-0.61 m, B-1.28 m.
- *Topology Precision Versus Length of Side*
  4 BSs are placed as a square , length of side ranging from 10 to 50 m. Topology construction is completed by location algorithm and each topology is being tested 100 times.
- *Topology Precision Versus Topology Shape*
  This test designed several different display methods randomly, length of side of each parallelogram is 40 m, average errors of topology is: a-0.34 m, b-0.41 m, c-0.36 m, d-0.35 m.

It turns out that under the condition the length of side of topology are similar to each other, there is no significantly difference in average location error. So we can think the topological shape has nothing to do with anchor positioning accuracy and this algorithm can meet the demand of anchor randomly arranged (Fig. 23.9).

(2). TAG Positioning Accuracy Test

TAG's location accuracy is closely related to the geometric relationship between TAG and Anchor. The accuracy is higher when edge-deployed than that when center-deployed. And the accuracy increases when TAG moves nearer to the center of topology.

- *Positioning Accuracy in Edge-deployed Topology*
  Test in edge-deployed topology shown as Topology B. Randomly pick up 10 positions within the location area, get 2,940 groups of effective data, based on which the average location accuracy of topology B is 1.32 m.

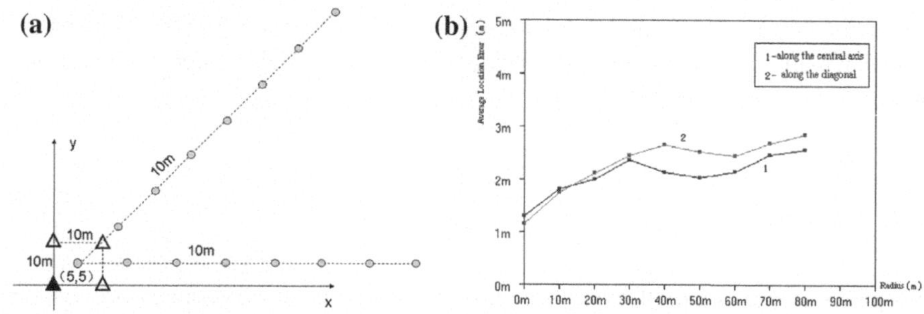

Fig. 23.10 TAG positioning accuracy test. **a** Test points, **b** Location error of topology A

- *Positioning Accuracy in Center-deployed Topology*
  Although having lower accuracy in center-deployed topology, it's very useful in practice work because it can cover larger areas with less anchors. Get 200 groups of data at each test point showing the relationship between location error of TAG and the distance between TAG and center of topology.

The chosen 15 test points are typical due to the symmetrical structure.

Conclusions can be drawn that with the increase of the distance between target nodes and the center of topology, the positioning error is generally on the rise and they're nearly the same at all directions. Generally, average location error is less than 3 m, which can meet general outdoor applications (Fig. 23.10).

23.7.3 Test of Multiple Service Fusion

(1). Test of Location Success Rate

The standard of location success is defined as getting the certain location coordinates. Two testers move in the location area taking 4 TAGs, and after 1,000 cycles the statistical result is: success rate of topology A is 92.15 %, and B 94.21 %.

(2). Voice Communication and Short Message Communication

The standard of voice communication success is defined that voice can be heard clearly and intelligibly. Test sample is made up of 100 digitals. In the choosing spots, the tester with TAG A read the sample and the tester with TAG B counted the number of samples that can be heard clearly.

One sample of short message test contains a packet of 50 bytes. TAG A continually sent 100 packets test samples to TAG B. And TAG B counted the packets received successfully and worked out the transmission success rate.

TAG A and B do voice and short message communication tests respectively in the spots away from BS 20, 40, 60 and 80 m , each of situation repeated 10 times. The result shows that: within the location area coverage, the averagevoice communication successful rate is 97.5 % and short message communication 99.25 %. The results show that MLOC supports multiple service fusion and can highly improve the service quality.

23.8 Conclusion

This paper proposes a multiple service fusion self-organization geolocation system: MLOC, and a detailed system solution is given from aspects of system architecture, network topology, hardware design and location algorithm, etc.

In the aspect of multiple service fusion, ensure tasks' real-time via a preemptive priority scheduling mechanism; realize the mergence of periodic and paroxysmal services. In the aspect of self-organization geolocation, we improved the maximum likelihood of centroid localization algorithm in [12], making it more suitable for self-organization geolocation system and proposed a topology construction algorithm with high accuracy.

The test results of MLOC prototype system show that MLOC can realize the fusion of voice, short message and other services with quality ensurance, is of highly use value in practice.

Whether MLOC localization algorithm is of applicability and high localization precision in the indoor environment needs further verification. Next step work is to test MLOC localization algorithm in indoor environment. In addition, networking scheme aimed at large-scale network will be the focus in the future work of MLOC.

References

1. He J, Geng Y, Wan Y, Li S, Pahlavan K (2013) A cyber physical test-bed for virtualization of RF access environment for body sensor network, p 1
2. He J, Li S, Pahlavan K, Wang Q (2012) A real-time testbed for performance evaluation of indoor TOA location system. In: IEEE international conference on communications (ICC), Ottawa, Canada, 10–15 June 2012
3. Orr JA, Cyganski D (2005) Fire fighter location tracking and status monitoring performance requirements. Technology report. Electrical and Computer Engineering Department, Worcester Polytechnic Institute, Worcester
4. Zhongmin P, Zhidong D, Xu S, Xu X (2009) Anchor-free localization method for mobile targets in coal mine wireless sensor networks. Sensors 9(4):2836–2850
5. Hui Q, Stephen W (2006) Anchor-free localization in rapidly-deployed wireless sensor networks. In: 2006 IEEE international conference on mobile ad hoc and sensor systems, MASS 2006, pp 627–632
6. Priyanrha N, Balakrishnan H, Demaine E et al (2003) Anchor-free distributed localization in sensor net-works. Technical Report TR-892, MIT laboratory for computer science, April 2003
7. Fang L, Du W, Ning P (2005) A beacon-less location discov-ery scheme for wireless sensor networks. In: IEEE conference on computer communications (INFOCOM)
8. Savarese C, Rabaey J, Beutel J (2001) Locationing in distributed ad hoc wireless sensor networks. In: Proceeding of ICASSP, 2037–2040 May 2001

9. nanoLOC TRX Transceiver (NA5TR1) Datasheet, Nanotron
10. IEEE 802.15.4a[OL]. http://www.ieee802.org/15/pub/TG4a.html
11. He J, Wang Q, Zhang Q, Yu Y, Liu B (2013) Modelling ranging error of indoor TOA-based position system based on receive signal strength. Int J Sens Netw 13(3):162–172
12. He J, Wang Q, Yong F (2009) A mass-target real time localization system. In: International conference on wireless communications and signal processing, WCSP 2009

Chapter 24
Study on Architecture and Topology of Mobile Underwater Wireless Sensor Networks

Ming He, Wenhui Liang, Qiuli Chen and Qingbing Zou

Abstract Mobile underwater wireless sensor networks (MUWSNs) is becoming an active research field. Owing to the factors that complex underwater conditions and the mobility of the underwater sensors have made MUWSNs to evolve dynamically, it is important to study the architecture and topology of MUWSNs. First, the research status of MUWSNs is clarified. Second, the features and architecture are analyzed. Third, the topological characteristics of MUWSNs are discussed. Finally, the directions of the future study are put forward.

Keywords MUWSNs · Autonomous underwater vehicle · Sensor deployment · Coverage efficiency

24.1 Introduction

Before Wireless Sensor Networks (WSNs) appeared, underwater data were perceived and collected by wired network, which was costly and needed the assistance of various engineering technology as well [1]. More and more researchers focus on underwater wireless sensor networks (UWSNs) with the rapid development of wireless sensor networks and increasing application requirements in the underwater environment. Plenty of sensors applying to underwater environment were arranged. Although the acoustic communication between nodes was flexible, dynamic networking was infeasible and monitoring of the blind area or even network failure was easy to form because of the influence of the current [2]. Then mobile underwater wireless sensor

This work is supported by the Natural Science Foundation of Jiangsu Province, China (Foundation No. BK2011124, BK2010129).

M. He · W. Liang (✉) · Q. Chen · Q. Zou
College of Command Information Systems, PLA Science and Technology University, Nanjing 210007, China
e-mail: liangwenhui2008@hotmail.com; review_paper@126.com
http://www.springer.com/lncs

X. Wang et al. (eds.), *Advanced Technologies in Ad Hoc and Sensor Networks*, Lecture Notes in Electrical Engineering 295, DOI: 10.1007/978-3-642-54174-2_24,

networks (MUWSNs) which combine underwater robots with underwater wireless sensor networks tightly have appeared. A number of ordinary nodes and autonomous underwater vehicles (AUV) constituted the underwater acoustic communication network, which provides a favorable technological means to underwater environment monitoring, ocean data collecting, submarine resources exploiting, pollution monitoring, disaster forecasting, and beach rescuing, etc. [3].

A type of OSI hierarchical structure was mentioned in [4] to elaborate details in each layer of underwater wireless sensor network, which supplied a clear idea for researches on underwater wireless sensor network, but AUVs were not considered. The framework of MUWSNs was put forward in [5], and analyzed the difference between MUWSNs and common UWSNs while involving little about the moving characteristic of nodes and methods of dynamic networking. Reference [4–6] more sufficiently considered the moving characteristics of MUWSNs. It expounded the design of underwater nodes, nodes connection, and dynamic networking of MUWSNs according to the views of node, line, and plane, but it did not concern about healing and optimization of network topology. The security issues with which UWSNs are faced and the security technologies are discussed in [7], but it does not consider AUVs and the methods of healing and optimization for network topology as well. All the research work mentioned above does not systematically study the topology of MUWSNs. Locating at underwater 3D environment, the network topology of MUWSNs possesses the feature of dynamic evolution. Thus, it is important to study the network topology in order to ensure high reliability and robustness of MUWSNs.

In this paper, the MUWSN topology is described systematically. It has a guiding significance for further study on topology control algorithm and reliability of MUWSNs. Section 24.2 introduces research status of MUWSNs. Section 24.3 dissects the characters and architecture of MUWSNs based on the mobility of AUVs. Section 24.4 extracts three scientific problems about MUWSNs' topology based on Sects. 24.2 and 24.3, and it analyzes them in detail. Finally, the directions of further research are indicated in Sect. 24.5.

24.2 Research Status

In 1993, the theory for underwater acoustic network application was first set forth by the American marine research agency and MIT. They developed an intelligent underwater collection network named AOSNs, using AUVs as the platforms of mobile sensor. The plan *Seaweb* from Office of Naval Research was a relatively successful underwater network from 1999 to 2004. Meanwhile, various universities set up research groups to conduct studies on underwater sensor network. In 1994, for the first time ever, the robot lab of MIT manufactured a mechanical tuna with bionic propelling system successfully, and other underwater mobile platforms with bionic propelling systems have come out later [8]. In [5], the Underwater Sensor Network Lab of the University of Connecticut conducted a systematic study of underwater

Table 24.1 The architecture of MUWSNs

Project title	Description
LOTUS	UASNs for long distance communication [11]
ROBLINKS	Communication algorithm of long distance, high data rate, and robustness for UASNs in shallow waters [12]
SWAN	Communication protocols of UACNs in shallow waters
ESONET	Multidisciplinary and multifactors integrated study of ocean through observation sensor networks

sensor network. In [9, 10], the John Heidemann group put forward the idea of an intensive underwater sensor network and designed an underwater acoustic communication system that possesses functions of receiving signal energy detection and power consumption control. With the support of MAST III, the EU expands a series of plans and projects on Underwater Acoustic Sensor Networks (UASNs), as shown in Table 24.1.

In addition, the National University of Singapore carries out research on the interaction of multirobots underwater and networking based on underwater acoustic communication, and structures the architecture of underwater sensor networks with the combination of wired and wireless network as well as the static nodes and AUVs [13]. At present, the researches of various countries obtain achievements mainly on routing strategy and energy-saving efficiency of UWSNs [14–16], and have developed MUWSNs test bed for practice verification [17].

24.3 Characteristic and Architecture of MUWSNs

24.3.1 Characteristic of MUWSNs

Different from traditional wireless sensor network, MUWSNs have the following features because of different application environment and communication mode [18, 19]: *a. Complexity of application environment.* The complex and severe underwater environment corrodes and affects the sensitivity of sensor node and it is difficult to maintain the regular operation. Thus, in order to adapt such an environment, the nodes should be specific.

b. Limitation of sensor energy. Because sensor nodes are arranged underwater, it is hard to supply energy and unable to displace. Therefore, energy consumption is one of the major problems to consider while sensor designing. Besides, energy consumption among nodes should be seen as one important evaluation indicator for designing shift algorithm.

c. Inefficiency of communication mode. The underwater sensor network mainly uses underwater acoustic communication for the restriction of underwater environment. However, some bugs lead to a much more complicated design of communication protocol, such as the low bandwidth of underwater acoustic channel, huge signal attenuation, big and dynamic changes of time delay, severe multipath effect, high error rate, etc. The communication efficiency is obviously inferior to wireless sensor network on land.

d. Dynamics of network topology. The flow and underwater creatures often make network topology change. It may bring about intermittent acoustic communication links and disabled local topology, etc. Therefore, MUWSNs should have refactoring capability.

24.3.2 The Architecture of MUWSNs

The form of UWSN architecture is various. Referring to the OSI hierarchy model, [4] establishes the architecture of UWSNs according to the difference in the level's structure and elaborates contents of each level, respectively. In [20], the architecture of UWSNs is introduced according to the difference in the monitoring area, and it also expounds the network protocol stack referring to the OSI hierarchical structure. The requirements for the UWSN security system architecture are analyzed in [7]. All these studies are directed at ordinary underwater sensor networks without considering mobility. In this section, we divide the architecture of MUWSNs into three layers according to the different positions of sensors, as shown in Fig. 24.1.

a. Surveillance network attached to the water bottom. It is constituted by sensor nodes anchored to the water bottom and there are possibly wireless sensor nodes through underwater acoustic communication as well as those combining with each other through wired connection. Monitoring information will be collected by AUV nodes or delivered to sink nodes at regular time, then they are sent to the surface station. Finally, users will obtain information on the water bottom timely by communicating with satellites, ships, or onshore stations.

b. Surveillance network in water. It consists of sensor nodes suspended in water, which should be divided into two types: first, the common sensor node which cannot move autonomously, being equipped with air bag and anchored to the water bottom by chains, and second, making use of buoys on the surface to control nodes at different depths. They are used for monitoring information about the underwater space. Another type is AUVs, which are actually underwater robots that can move autonomously and reflect the mobility of MUWSNs. They can gather monitoring information from common nodes and move to fault location to heal the network topology.

c. Surveillance network on the surface. It consists of wireless sensor nodes over the water which contains common nodes and sink nodes. The common nodes are used to perceive the information about the 2D water surface. The sink nodes collect monitoring information from underwater sensor nodes and AUVs, and deliver the information to users through satellites, ships, and onshore stations.

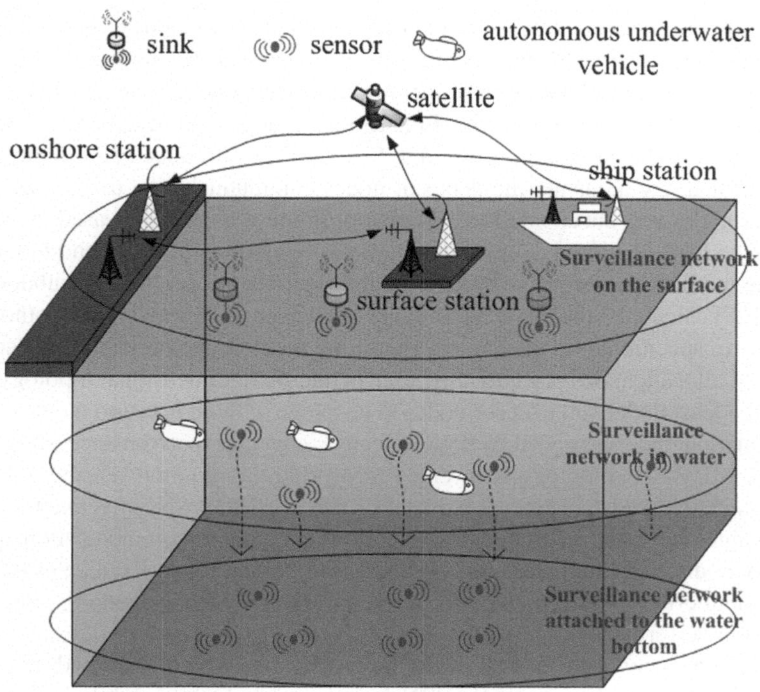

Fig. 24.1 The architecture of MUWSNs

24.4 MUWSNs Topology

The research on MUWSNs topology is an open field. The special underwater condition causes challenges for not only sensor nodes' arrangement but also topology generation, and it makes MUWSNs evolve dynamically, impacting on the reliability of networks. Therefore, topology is the key to MUWSNs research [21–23]. Analyzing MUWSNs features, architecture, and research status, three scientific problems of MUWSNs topology are summarized: how to generate three-dimensional MUWANs topology; how to heal topology ineffectiveness; and how to optimize topology.

24.4.1 Network Topology Generation Method

Topology generation is the whole process of generating effective overlay networks by putting the sensors into water. In this process, the initial energy of each sensor nodes, the communication link between them, and the coverage of whole networks are fixed, which has an important influence on the networks' coverage efficiency and their living time. In [24], Sandra studied topology control in the area of power saving and energy optimization, but she did not consider the sensor nodes' ineffectiveness and other factors. In [25], the special sensor nodes that are similar to AUV are introduced to optimize topology to improve connectedness and coverage. In [26], for the first time, Pompili studied three-dimensional arrangement of underwater sensor nodes and introduced Bottom-grid, and Akkaya [27] introduced self-deployment based on that. By adjusting the depth of nodes continuously, the overlap between adjacent nodes is reduced and the coverage of monitor area is increased, but these two methods need all nodes' information. By solving the maximum and minimum of game matrix, and according to the probability of occurrence in each subdomain, Golen [28] calculated the numbers of nodes in each area. However, this method did not involve specific nodes allocation. Therefore, the first key problem to be solved is how to allocate nodes scientifically, to generate three-dimensional topology, and to use the least underwater sensor nodes to coverage without aperture monitor area.

Evaluation indexes of MUWSNs' effective coverage performance should be defined. MUWSNs topology is invented to overlay the required sensing area and communication area, however, as a matter of fact, the MUWSNs effective topology area changes due to the restrictions on water flow and acoustic communication, which leads to monitoring blind area. As a result, it is important to work out an evaluation index about effective coverage performance of MUWSNs to evaluate the topology generation performance.

Three-dimensional MUWSNs evolution model should be constructed. In accordance with the complexity of MUWSNs' working conditions, and diverse mobile factors such as impact of the waves, ocean current, and ocean's irregular movement, an MUWANs evolution model can be constructed. With the ocean current's model imitated by the ocean flow data, the sensors arrangement can be improved. And the mobility evolution situation of networks can be predicted by the emulation techniques of constructing model and with MUWANs evolution model studied. The MUWSNs evolution model includes three submodels: nodes deployment model; energy consumption model of nodes mobility; and AUVs mobile state model.

MUWSNs' three-dimensional topology generation algorithm should come out. The three-dimensional topology generation algorithm is the key to create the MUWSN topology. The proper and high-efficient topology generation algorithm should have the following features:

- Low nodes removing energy consumption. Underwater sensor nodes could take only limited energy, and it is inconvenient to replenish. Therefore, it is necessary to reduce energy consumption and prolong nodes' service life-span during the process of topology generation.
- Small link communication delay. Some restricted factors such as underwater acoustic communication bandwidth lead to short communication distance and long delay. Thus, the network topology generated by 3D topology generation algorithm should decrease the distance or route between two sensor nodes so as to reduce communication delay.
- High efficiency of overlay network. Because of the restricted monitor overlay area, the 3D topology generation algorithm should use the least sensor nodes to cover

all the required sensing range and communication area. It is not allowed to have any monitoring blind area.
- High reliability of network topology. The complex underwater conditions have a lot of influence on MUWSNs, which lead to tremendous changes in MUWSNs. The MUWSNs generated by 3D topology algorithm should be as reliable as they can deal with the changes caused by network topology promptly.

24.4.2 The Self-healing Mechanism of MUWSN Topology

MUWSN is in a complex environment. Various external factors, such as water flow and underwater creatures, cause the shift of sensor nodes, which change the original network topology, and may lead to ineffectiveness of parts of topology, network segmentation, and even paralysis. Therefore, when partial topology inefficiency happens, it is a critical issue in the MUWSN topology study on how to heal the topology effectively.

The AUVs' position choice algorithm is necessary. Nowadays, increasing number of researches are based on AUVs topology healing [29]. Due to the mobility of AUVs in MUWSNs, mobile strategy can be created to move to the designed position spontaneously and finish the sense and communication tasks instead of ineffective nodes. This can heal the network topology fast and resume the whole connectivity. However, as a substitute for ineffective nodes, AUVs are not invented to shift to the original position, but to consider topology's communication delay and energy consumption and reduce them as much as possible. In terms of this problem, the position choice of AUVs as the substitute is of great concern, which can not only heal ineffective Internet, but improve topology's features of delay and energy consumption. Therefore, it is necessary to judge and choose a proper position with the AUVs' position choice algorithm instead of moving AUVs to the original position. Nowadays, there are only few studies on AUVs' position choice algorithm, but many on the AUVs' project.

The fictitious force for AUVs' motion should be directed. After figuring out the movement position with AUVs' position choice algorithm, another important problem arises: how to shift the AUVs to the accurate position and reduce energy consumption, as there are various forces affecting AUVs in the underwater environment, such as gravity, buoyant force, water flow force, attractive force, etc. It is important to take advantage of the relation of these forces and shift AUVs to the expected position with the least energy consumption. There are four fictitious forces mentioned in [30], with which AUVs can shift to the ineffective topology position. The mentioned fictitious forces can be solved in theory, but it is difficult in the real situation. Due to the difficulty of underwater energy supply, proper force becomes the key to save AUVs' energy and prolong their service life.

24.4.3 Topology Optimization Strategy

Topology optimization means reducing mostly network communication delay and energy consumption and making it have strong survivability and robustness with guarantee of network overlay on the whole monitoring area. For traditional wireless sensor networks, there are already many topology optimization algorithms, but they are not suitable for the 3D underwater environment. MUWSN topology optimization is an NP problem, and its research achievements are still limited.

Because MUWSNs are influenced by complex underwater conditions, the network topology changes a lot, and the communication delay and energy consumption are also affected greatly. In order to guarantee the high efficiency and reliability of MUWSNs, it is necessary to carry out the topology optimization. There are two ways of topology optimization: event trigger optimization and preventive optimization.

- The event trigger optimization means that the trigger optimization algorithm would optimize the network topology when MUWSNs' topology is monitored to change or fail. This method is combined with the topology of self-healing strategy, and often apply to the whole optimization through using the partial optimization algorithm. When the topology changes because of the nodes failure, the self-healing algorithm optimizes ineffective part network, and carries out the topology optimization. As this method is based on the awareness of topology's change, its timeliness is relatively low.
- The preventive optimization means that the optimization strategy is carried out to optimize network topology regularly no matter whether MUWSNs' topology changes or not. There are many key sensor nodes in MUWSNs whose degree is relatively high. In order to improve MUWSNs' reliability and survivability, it is important to avoid the long-term existence of these nodes. When the preventive optimization is adopted, the key nodes of MUWSNs could be exchanged by changing routes. As a result, MUWSNs' efficiency is optimized largely by reducing the stationarity and long-term existence of key nodes. The preventive optimization is able to ensure the networks' real-time communication, but it is a whole optimization strategy, which needs numerous network information. As a result, it could consume many network resources.

24.5 Conclusion

With the deepening application of the underwater priority areas, requirements of the reliability for MUWSNs' topology are constantly improving. Topology generation, healing, and optimization are key links of MUWSNs' network reliability, and are also the key problems of whether MUWSNs can be widely used or not. At present, the result of MUWSNs' topology research is less, especially in respect to the topology healing and optimization. Therefore, based on basic network topology, in cases of network node or link failure, by combining topology control, topology coverage, and topological reconstruction, the optimization of multiobjective combination (energy consumption, delay, coverage, and healing) will be the focus of the future research direction in this field to improve MUWSNs reliability. The next steps in the work are as follows: allow MUWSNs network topology to be the breakthrough point; establish MUWSNs network evolution model; further research to design high performance topology healing algorithm; looking for efficient topological global optimization strategy combining with swarm intelligence algorithm; simulation experiment can be made for MUWSNs-related research results by applying simulation tools such as OMNet++, NS2, and meanwhile, simulation process can be modified according to the sea trials.

References

1. Liu K, Yang Z, Li M et al (2010) Ocean-sense monitoring the sea with wireless sensor networks. ACM SIGMOBILE Mob Comput Commun Rev 14(2):7–9
2. Casari P, Zorzi M (2011) Protocol design issues in underwater acoustic networks. Comput Commun 34:2013–2025
3. Shusta J (2010) Acoustic network architecture. In: Proceedings of 5th ACM international workshop on underwater networks (WUWNet). Woods Hole, MAACM, pp 28–34
4. Akyildiz IF, Pompili D, Melodia T (2005) Underwater acoustic sensor network: research challenges. Ad Hoc Netw 3(3):257–279
5. Cui J-H, Kong J, Gerla M et al (2006) ChallengesBuilding saclable mobile underwater wireless sensor networks for aquatic applications. IEEE Netw Spec Issue Wirel Sens Netw 20(3):12–18
6. Chao L, Shuo W, Min T (2009) Survey on mobile underwater wireless sensor networks. Control Decis 24(6):801–812
7. Wei Z, Yang G, Cong Y (2012) Security of underwater sensor networks. Chin J Comput 35(8):1594–1606
8. Lu F, Mirza D, Schurgers C (2010) D-Sync Doppler-based time synchronization for mobile underwater sensor networks. In: Proceedings of 5th ACM international workshop on underwater networks (WUWNet), woods Hole, ACM, pp 31–38
9. Heidemann J, Ye W, Wills J et al (2006) Research challenges and applications for underwater sensor networking. In: Proceedings of IEEE wireless communication and networking conference, Las Vegas, pp 228–235
10. Heidemann J, Stojanovic M, Zorzi M (2012) Underwater sensor networks applications, advances and challenges. Philos Trans R Soc A 370:158–175
11. Li B, Zhou S, Stojanovic M et al (2008) Multi-carrier communication over underwater acoustic channels with non-uniform doppler shifts. IEEE J Ocean Eng 33(2):198–209
12. Liu L, Zhou S, Cui J-H (2008) Prospects and peoblems of wirless communications for underwater sensor networks. Wirel Commun Mob Comput 8(8):977–994
13. Junfeng X, Li K, Min G (2012) Reliable and energy-efficient multipath communications in underwater sensor networks. IEEE Trans Parallel Distrib Syst 23(7):1326–1335
14. Tan HP, Diamant R, Seah WK, Waldmeyer M (2011) A survey of techniques and challenges in underwater localization. Ocean Eng 38(14):1663–1676
15. Erol-Kantarci M, Mouftah HT, Oktug S (2011) A survey of architectures and localization techniques for underwater acoustic sensor networks. IEEE Commun Surv Tutorials (COMSUR) 13(3):487–502
16. Sanchez A, Blanc S, Yuste P, Piqueras I, Serrano JJ (2012) Advanced acoustic wake-up system for underwater sensor networks. Commun Inf Sci Manag Eng 2(2):1–10

17. Goodney A, Cho Y, Heidemann J, Wroclawski J (2010) An underwater communication and sensing testbed in Marina del Rey (poster abstract). In: Proceedings of 5th ACM international workshop on underwater networks (WUWNet), ACM, Woods Hole
18. Luo H, Guo Z, Dong W et al (2010) LDBLocalization with directional beacons for sparse 3D underwater acoustic sensor networks. J Netw 5(1):28–38
19. Caruso A, Paparella, Vieira LFM, et al (2008) The meandering current mobility model and its impact on underwater mobile sensor networks. In: Proc of INFOCOM, Piscataway, NJIEEE, pp 221–229
20. Guo Z, Luo H, Hong F et al (2010) Current progress and research issue in underwater sensor networks. J Comput Res Dev 47(3):377–389
21. Erol-Kantarci M, Oktug S (2011) A survey of architectures and localization techniques for underwater acoustic sensor networks. IEEE Commun Surv Tutor 13(3):487–502
22. Zhou RZ, Peng JZ, Cui J-H et al (2011) Scalable localization with mobility prediction for underwater sensor networks. IEEE Trans Mob Comput-TMC 10(3):335–348
23. Climent S, Capella JV, Meratnia N, Serrano JJ (2012) Underwater sensor networks a new energy efficient and robust architecture. Sensors 12(1):704–731
24. Sendra Sandra, Lloret Jaime, García Miguel, Toledo José F (2011) Power saving and energy optimization techniques for wireless sensor networks. J Commun 6(6):439–459
25. Yoon S, Azad AK, Oh H, Kim S (2012) AURP an AUV-aided underwater routing protocol for underwater acoustic sensor networks. Sensors 12:1827–1845
26. Pompili D, Melodia T, Akyildiz IF (2009) Three-dimensional and two-dimensional deployment analysis for underwater acoustic sensor networks. Ad Hoc Netw 7(4):778–790
27. Akkaya K, Newell A (2009) Self-deployment of sensors for maximized coverage in underwater acoustic sensor networks. Comput Commun 32(7–10):1233–1244
28. Golen EF, Mishra S, Shenoy N (2010) An underwater sensor allocation scheme for a range dependent environment. Comput Netw 54(3):404–415
29. Erol M, Vieira LFM, Gerla M (2007) Auv-aided localization for underwater sensor neteorks. In: Proceeding of the international conference on wireless algorithms systems and applications(WASA2007), Piscataway, NJIEEE, pp 44–54
30. Liu Jun Yu, Geng Zhang Huipeng (2011) A topology reconfiguration algorithm through node position control for space information networks. Acta Electronica Sinica 39(8):1837–1844

Author Biographies

Ming He was born in 1978. He is a Ph.D. and an associated professor in PLA Science and Technology University. His research interests are reliability of sensor network.

Wenhui Liang was born in 1986. He is a M.S. candidate in PLA Science and Technology University. His research interests are reliability of sensor network.

Qiuli Chen was born in 1990. She is a M.S. candidate in PLA Science and Technology University. Her research interests are network security.

Qingbing Zou was born in 1989. He is a M.S. candidate in PLA Science and Technology University. His research interests are network security and Internet of Things.

Chapter 25
The Design and Simulation Based on Ultrasonic and CSMA/CD Alike Tracking Method for Wireless Sensor Networks

Changliu Niu, Xiwei He and Shouli Zhou

Abstract To address the defects of existing tracking method in some industrial applications, a new wireless sensor network tracking method is put forward based on ultrasonic and CSMA/CD mechanism: prior to each positioning, unknown node need to query the current ultrasonic channel state to determine whether the next step is to locate or to wait. After the completion of the positioning, the unknown node will broadcast positioning ended signal to release the channel. The results of MATLAB simulation of existing method and CSMA/CD alike method show that in a particular application, the new method can reduce the tracking error and extend the nodes' lifespan.

Keywords Ultrasonic · CSMA/CD · Wireless sensor networks · Position tracking

25.1 Introduction

The target tracking in wireless sensor network based on ultrasonic is increasingly applied in industrial production, indoor tracking, and so on [1, 2]. In the situation where multiple unknown nodes need tracking, generally there are two modes: active

Project supported by the Natural Science Foundation of Zhejiang Province (Grant No. LY12F04003).

C. Niu (✉) · X. He
College of Information Engineering, North China University of Technology, Beijing 100041, China
e-mail: chl_niu@126.com

X. He
e-mail: xiwei_he@foxmail.com

S. Zhou
College of information engineering, Zhejiang University of Technology, Hangzhou 310023 , China
e-mail: zhoushl@zjut.edu.cn

X. Wang et al. (eds.), *Advanced Technologies in Ad Hoc and Sensor Networks*, Lecture Notes in Electrical Engineering 295, DOI: 10.1007/978-3-642-54174-2_25,

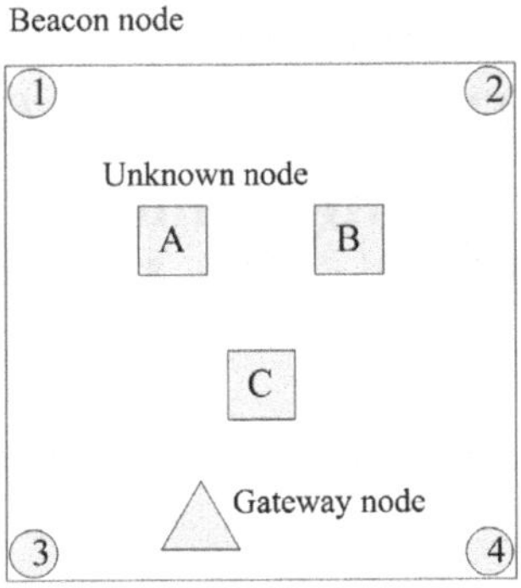

Fig. 25.1 Two-dimensional distribution of the node

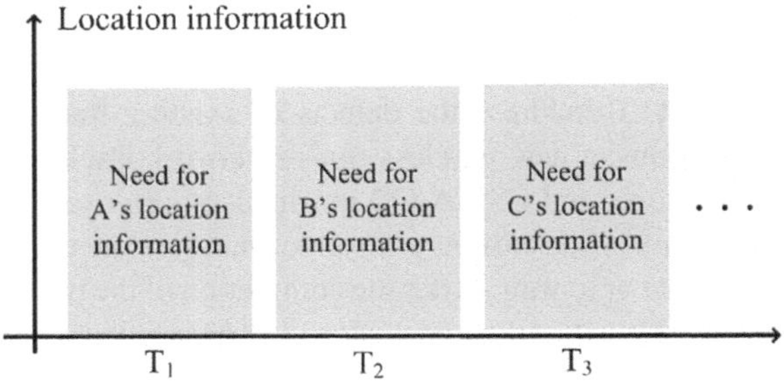

Fig. 25.2 Multiple selective tracking

mode [3] and passive mode [4, 5]. In the active mode, the unknown node carries ultrasonic launching module, the beacon node carries ultrasonic receiving module. While in the passive mode, the unknown node carries ultrasonic receiving module, the beacon node carries ultrasonic launching module. When the frequency of positioning is high enough, the tracking of unknown node will be realized.

The polling method, positioning in the order of A-B-C-A, is commonly adopted in active mode as shown in Fig. 25.1. This method can position or track multiple unknown nodes in the same time. However, in some industrial production applications, such as the AGV in manufacturing shop, a particular unknown node's (for example, node A, we call it current unknown node) location information is needed while the other unknown node B and C's location information is not needed in a period of time T1. And in another period of time T2, the unknown node B's (at this moment, the current unknown node is B) location information is needed while the other unknown node A and C's location information is not needed. In order to express it commodiously, we call this situation multiple selective tracking as shown in Fig. 25.2. In this situation, useless location information will be produced additionally if the polling method is adopted, with a negative impact on tracking error and energy of the nodes.

This article puts forward a CSMA/CD alike tracking method based on active mode, and compares this method with the polling method in the occasion of multiple selective tracking.

25.2 The Analysis of CSMA/CD Method

25.2.1 Workflow of the Polling Method

In the active mode, the unknown node carries ultrasonic launching module, the beacon node carries ultrasonic receiving module. The gateway node broadcasts the RF synchronizing signal at the interval of T_S. The unknown node A will launch ultrasonic immediately when it receives the first synchronizing signal. When the synchronizing signal arrives in each beacon node, they will start the timer at once. And the timer will be stopped if these beacon nodes receive the ultrasonic. This way, we can get distances from the node A to each beacon node: $d_{A1}, d_{A2}, d_{A3}, d_{A4}$. Similarly, the unknown node B will launch ultrasonic immediately when it receives the second synchronizing signal. The beacon nodes repeat the process above. In the same way, we can get $d_{B1}, d_{B2}, d_{B3}, d_{B4}$ and $d_{C1}, d_{C2}, d_{C3}, d_{C4}$. These data will be transmitted to the gateway node through RF signal and will be used to calculate the location. This is the polling method: calculating the location in the order of A-B-C-A as shown in Fig. 25.1.

25.2.2 Workflow of the CSMA/CD Alike Method

Refer to the CSMA/CD in Ethernet [6], a new method is put forward based on existing method. We introduce a variable called ultrasonic channel state. Its initial state is idle-state. For any unknown node which needs positioning, such as node A, its workflow is shown in Fig. 25.3. Figures 25.4 and 25.5 depict the workflow of beacon node and node B, C. The method's workflow is as follows:

(1) Prior to each positioning, node A inquires its ultrasonic channel state. If the state is busy-state, it will inquire again after a delay. If the state is idle-state, it will emit a RF positioning request signal. When gateway node receives the positioning request signal, it broadcasts a synchronizing signal.

(2) When node A receives the synchronizing signal broadcasted by gateway node, it starts to launch ultrasonic. When the other unknown nodes receive the synchronizing signal, their ultrasonic channel state will be set to busy-state. As soon as each beacon node receives the synchronizing signal, the timer is started. And the timer will be stopped if these beacon nodes receive the ultrasonic. This way, we can get distances from the node A to each beacon node: $d_{A1}, d_{A2}, d_{A3}, d_{A4}$. These data will be transmitted to the gateway node through RF signal and will be used to calculate the location.

(3) The node A broadcasts positioning ended signal. When other unknown nodes receive the signal, the ultrasonic channel state will be set to idle-state.

This method doesn't adopt the polling mechanism, invalid data is avoided. But it increases the time expenditure, including the time to transmit positioning request signal, positioning ended signal, and so on.

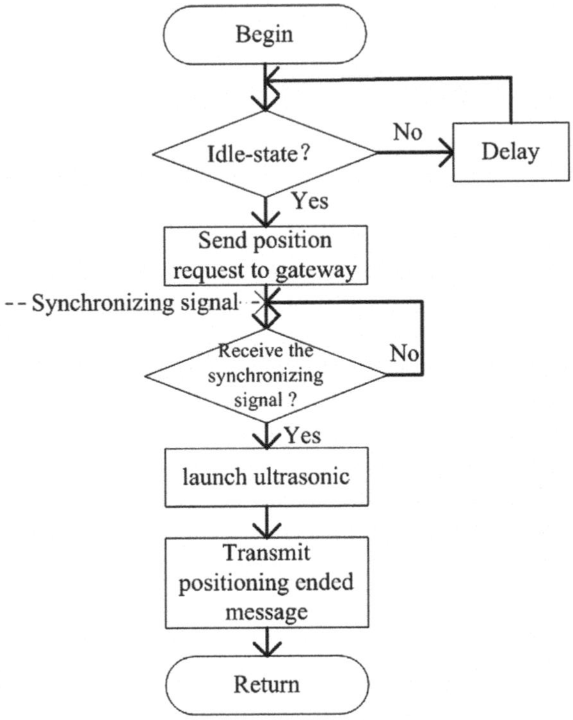

Fig. 25.3 Workflow of node A

25.2.3 Minimum Mean Square Error Estimation

The commonly used minimum mean square error estimation is adopted to calculate the two dimension coordinate of the unknown nodes. Figure 25.6 shows the three-dimensional distribution of the beacon nodes and the unknown node. In the xy-plane there are four beacon nodes: (x_1, y_1, z_1), (x_2, y_2, z_2), (x_3, y_3, z_3), (x_4, y_4, z_4). In the calculations, we assume that $z_1 = z_2 = z_3 = z_4 = 0$. The coordinate of unknown node A is (x, y, z). The distances from node A to beacon node 1, 2, 3, 4 are d_1, d_2, d_3, d_4. Then, we can get the following equations:

$$\begin{cases} (x-x_1)^2 + (y-y_1)^2 + (z-z_1)^2 = d_1^2 \\ (x-x_2)^2 + (y-y_2)^2 + (z-z_2)^2 = d_2^2 \\ (x-x_3)^2 + (y-y_3)^2 + (z-z_3)^2 = d_3^2 \\ (x-x_4)^2 + (y-y_4)^2 + (z-z_4)^2 = d_4^2. \end{cases}$$

The first three equations subtract the last equation respectively. We have

$$\begin{cases} x_1^2 - x_4^2 - 2(x_1-x_4)x + y_1^2 - y_4^2 - 2(y_1-y_4)y = d_1^2 - d_4^2 \\ x_2^2 - x_4^2 - 2(x_2-x_4)x + y_2^2 - y_4^2 - 2(y_2-y_4)y = d_2^2 - d_4^2 \\ x_3^2 - x_4^2 - 2(x_3-x_4)x + y_3^2 - y_4^2 - 2(y_3-y_4)y = d_3^2 - d_4^2. \end{cases}$$

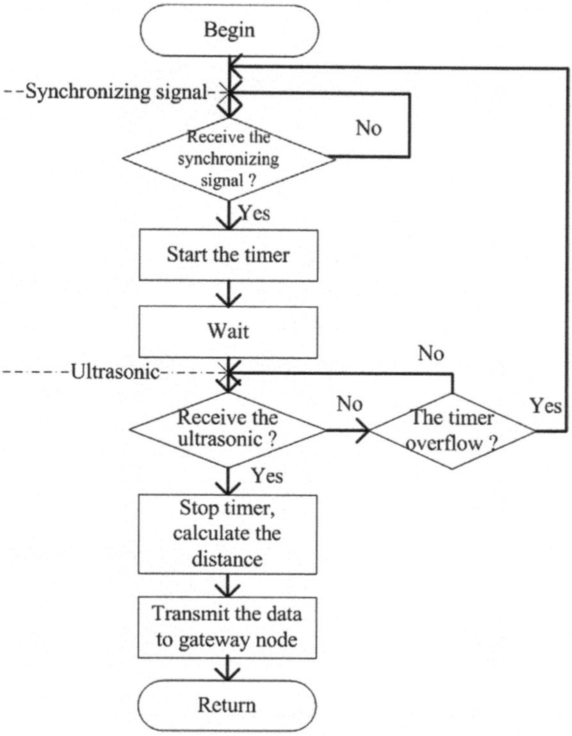

Fig. 25.4 Workflow of beacon node

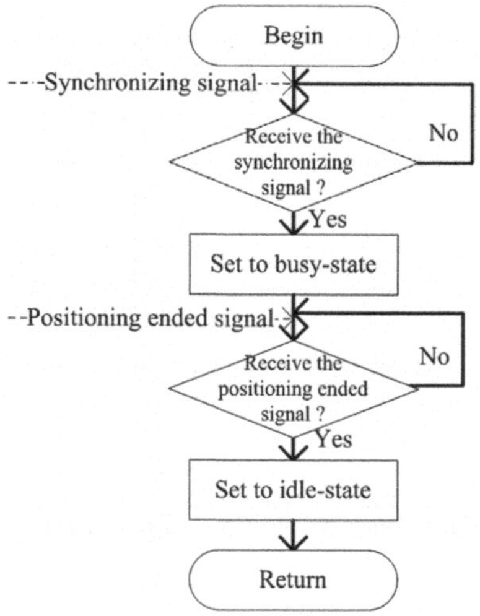

Fig. 25.5 Workflow of node B, C

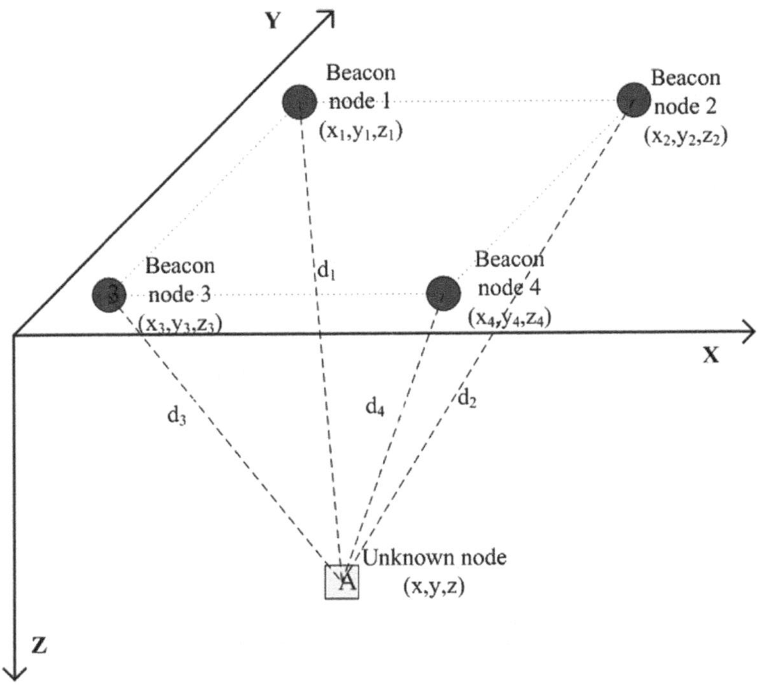

Fig. 25.6 Three-dimensional distribution of the nodes

The equations above can be expressed in the form of $AX=b$, where

$$A = \begin{bmatrix} 2(x_1 - x_4) & 2(y_1 - y_4) \\ 2(x_2 - x_4) & 2(y_2 - y_4) \\ 2(x_3 - x_4) & 2(y_3 - y_4) \end{bmatrix}, \qquad X = \begin{bmatrix} x \\ y \end{bmatrix},$$

$$b = \begin{bmatrix} x_1^2 - x_4^2 + y_1^2 - y_4^2 + d_4^2 - d_1^2 \\ x_2^2 - x_4^2 + y_2^2 - y_4^2 + d_4^2 - d_2^2 \\ x_3^2 - x_4^2 + y_3^2 - y_4^2 + d_4^2 - d_3^2 \end{bmatrix}.$$

According to the minimum mean square error estimation, we have

$$X = (A^T A)^{-1} A^T b.$$

Then the projected coordinate in the xy-plane of unknown node can be calculated [7].

25.3 The Experiment Process and Analysis

In order to test and verify the CSMA/CD alike method, the following experiments are designed. First, CSMA/CD alike method and polling method's elapsed time of a single positioning T_d is measured using the C51RF-CC2431 development board of

Table 25.1 The change of T_d as N varied

The number of unknown nodes N	1	2	3	4
Elapsed time of a single positioning in CSMA/CD alike method T_{d1} (ms)	71.4	70.0	73.1	71.2
Elapsed time of a single positioning in polling method T_{d2} (ms)	47.6	90.9	142.8	201.1

WXL company. Then, design a tracking experiment; compare the tracking error of the two methods based on MATLAB calculation.

25.3.1 The Elapsed Time of a Single Positioning

By the actual measurement of C51RF-CC2431 development board, we can get CSMA/CD alike method and polling method's elapsed time of a single positioning T_d, as shown in Table 25.1.

The CSMA/CD alike method increases the time expenditure, because it has to transmit positioning request signal, positioning ended signal, and so on. When $N = 1$, T_{d1} of CSMA/CD alike method is 23.8 ms longer than the T_{d2} of polling method. This is the time expenditure increased. The T_{d2} will increase obviously with N increasing. While in the CSMA/CD alike method, only the node that needs its own location information is being positioned, and the other $N - 1$ unknown nodes are neglected. So, T_{d1} can hold steady. This is equivalent to using the increased time expenditure of 23.8 ms to substitute for the effect of merely positioning the current unknown node in system when $N > 1$.

25.3.2 Tracking Experiment

Assume that there is a 5×5 m monitoring area. The four beacon nodes are placed on the four corners in the monitoring area. There are N unknown nodes waiting to be positioned in the monitoring area. The motion velocity of unknown nodes is 5m/s, and the motion trail is a sine curve as shown in Fig. 25.12. The distance from point A to point B is 7.81 m. The elapsed time T is 7.81 m/5 m/s = 1.562 s, and the positioning times from point A to point B is $M = T/T_d$. Table 25.2 shows the two method's M, which have been rounded.

Using MATLAB, the tracking polygonal line in the two methods can be draw out. In polling method, the current unknown node's positioning times from point A to point B decreases with N increasing. When the number of the unknown node N increases from 1 to 4, the positioning times are 33, 17, 11, 8 times. Figures 25.7, 25.8, 25.9, 25.10 are the illustrations of the corresponding polygonal line.

Table 25.2 The positioning times from point A to point B in the two methods

The number of unknown nodes N	1	2	3	4
The positioning times in CSMA/CD alike method M_1	22	22	22	22
The positioning times in polling method M_2	33	17	11	8

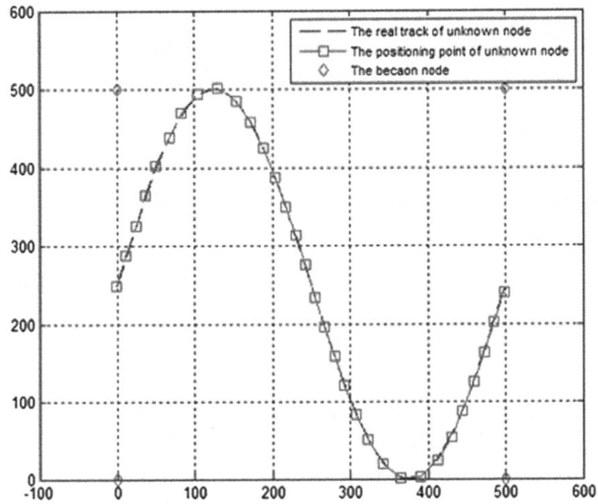

Fig. 25.7 The tracking polygonal line in polling method when $N = 1$

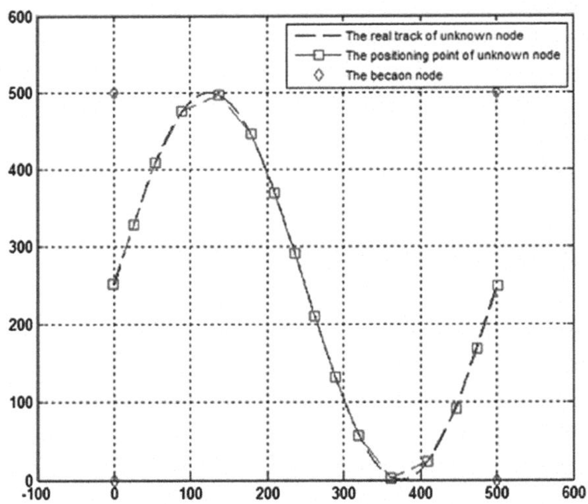

Fig. 25.8 The tracking polygonal line in polling method when $N = 2$

In the CSMA/CD alike method, the current unknown node's positioning times from point A to point B is 22 times, regardless of the number of the unknown node. Figure 25.11 shows the tracking polygonal line in CSMA/CD alike method (Fig. 25.12).

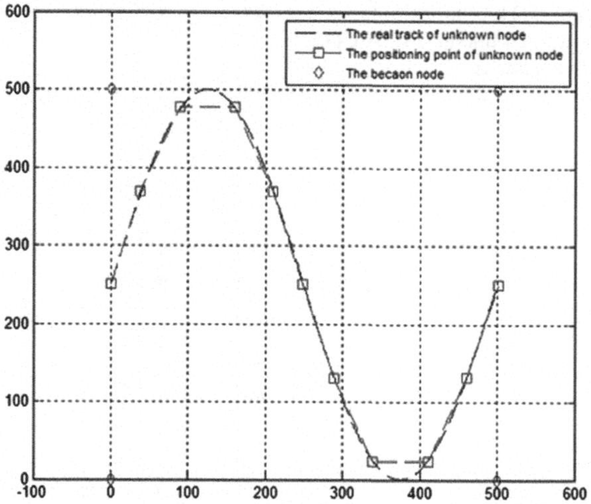

Fig. 25.9 The tracking polygonal line in polling method when $N = 3$

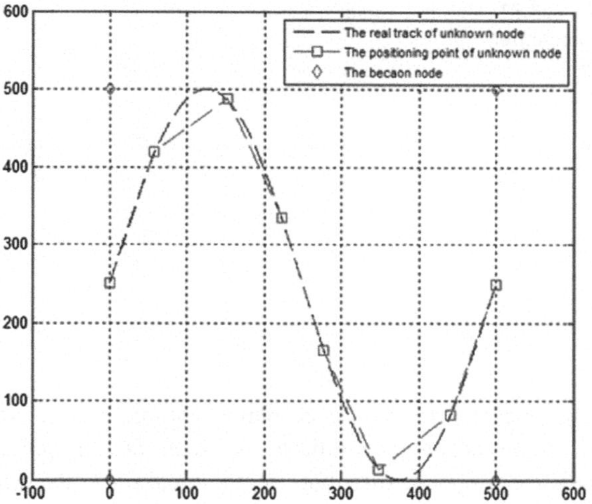

Fig. 25.10 The tracking polygonal line in polling method when $N = 4$

The Fig. 25.11 show that, with N increasing, the similarity between tracking polygonal line and real track decreases gradually in the polling method. While in the CAMS/CD alike method the similarity can be kept in an ideal level.

25.3.3 Average Tracking Error

On the unknown node's real track, select 30 compare points uniformly according to x-axis. To every compare point, fix its x-position and get a y-position on the tracking polygonal line, then we can get an estimate positioning point as shown in Fig. 25.13.

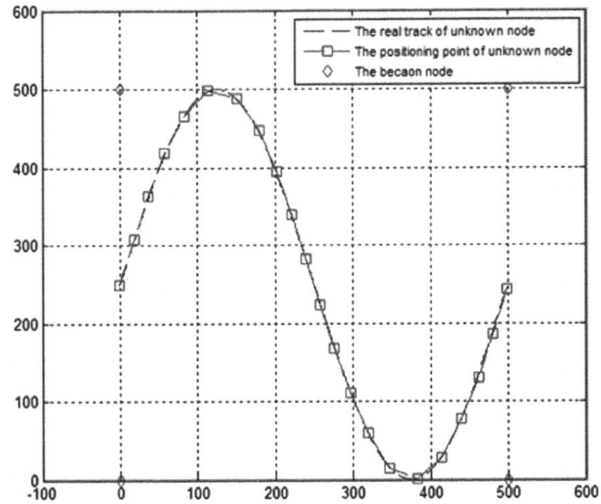

Fig. 25.11 The tracking polygonal line in CSMA/CD alike method

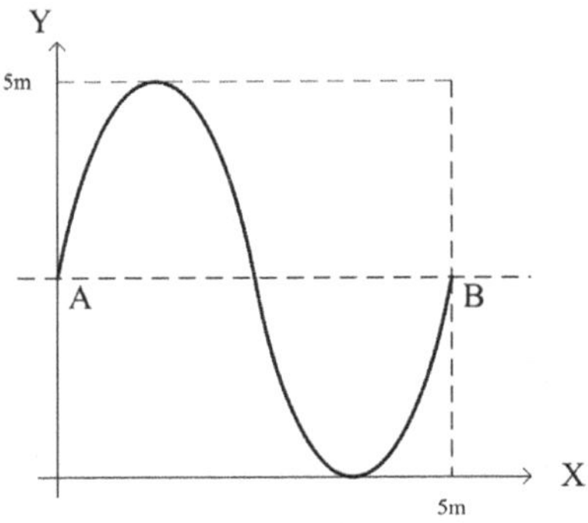

Fig. 25.12 Unknown node's motion trail

The distance between each pair of compare point and estimate positioning point d_i, can indicates the similarity between tracking polygonal line and real track. The positioning points distribute near the real track, the error can be limited to 5 cm. We define $W = (d_1 + d_2 + d_3 + \cdots + d_{29} + d_{30})/30$ as the average tracking error.

According to the data in Table 25.2, the polygonal line that shows the changes of average tracking error W as N varied can be calculated.The result is shown in Fig. 25.14. It can be seen from Fig. 25.14 that in the multiple selective tracking occasion, as the number of unknown nodes N increases, the W in the polling method increases and the W in the CSMA/CD alike method hold steady in a low level. The conclusion is that the CSMA/CD alike method has certain advantages when $N = 2$, and when $N > 2$ the method has definite advantages.

Furthermore, in the practical use, the other unknown nodes can work in low power mode because they are not involved in the positioning process fully, which can extend their lifespan.

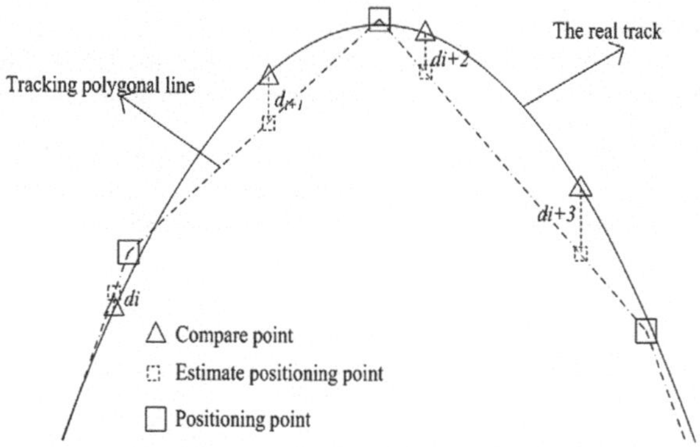

Fig. 25.13 Calculate the average tracking error

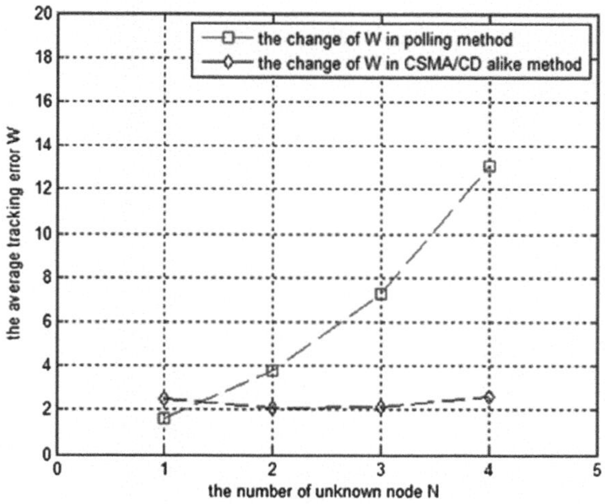

Fig. 25.14 The change of *W* as *N* varied

25.4 Conclusions

In this paper, to address the defects of existing polling method in some workshop transportation applications, the CSMA/CD alike method is put forward based on CSMA/CD mechanism in the Ethernet. The new method has definite advantage in the multiple selective tracking occasions, which can decrease the average tracking error and extend the nodes' lifespan.

References

1. Yang MH, Li LJ, Li CH (2004) Application of ultrasonic three-dimensional localization in virtual reality system. Comput Simul 21(11):145–151
2. An WP, Shen ZG, Pan YF (2009) Research on mining tracking algorithm for wireless sensor networks. Microcomput Inf 25(12–1):79–81
3. Han S, Luo HY, Chen Y, Ding YZ (2010) The design and implementation of a TDOA-based ultrasonic indoor localizing system. Chin J Sens Actuators 23(3):347–353

4. Kang LJ, Li M, Yang JH, Liu J, Dong YF (2010) A high-precision positioning method for wireless sensor networks based on ultrasonic and TDOA. Microcomput Inf 26(4–1):79–84
5. Sun YY (2011) Robot path planning based on ultrasonic location system. Ocean University of China, Qingdao
6. Xie XR (2007) Computer networks. Publishing House of Electronics Industry, Beijing, pp 79–83
7. Zhang F (2007) In: Design and implement of indoor localization system based on ultrasonic technology in WSNs. Wu Han University of Technology

Chapter 26
Security Analysis of WIA-PA Protocol

Yue Qi, Wenjuan Li, Xinqiang Luo and Qin Wang

Abstract The WIA-PA protocol is a secure and reliable communication standard for industrial process automation. WIA-PA standard defined on the security, but the description of the key management is not in detail, and there are security limitations in the kind of key management and authorization process. In this paper, we analyze the provided security mechanisms against well-known threats in wireless medium, and propose recommendations to mitigate the shortcomings. Finally, it summarizes concluded WIA-PA network security issues and the limitations of the security mechanism. This paper provides a reference for the WIA-PA network designer in security.

Keywords WIA-PA · Security consideration · Threat analysis

26.1 Introduction

Wireless Networks for Industrial Automation-Process Automation (WIA-PA) is a wireless network standard suitable for industrial process measurement, monitoring and control systems. It became the IEC standard [2] in 2011. WIA-PA, ISA100.11a and WirelessHART are the three standards in the field of industrial wireless network.

WIA-PA applies in process and equipment monitoring, environment and energy monitoring, asset management and advanced diagnostics, etc. WIA-PA network consists of Network Manager, Security Manager, Gateway, Router Device, Field Device, and Handheld Device. Gateway and Router form a mesh network. Router as cluster head, field device and handheld (if there are any) constitute a star network. Network Manager is responsible for resource management and monitoring of the entire network. Security Manager is in charge of key management and secure authentication of gateway, router, field device, and handheld device [1].

Y. Qi (✉) · W. Li · X. Luo · Q. Wang
School of Computer and Communication Engineering, University of Science and Technology Beijing, Beijing, China
e-mail: qiyuee@ustb.edu.cn

X. Wang et al. (eds.), *Advanced Technologies in Ad Hoc and Sensor Networks*, Lecture Notes in Electrical Engineering 295, DOI: 10.1007/978-3-642-54174-2_26,

WIA-PA standard has specified chapter to define security. Key management, secure access authentication, secure communication, and frequency hopping in this standard can surely resist most of the security threats in the industrial wireless environment. However, its lack of a more detailed explanation for key management. In key management mechanism, there is no broadcast key. Some security threats during access authentication process still exist. All the issues mentioned before must bring some security threats, which could not be solved by the security mechanism provided by the standard. Preciously research on WIA-PA security mainly focus on industrial wireless secure architecture design [3], WIA-PA authentication, key update or secure communication protocol stack research and so on. There has no systematic analysis on the security of WIA-PA standard according to the security threats, which exist in the industrial wireless environment. Designer or engineer of WIA-PA product always spend a lot of time on establishing security architecture of system, defining security module for each layer, or design of intrusion detection scheme. On the basis of the security mechanism of WIA-PA, this article analyzes advantages and disadvantages of security mechanism in the terms of security threats, and provides a reference for the WIA-PA network designer in security.

The rest of the paper is organized as follows: In Sect. 26.2, security scheme provide by WIA-PA standard is introduced. An analysis advantages and disadvantages of security scheme in the terms of security threats and some advices on the shortcomings of WIA-PA standard are given in Sect. 26.3. In Sect. 26.4, a summary of limitations of WIA-PA network security scheme is introduced. The paper is concluded in Sect. 26.5, give some possible research tendency in the future.

26.2 WIA-PA Security

The WIA-PA protocol is a secure and reliable wireless standard for industrial process automation, security is an optional part. Field device collects the data produced during the process of industrial automation and transport them to upper device in a safe way. When security is enable, process data along with network resource allocation information, secure key, time information, etc., are required to be transported in the form of security. In a word, all data transported in WIA-PA network should meet security requirements such as confidentiality, integrity, and authentication. The MAC layer security in the WIA-PA network is based on the IEEE STD 802.15.4-2006. It provides two levels of security service above the MAC layer [1]: end-to-end security in application sublayer and point-to-point security in data link sub layer (DLSL). Besides, WIA-PA provides secure access authentication mechanism of whole network.

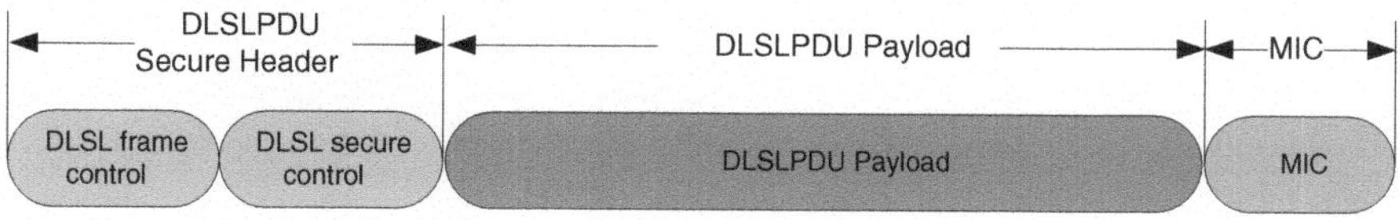

Fig. 26.1 WIA-PA DLSLPDU frame format

26.2.1 Point-to-Point Transportation Security

The DLSL uses data encryption and Message Integrity Code (MIC) to guarantee the point-to-point data security. MIC is computed over the DLSL keys and DLSLPDU payload and used for implementation of message integrity check. Point-to-point transportation security can defend against devices which try to access the network without authentication or eavesdrop data transported in wireless medium. MAC layer security can also guarantee point-to-point transportation security. Therefore, we can choose either MAC security or DLSL security to implement the point to point security.

DLSLPDU frame format is shown as Fig. 26.1. There is no specification for which encryption algorithm to use in WIA-PA standard. In IEEE STD 802.15.4:2006, it uses AES CCM* algorithm to implement the encryption and integrity check. The advantage of this method is that the receiver uses the same algorithm as the sender to decrypt data without another decryption algorithm, which can cut the amount of code in halves and then largely save storage and computing resources. This method is very suitable for resource-limited wireless sensor networks. Four parameters in AES CCM* algorithm are DLSL key, DLSL frame header, DLSL payload and nonce. DLSL security guarantees integrity and confidentiality between two neighbor devices.

AES CCM* method implements the encryption and integrity check. Counter mode is responsible for encryption and Cipher Block Chaining-Message Authentication Code (CBC-MAC) mode can provide integrity check.

(1) CBC-MAC mode

In CBC-MAC mode, the string which includes key, DLPDU header (a), DLPDU payload (m) and 13-byte nonce is divide into several 16-byte string, B0, B1, Bi. The 16-byte sub string Bi and key are set as the input of AES to compute MIC. The flow chart of CBC-MAC is shown in Fig. 26.2. CBC-MAC algorithm is shown in detail in reference [4].

(2) CTR mode

In CTR mode, the same 16-byte sub string as mentioned in CBC-MAC mode and the encrypted counter are the input to compute the cipher text. The flow char of CTR mode is shown in Fig. 26.3.

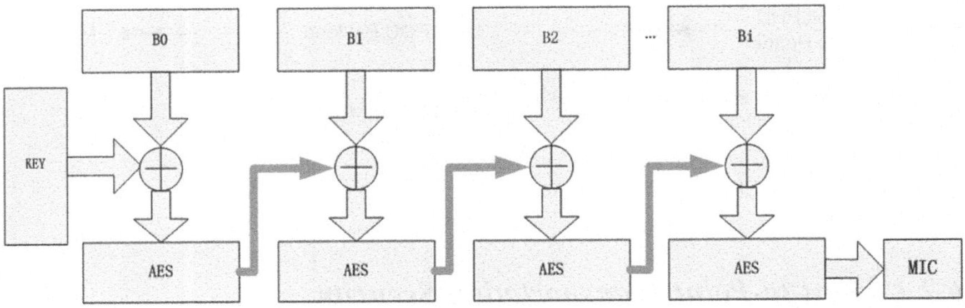

Fig. 26.2 Flow chart of CBC-MAC mode

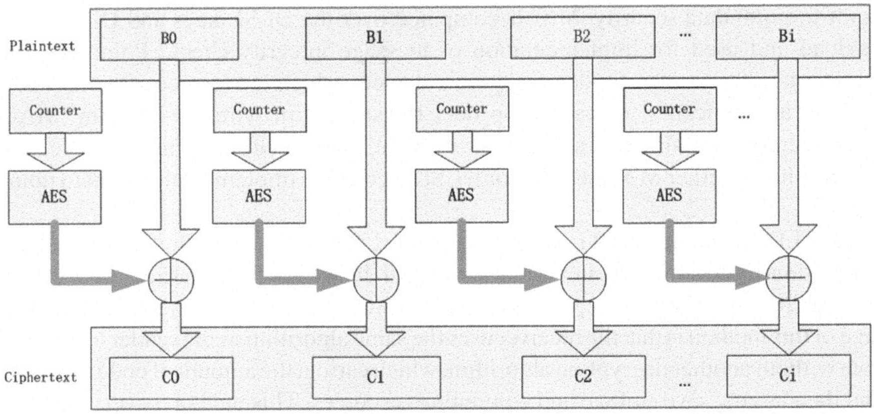

Fig. 26.3 Flow chart of CTR counter mode

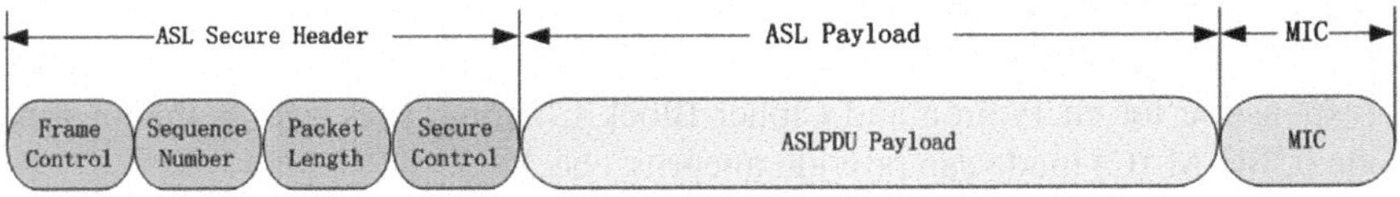

Fig. 26.4 WIA-PA ASL packet format

26.2.2 End-to-End Transportation Security

Application sublayer uses data encryption and MIC to secure the communication between the source and destination. MIC is computed over ASL key and ASL payload to implement integrity check. The format of ASLPDU is shown in Fig. 26.4. The secure control field defines the security level. If we choose both encryption and integrity check, encryption comes first on the sender side. On the receiver side, the sequence is inverted. If integrity check is failed, there is no need to execute decryption operation. WIA-PA protocol does not define the encryption algorithm at application sub layer. In order to save network resources, we can adopt the AES CCM* method as preciously mentioned.

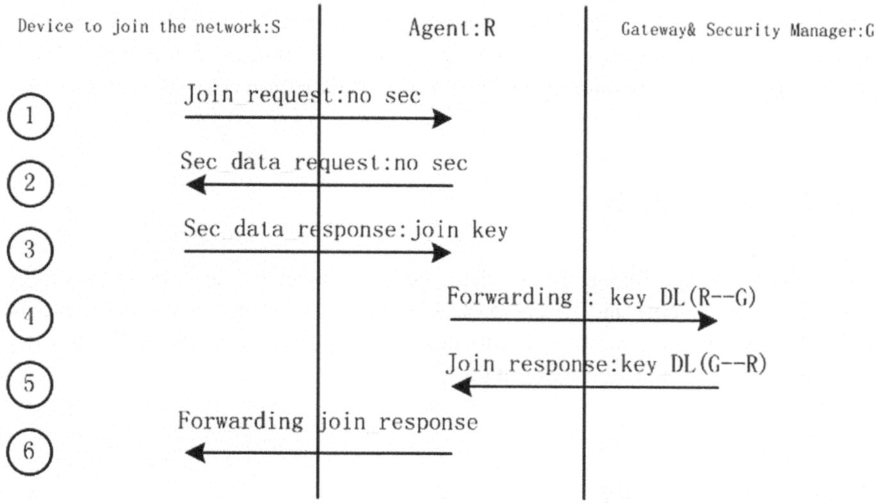

Fig. 26.5 Process diagram of secure identity authentication

26.2.3 Secure Identity Authentication Scheme in WIA-PA

Before secure access to WIA-PA network, join key and other security configuration information are written into devices in a wired way via handheld device. Join key is shared between device and security manager and used for access authentication.

Security authentication process is shown in Fig. 26.5.

Step 1: Device to join the network (S) sends join request to Agent R, which is an ordinary join request without any secure behavior.

Step 2: Agent R send secure information request to S, which is used to verify S legality. Because S is not yet a member of the network, there is no available key between S and R. There is no secure behavior in this request.

Step 3 to 5: S uses join key to encrypt unique 64-bit long address and then get secure information. S sends secure information to G which needs R to forward. On the path of R-G, DLSL security is used for secure transportation. Once received by Gateway, secure information is validated by security manager. If secure information is right, G sends secure information response to R. On the path of G-R, we use DLSL security to guarantee point-to-point transportation security.

Step 6: When received the response from G, R will forward the response to S. During this process, there is no secure behavior.

According to the analysis of secure authentication process, gateway can authorize the device which tries to access to this network. On the device side, it cannot verify whether sender of the join response is legal or not. In a word, it belongs to one-way authentication.

In a word, WIA-PA protocol provides confidentiality, integrity, identity authentication and availability (frequency hopping and time slotting mechanism [1]), but it does not support nonrepudiation, audit or intrusion detection at the moment.

26.3 Threat Analysis

Due to the broadcast feature of wireless medium, attacker can eavesdrop, replay packets, capture nodes, and so on [5–8, 11]. This section lists security threats to WIA-PA network and discusses threats that can be prevented by WIA-PA security mechanism and that must be focused on.

26.3.1 Eavesdropping

Wireless signal in the physical space is exposed, an attacker can easily monitor the channel, active or passive analyze traffic, launch jamming attacks or replay the previous data.

Eavesdropping occurs when an attacker surreptitiously monitor network and get information in an illegal way. WIA-PA confidentiality services can defend against eavesdropping attack. In WIA-PA protocol, application sublayer and DLSL use AES symmetric encryption mechanism. Despite some attacks launched against AES [9, 10], they can not destroy availability of algorithm, AES is still a NIST USA recommendation. Because the key information exists in the network will be regularly updated, it is difficult for an attacker to master the material. Moreover, frequency hopping mechanism makes it difficult for an attacker to intercept the signal, unless the attacker can get the network hopping sequence. In short, eavesdropping in WIA-PA network is not easy to implement.

26.3.2 Traffic Analysis

In wired network, the attacker need to physically connect to the network in order to analyze the network traffic and initiate subsequent attacks. However, in wireless networks, an attacker exploits the broadcast nature of wireless signals which means one does not need to actually connect to the network to analyze traffic.

If the network does not use security mechanism, then the network layer packet's source address, destination address, routing information and other information can be sent in plain text. By traffic analysis or packet analysis, an attacker can speculate network topology, used routing algorithm, used device type, and other information. Furthermore, he can launch subsequent attacks based on that information. On the DLSL, WIA-PA encrypts data and check message integrity. This security mechanism can defend against such traffic analysis. However, this means that router must decrypt the packets at the data link layer to get the destination address, routing information, re-encrypt packet and then send it to the destination address. It can bring additional time overhead. If we select AES or CCM built in radio hardware module to complete this task, it can save a lot of time overhead. To wireless network nodes which have limited energy, limited computing power and limited storage resources, it requires a compromise between security and system performance, which can greatly increase traffic analysis threats.

26.3.3 Selective Forwarding

Selective forwarding refers that the malicious node selectively drops the received packet or completely discard the received packet without forwarding. Certain probability of packet loss behavior can be mistaken for legal behavior which is not easy to find and easily regarded as a follow-up attack of traffic analysis.

WIA-PA network manager is in charge of monitoring the entire network and regularly collect network device status reports. According to these regular reports, assessment and diagnosis of network performance can detect and eliminate this attack to some extent.

26.3.4 Interference

Interference is not launched deliberately. Some device may emit exactly the same frequency channel, cover the original signal and impede the normal communication.

WIA-PA works in the 2.4 GHz band, including 16 channels. There are also other devices that use the same band such as Wi-Fi, Bluetooth, ZigBee, WirelessHART, ISA100.11a, and so on. WIA-PA network data link sublayer and network managers can implement coexistence of WIA-PA network and other wireless RF systems. WIA-PA network data link sublayer provides coexistence strategies, for example, slot transmission, low duty cycle, multichannel, adaptive frequency hopping and collision avoidance mechanism [1].

WIA-PA network introduces FHSS method on the basis of DSSS. Adaptive frequency hopping and other mechanisms can effectively suppress sudden interference and eliminate the frequency selective fading.

WIA-PA uses three kind of hopping mechanisms, namely Adaptive Frequency Switch (AFS), Adaptive Frequency Hopping (AFH) and Timeslot Hopping (TH). They can minimize the interference on the network in a better way. In order to support three frequency hopping mechanisms, related device needs to maintain a list of channels, including premium channel tables, poor channel tables, and recycling channel tables. Through the channel quality testing indicators (RSSI / LQI / PRR) we can monitor the current used channel for real-time and select an available channel from premium channel list. When channel quality drops below threshold, we throw it into the poor channel table. Then periodically update poor channel table and take better channel to the recycling channel table. When there is no available channels, select channel from recycling table and promote it to premium channel quality channel [12].

However, industrial process automation system requires high reliability. With more and more use of Wi-Fi, Bluetooth-enabled devices, mechanism to prevent interference need to be further improved.

26.3.5 Jamming

Compared with interference, jamming has different motivation. Attackers actively launch this attack by using Wi-Fi, Bluetooth mobile electronic device. Thus, it affects network performance. Although WIA-PA uses AFH mechanism to defense interference, an active attacker who mastered the hopping sequence can still initiate Jamming attack, affect WIA-PA network data transmission and consume node energy.

26.3.6 Collision

When two or more devices attempt to transport data in the same channel at the same time, collisions occur. WIA-PA network uses TDMA, AFH mechanism and the CRC check to detect and avoid collisions.

26.3.7 Tamper Attack

Tamper attack refers to modify data transmitted over the network. If message is only protected by the CRC checksum or digest algorithm, an attacker can modify data and then re-calculate CRC or summary. An attacker can get key material after node capture, and then further modify packet. In WIA-PA protocol, the application sublayer and the DLSL use MIC check to ensure data integrity which can effectively defend against data tampering attack. Generally speaking, in case of no access to key information, it is very difficult for an attacker to initiate tamper attack. Once an attacker masters key materials, he can easily modify packets and rebuild MIC. Periodically key refreshment is an effective way. Improving the physical protection method can reduce the opportunity to obtain key materials.

26.3.8 Wormhole

In a wormhole [13, 14] attack, an attacker establishes a tunnel between two nodes through wireless or wired link. Multi-hop routing nodes think that there is a single hop between them. Thereby, they attract nearby nodes to send packet via this wrong path and destroy the network. In WIA-PA network, wormhole attack is easy to be launched by an adapter and wired devices connected with WIA-PA network. If an attacker masters key information, he can establish between the two routing devices using wireless tunnels which have stronger receiving and sending power. Wormhole attack is related to routing algorithm used in network. If you use the map route which means that network has redundant paths, it may suffer wormhole attack. If you use tree routing, i.e., packet is sent from the source address to destination address in strict accordance with allocation, this can reduce the wormhole attack probability. However, it is prone to packet loss and reduces the reliability of the network. Reference [15] proposed a scheme which can defend wormhole attack. It adds the longest transportation distance to the packet. This scheme can defense wormhole attack.

WIA-PA network does not provide details of the routing algorithms, but if we use a centralized resource allocation, all nodes on the route are authorized nodes which can resist wormhole attacks. However, there are security risks during WIA-PA network secure authentication process. First, WIA-PA network lacks broadcast key. Device which tries to join the network may choose a malicious node as its proxy agent to join network. Second, join request is nonencrypted, as shown in Fig. 26.5 as step 1 and 2. If a malicious node gives join response, node may join an illegal network.

26.3.9 DOS Attack

DOS [16] is a kind of common network attack. It makes network be contrary to the availability requirement of network. There are two ways to launch DOS attack in WIA-PA network,

(1) When device tries to join the network, first initiates a nonencrypted join request, as shown in Fig. 26.5 as Step 1. Then the proxy node sends secure information request. The process after that is safe, but the nonencrypted request can make it possible that malicious node refuses to send join response.

(2) An attacker can tamper DLPDU, recalculate CRC and then keep sending this kind of packets. On the receiver side, the receiver checks the integrity of messages. Receiver may spend time on checking the wrong MIC, dropping the wrong packet and then sending request for packet retransmission. DOS attacks severely consume network traffic and prevent network from providing normal service.

WIA-PA network requires auxiliary DOS attack detection mechanism to defend against such attack.

26.3.10 Sybil

Sybil [17] attack occurs when an attacker masquerades as nodes which have multiple identities. Some network which is lack of authentication mechanism network is vulnerable to Sybil attack. Because the enemy can easily masquerade as multiple identities. However, in WIA-PA network, when devices try to join the network, secure authentication is required which makes it difficult to launch Sybil attack to WIA-PA network. However, during the nonencrypted join request phase (shown in Fig. 26.5 as step1 and 2), malicious nodes can disguise as a proxy node and send join response to device which tries to join the network. It makes it possible to join the network and destroy the network.

In conclusion, security mechanism provided by WIA-PA protocol can defend against most of the attacked mentioned above. However, we still need to take further measure to defend against traffic analysis, jamming, wormhole, DOS and Sybil attack. Besides these attacks, node physical protection is also very important. When captured by attackers, nodes need to execute self-destruction to avoid node cloning or further attack. When nodes leave network, we need to timely erase the memory.

26.4 Limitations of WIA-PA Standard

WIA-PA standard is a secure and reliable communication standard for industrial process automation. There are still some limitations for its security mechanism:

WIA-PA protocol does not support public key encryption algorithm. It does not provide nonrepudiation or other secure services.

Audit, intrusion detection and intrusion prevention mechanism are not specified in WIA-PA protocol.

Key management is not explained in detail. Because of lack of broadcast key, the WIA-PA network is vulnerable to Sybil attack during access process.

At the initial stage of secure authentication, the request is not encrypted. This security flaw can make the WIA-PA network vulnerable to DOS attack and Sybil attack.

The framework of security manager and its interface with network manager is not defined in detail.

26.5 Conclusion

This paper discusses the security features of WIA-PA protocol and analyzes advantages and disadvantages of security mechanism in the terms of security threats. Meanwhile, it points out some security limitations exist in WIA-PA. Physical protection of WIA-PA nodes is also one of the important parts in WIA-PA security, which can avoid attackers from node cloning or further destruction when they capture the nodes. Moreover, its of great significance to specify the monitoring of network performance which is in the charge of network manager. Secure access to wired network is not defined in WIA-PA protocol, which is also an essential part to guarantee the secure operation of network. WIA-PA provides core secure services, such as confidentiality, integrity, identity authentication and availability, but there are still some issues to be solved among them. Lack of broadcast key and security flaw exists in secure authentication make WIA-PA network vulnerable to malicious attacks at the access stage. Besides, WIA-PA does not provide some other key secure services such as nonrepudiation, audit, intrusion detection, etc.

Although we avoid using Public Key Infrastructure (PKI) in nodes, but the credible certification center (network manager or security manager) and relatively high processing power and energy resource make WIA-PA device different from the traditional sensor device. Research on implementation of ECC and RSA algorithm on sensor nodes implies the possibility of using PKI in wireless sensor network [18]. Ways to optimize the WIA-PA security mechanism include improve the existing security flaws, classify the security services in a comprehensive way, define the service such as audit, nonrepudiation, and intrusion detection and provide solutions to those services.

Acknowledgments This work was supported by National High-Tech Research and Development Program of China under Grant No.2011AA040101.

References

1. Industrial Wireless Networks WIA Specification (2001) Part I: WIA system architecture and communication specification for process automation (WIA-PA)
2. WIA-PA Wireless Industry Automation Specifications for Process Automation, IEC62601. http://www.industrialwireless.cn
3. Min Wei, Ping Wang, Quan Wang (2009) Research and implementation of security strategy for wireless industry control network. Chin J Sci Instrum 2009(04):679–684
4. Dworkin MJ (2004) Recommendation for block cipher modes of operation: the CCM mode for authentication and confidentiality, SP 800–38C
5. Raza S, Slabbert A, Voigt T et al (2009) Security considerations for the wirelesshart protocol. In: IEEE conference on emerging technologies and factory Automation, ETFA, vol 2009, pp 1–8
6. Roosta T, Shieh S, Sastry S (2006) Taxonomy of security attacks in sensor networks and countermeasures. In: The first IEEE international conference on system integration and reliability improvements, vol 25, p 94
7. Welch D, Lathrop S (2003) Wireless security threat taxonomy. In: Information assurance workshop, IEEE systems, man and cybernetics society, vol 2003, pp 76–83
8. Giannetsos T, Dimitriou T, Prasad NR (2010) Weaponizing wireless networks: an attack tool for launching attacks against sensor networks. In: Black Hat technical security conference
9. Bogdanov A (2008) Multiple-differential side-channel collision attacks on AES. In: Cryptographic hardware and embedded systems CCHES 2008, Springer, Berlin, Heidelberg, pp 30–44
10. Phan RCW (2004) Impossible differential cryptanalysis of 7-round advanced encryption standard AES. Inf process lett 91(1):33–38
11. Ukil A (2010) Security and privacy in wireless sensor networks. Smart Wirel Sens Netw Intechweb Croatia 2010:395–418
12. Xu W, Liang W, Feng C (2011) Design and implementation of key technologies in WIA-PA networks. Appl Res Comput 2011(06):2265–2270
13. Nagrath P, Gupta B (2011) Wormhole attacks in wireless ad hoc networks and their counter measurements: a survey. In: IEEE 3rd international conference on electronics computer technology (ICECT), vol 6, pp 245–250

14. Butty L, Hubaux JP (2007) Security and cooperation in wireless networks, Cambridge University Press, Cambridge
15. Hu YC, Perrig A, Johnson DB (2003) Packet leashes: a defense against wormhole attacks in wireless networks. In; INFOCOM 2003, twenty-second annual joint conferences of the IEEE computer and communications, IEEE societies, vol 3, pp 1976–1986
16. Raymond DR, Midkiff SF (2008) Denial-of-service in wireless sensor networks: attacks and defenses. IEEE Pervasive Comput 7(1):74–81
17. Newsome J, Shi E, Song D et al (2004) The Sybil attack in sensor networks: analysis and defenses. In: Proceedings of the 3rd international symposium on information processing in sensor networks. ACM, pp 259–268
18. Watro R, Kong D, Cuti S et al (2004) TinyPK: securing sensor networks with public key technology. In: Proceedings of the 2nd ACM workshop on security of ad hoc and sensor networks, ACM, pp 59–64

Chapter 27
Design and Implementation of IoT Security System Towards Campus Safety

Weijun Qin, Bo Li, Jiadi Zhang, Shanyan Gao and Yunhua He

Abstract Campus security has a personal relationship with all the teachers, students, parents and society. Campus's personal assault, crowded stampede, the safety of school bus and emergency treatment are all to be solved right now. In order to fuse and mine the data of security system, this article, first of all, presents a display method based on the massive data storage and browser. Then, it designs and develops a campus network security platform and the prototype system. Finally, experiment is carried out to prove the feasibility of campus network security platform in logic and performance.

Keywords Internet of things · Campus safety · Security system

W. Qin (✉)
State Key Laboratory of Information Security, Institute of Information Engineering, Chinese Academy of Sciences, Beijing 100093, China
e-mail: qinweijun@iie.ac.cn

B. Li · J. Zhang · S. Gao
School of Software and Microelectronics, Peking University, Beijing 102600, China

B. Li
e-mail: libo1989@pku.edu.cn

J. Zhang
e-mail: 514086918@qq.com

S. Gao
e-mail: 358245141@qq.com

Y. He
Xidian University, Xi'An 710126, China
e-mail: heyunhua610@163.com

X. Wang et al. (eds.), *Advanced Technologies in Ad Hoc and Sensor Networks*, Lecture Notes in Electrical Engineering 295, DOI: 10.1007/978-3-642-54174-2_27,

27.1 Introduction

Campus security has a personal relationship with each of the teachers, students, parents and the society. Not only is it related to the safety and healthy growth of students, but also it attaches relationship to the family's well-being and social stability. According to the report of WHO in 1990, most countries in the world, accidental injury is the main cause of injury, disability in children and adolescents [1]. In our country, the majority of children's accidents happened on the way to school or in the school. Therefore, personal assault, the stairs collapse stampedes, social hooligans nuisance of teachers and students, driving safety of school bus and emergency treatment and other events should be attached great importance to and solved right now.

In the late October 25, 2006, crowed stampede accident happened when the students from the fourth grade to the sixth grade of Guang Na primary school in the down stairs, which result in 8 dead and 45 injured students. In May 12, 2010, murder happened in the holy water kindergarten, Nanzheng County, Shanxi Province. The suspect Wu Huanmin holding a kitchen knife, cut into the kindergarten, which caused serious consequences of 9 dead 11 injured. In April 9, 2012, Jinlong Yucai kindergarten school bus was crashed by a big truck from the side at an intersection of Yangchun, Guangdong, which killed three children. According to the official statement, that bus is required to load 11 people, but the true number is 19 in fact, including 17 students of the kindergarten. In foreign countries, the campus safety accidents also occur frequently. In December 14, 2012, a 20-year-old Adam Lanza made the world "heartbreaking" bloody shootings in the Sandy Hooke primary school of United States. After the suspect had killed 20 children whose age is from 5 to 10 and 6 adult, he took his own life. Therefore, strengthening the management of campus safety requires the measures which should have high security, high real-time and standardization. In nowadays, with the rapid development of network in information and technology, applications in campus security field which combine the two aspects to achieve a functional perfect, wide usage, high real-time and high concurrency will have a great value [2].

At present, the Internet of things has been applied in some security system, such as video surveillance, burglar alarm. Each system plays its unique function in the field of security. However, the system is relatively independent. Not only did the existence of information islands have brought users many changes, but also it did not reflect the best performance of overall linkage in an emergency situation. There are many problems in the traditional security system of internet of things. Most video in video surveillance system and display device are not symmetric. Massive Information is independent and difficult to make the full use. If users want to get the information about an accident, they have to use the tracing method because it cannot give advance warning. Consequently, how the multiple sensor data of the internet of things carry out intelligent analysis, to provide early warning information in real time, is the core problem in network security field [3].

With the combination of Internet technology and security system, the key technology and the theory of multiple sensor information gathering and visualization, has important application prospect and theoretical value. According to the related researches at home and abroad, the preliminary research achievements in technology has been made, but there are still many problems. Being aimed at the multiple information sources and types of heterogeneous problem in the network security, the existing research can combine data fusion and display, but lack the deep information mining. As for the streaming media data display and interactive, the existing research can only show the primary interaction and cannot meet the display of three-dimensional and highly interactive requirements. As for analysis on the tendency of events, the existing research is also unable to achieve accurate predictions and is lack of chain display about the relationship of the people, cars, etc. Therefore, there are many technical challenges in this field and it is necessary to study related key theory and technologies deep.

In order to solve those problems, it is necessary to establish a unified data platform for data fusion of each security system. The platform should be easy to access. Not only can the Users monitor the security situation at any time and any place, but also it can carry out mining massive data and present users directly. In recent years, most of the security systems are based on C/S and rarely on B/S, because of the analytical efficiency of browser and the expression ability of data. With the development of science and technology, browser's performance is no longer the bottleneck of B/S architecture, and the development of Javascript and CSS make a great contribution to the browser's display ability. Zero maintenance and high universality of B/S is the targets in network security platform in this article.

Section 27.1, firstly, proposed the mass data storage and display method, which are all based on B/S architecture. And then gave the design and implementation of the security platform and massive data. Secondly, it presented the design and implementation of the prototype system. In the end, experiments were carried out to prove the feasibility of campus network security platform in logic and performance. The Sect. 27.2 briefly described the related work at home and abroad. The Sect. 27.3 described the campus security requirement, the concept and the process. The Sect. 27.4 described the design and implementation of the security platform based on B/S architecture. The Sect. 27.5 described the experimental scene, methods, results and conclusions and did some experiments combined with the campus security scene. At the end of the article, summaries of the current work were made and it provided the further research direction and content.

27.2 Related Works

At present, there are many mature campus security management systems.

In order to ensure personal safety and improve kindergarten children's safety coefficient, an application were launched for the kindergarten daily shuttle work and the safety management measures at home. This system was integrated with the safe transfer, voice calls, teacher attendance and access control functions. With the core concept of "safe campus", Jieshun technology designed a set of "Jieshun JSE

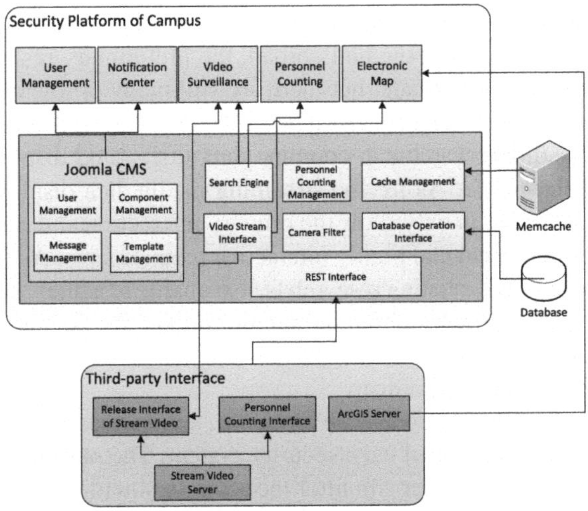

Fig. 27.1 System architecture

security management platform" solutions as the core of campus safety security. This system could make a centralized management, unified control and intelligent linkage for the access control, video surveillance, vehicles, patrol, radio broadcast, alarm and information management and provided a unified login, operation management platform for customers.

During the procession of constructing campus safety environment, there are not only software applications, but the fields in technology research, comprehensive management, standard setting, also have great achievements. In order to solve the campus security in The United States, many schools employ local polices or get community support to help to develop campus safety plans, solve the campus safety accident and build a strong community safety network. In addition, there are some legislations such as *no gan in campus*, *improve the campus environment* and so on in the United States. In the *education target of United States in 2000* which was went through in 1994, the seventh item is about "safety campus", "drugs, unauthorized guns and wine are not permitted within the campus".

Although campus security system at home and abroad are more perfect and mature in their respective functions, single function, lack of overall monitoring management, real-time difference, unable to handle concurrent data are their common shortcomings. Most of systems separate the campus personal management from the school bus, so they cannot make a real-time warning and prompt response under the unexpected situation [4, 5].

27.3 System Architecture

With the campus security application scenarios and the wireless sensor network, this article designed a network security information platform based on B/S architecture. As is shown in Fig. 27.1, this system consisted of three parts, which are a display subsystem, data processing subsystem and the data source subsystem.

The display subsystem is responsible for the visualization of data and provides the interface between users and platform. Data visualization includes real-time video surveillance, personnel counting, face recognition [6, 7], vehicle trajectory, vehicle license plate recognition and so on. With the electronic maps and data statistics, this system has realized the visualization of data.

Data processing subsystem includes data storage, caching, and clustering analysis and retrieval of sensor network data. This system adopts data mining and search engine technology from massive data to extract useful information and provides efficient retrieval methods. The raw data obtained from the third-party system, such as including personal technical data, the license plate recognition data, vehicle trajectory data and so on will be stored in the database, and then the thread running in background will make an analysis about them, extract useful information, such as the average number of students in campus daily and in the end, the analysis results will be stored in the database in order to provide data sets show. Cache management module provides the cache of frequently accessed data and access state of the user. If user wants to access the specific data, the system will search them in the cache. If failed, it will get the data from the database and put them into cache. Using cache mechanism reduces the frequency of users' accessing to the database and improves the performance of the system. The search engine module uses segmentation technology to create index and provides the data search interface in order to access data conveniently. The content management and user management are based on Joomla CMS system. The extensibility, flexibility and convenience for user authority management have provided the safeguard for the system.

The third party systems provide the platform with data source, which are consisted of personnel counting system, video steam management system, the ArcGIS map service, license plate recognition system, vehicle trajectory analysis system. They also present data communication interfaces based on the requirement, such as socket communication interface, REST interface and so on in order to acquire data conveniently.

Because of the B/S architecture, the platform has a good portability and scalability. User can access the system through browser without installing any client software and it also realizes cross platform. The platform provides access interfaces of the third party system, so it makes functions expansion easily. Platform released REST API and authorized the third party system or the client can access the data on the platform.

27.4 Design and Implementation

27.4.1 Visual Display of User Interface

Electronic Map Electronic Map is realized by Web ArcGIS [8–10] and gets source data through AJAX. Those data define the campus information. According to the specific data of different attributes of each campus, this map can display the student's number respectively. It can also be added to a given campus dynamically which should be added the latitude and longitude of the name attributes. Each of the campus icons has two buttons, "video monitoring" and "statistics", which can jump to the page respectively.

Data Chart Security platform uses Web chart for visual display. The historical records and the historical average records are displayed by bar chart and line chart, the passing frequency of vehicle by bar chart, the gather density by regional hot map. The implementation of web chart depends on Highchart library which is one kind of Javascript Library, which can show the statistical data directly.

Interactive Video Monitoring Interactive video surveillance can provide traditional view function and show the analysis results to users. The design of Interactive video surveillance is shown in Fig. 27.2.

The browser gets video streaming and video analysis data by querying the database. Flash Player is used to broadcast video stream and at the same time, it uses Javascript code to draw video analysis data.

Notification Center Safety warning is an important part of the security platform. In this platform, it adopts the way of notification center to achieve security warning. If the occurrence of hazardous event happens, such as, a suspicious vehicle in the bayonet, notification center will give an alarm by the way of flashing or pop-up window to tell users the event. Notification center is located in the status bar of the display page, which is convenient for users to use. To ensure the real-time of the event, message is delivered by the message queue, that is to say, after the third subsystem have put the emergency message into the message queue of the platform, the web monitoring thread of the platform will read the message from the queue and then push the message to the front page with Comet mechanism. After the event has been catched, the message will be stored in the database. This progress is shown in Fig. 27.3.

27.4.2 User and Content Management

Content Management System Content management system (Content Management System, referred to as CMS) refers to a mode of cooperation for the management of work flow, which is widely used in the storage and revised of articles and documents. This article has adopted the open source CMS Joomla. Joomla is required to use PHP

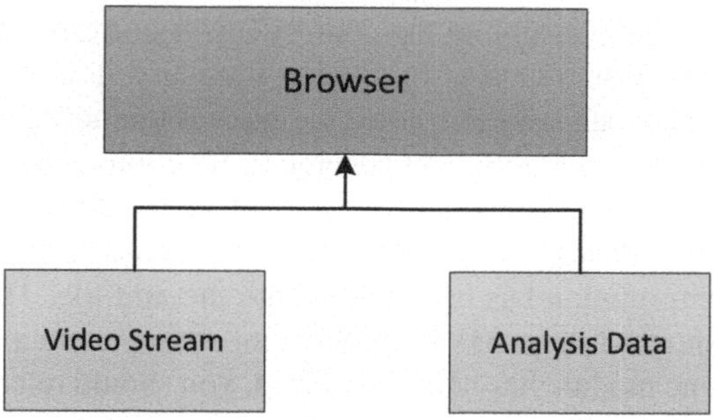

Fig. 27.2 Interactive video monitoring

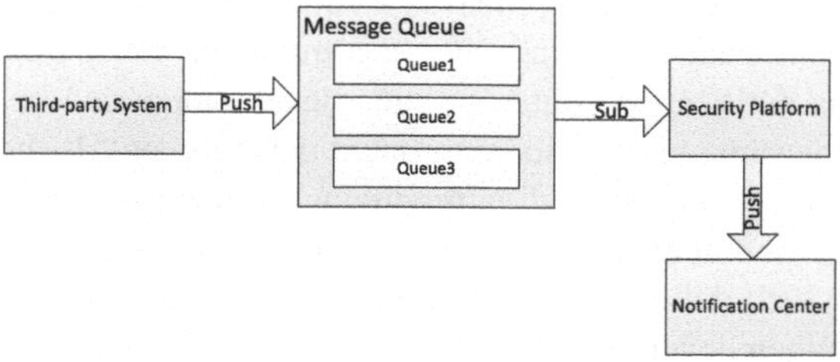

Fig. 27.3 Message pushing mechanism of notification center

programming language and MySql database and can run in different operating system such as Linux, Windows, MacOSX and so on.

According to the architecture of Joomla System, Joomla Framework and JApplication ensure the smooth running of the system. The upper layer extension is open to developers for secondary development, so developers can meet users' requirements at a minimum cost.

User Management User management is one of Joomla Modules and establishes a set of complete user management mechanism. Not only can the system administrator add users and put them into different group, but also he or she gives them different access authority. Joomla can automatically check the user access authority and display the authorized pages so that different authority pages can be accessed by different authority users. At the same time, during the development process, system can get the current users' group name and user authority with the interface built-in system.

The user groups of this platform are divided into three categories. The first category is visitor who has no access authority for core functions. The second category is the men who are responsible for campus security, such as security, teachers. They are permitted to get the campus security information and view the video stream data. However, they don't have the configuration authority. The third category is the administrator of this platform. Not only do they have the same access authority to the second category users, but also they have the configuration authority.

Function module development and system configuration Developments of Function modules are mainly centralized in the extended layer of Joomla and divided into three parts: component development, module development and template development.

The partition of the components is corresponding to all the function modules of this system such as map page, video surveillance, parameter configuration and so on. These pages are permitted to set access authority independently. When it is necessary for page jumping and generating functions, system will determine the user's access authority. Components development should be transferred into compression file in accordance with the document requirements of Joomla and ensure that installation information has been added into the xml file. The development process of Joomla should follow MVC architecture which contain model, view and controller. After one module has been developed, you should release it to the Joomla system with its built-in installation function.

Module development is a supplement for component development and module can be understood as a part of component. When a module needs to be displayed, such as notification center, module development can be used. Module development has the same process to component development. It should to be noted that you should change the responding module when adding installation information into xml file.

Template mainly defines the effect of the theme and interface of Joomla system, which defines the system pages in the background, layout and overall style sheets. When developing template, you should define the loading position of every components and modules and its installation information file should be identified.

After the components, modules and templates have been developed completely, you should login Joomla management system. First of all, configure required template. Then configure the loading positions for different modules. Finally, add functionality module page and configure the function module menu.

27.4.3 Data Mining and Retrieval

Multi-source sensor network raw data, mining useful information According to the multivariate statistical information and analysis of raw data, the platform can achieve aggregation of information available and trend analysis to monitor events. We select appropriate approach to data mining for a particular subsystem. Take examples of personnel counting and vehicles trajectory subsystem.

According to numerical statistics of every schools, the personnel counting subsystem will find out the average value at working and non-working day and based the average value, it can set thresholds for comparison with real numbers. Once the real number exceeds the threshold, the system will alarm to alert the managers the fact that the campus may be crowded.

Vehicle trajectory analysis subsystem based on the third party vehicle mount monitor database, provides specific vehicle historical trajectory queries, suspicious vehicle trajectory analysis for group activities and alarm for sensitive events in specific areas. According to the characteristics of vehicle and events, Systems can make clustering analysis storage, such as building suspicious vehicles databases based on the historical records of risk events, building databases of different models based on the type of vehicle, building regional databases based on frequently sensitive

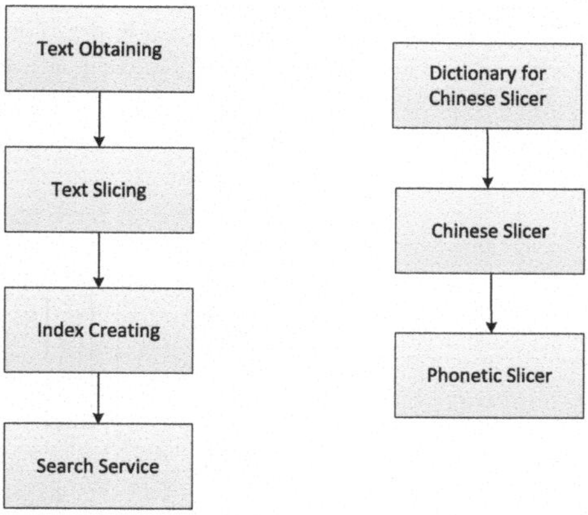

Fig. 27.4 Workflow of search engine

areas and combined with the needs of users, making data clustering analysis and visualization display.

Search Engine Search engine is used to create indexes for large amounts of data in the database and provide data searching and mining with efficient search service later. In this project, the search engine is based on Solr which is a high performance, supporting java programming language and full-text search server.

The workflow of Solr is as followed in Fig. 27.4. Obtaining text, creating indexes and providing search service can be realized by configuring Solr's environment, but segmentation of text requires intervention to achieve more satisfactory results.

Segmentation of text needs Chinese slicer and the workflow is as followed in Fig. 27.4. You should build a dictionary for special names which will be used by Chinese slicer. After the text has been sliced, phonetic slicer will hand over for the next processing. Thus, the search engine can not only support Chinese search, but it can also support the phonetic search.

27.4.4 Data Storage and Caching

Data storage The data of security platform is mainly from the sensor network node data, such as personnel counting data, the vehicle trajectory analysis data, license plate recognition data, and so on. Because of the prevalent heterogeneity of different data source and the scalability of third-party system, this system adopts NoSQL database MongoDB to store data of IOT. MongoDB is collection oriented and is easy to store the type of object data. The flexibility of its data model structure allows the system for the third-party's data with different structures easily accessible.

Taking the data storage of campus personnel counting as an example, Campus is the center of count and the smallest unit is a camera. Generally, each school requires multiple cameras for monitoring the school's gates. So it is necessary to combine the

Fig. 27.5 Data caching

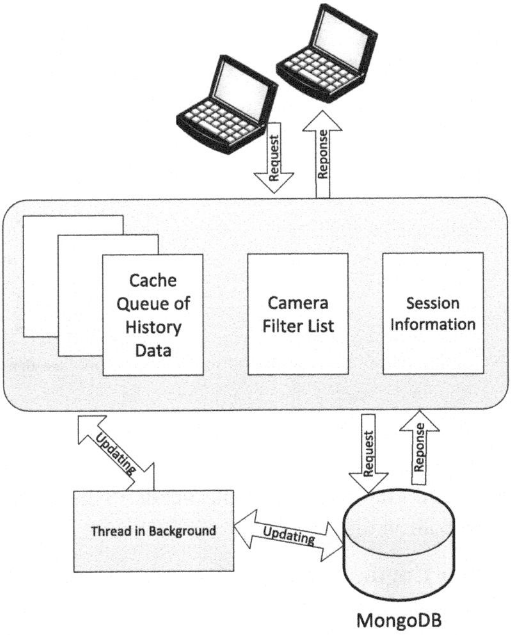

camera with the schools. That is to say, we can put all the cameras into the camera list which is a data type of school. For the different functions of the camera, we can set different tags attribute for every camera which represents the count or density label. Users can also add a custom label to a camera for searching and sorting easily.

The data of personnel counting is a type of log data which cameras in large detail generate and share the common characteristics that the size of a single data is small, the number of data is in large amount and the data structure is not fixed. Therefore, this system adopts the way of Key-value model to store massive data. A document of MongoDB represents a record and it contains id of this document, id of the last layer, timestamp and counting records. Id is unique in the database, which is consisted of the generating time, module type and a sequential number. The record way adopts dynamic list and it can change the result of counting dynamically. In addition, Id and timestamp should be built to improve the efficiency of retrieval.

The high scalability and better performance of MongoDB have provided multi-source heterogeneous data with a perfect solution and meet the data storage requirements and query performance.

Data caching As is shown in Fig. 27.5, the cache module was realized by Memcache. If the client wants to get data from platform, searching process will happen in the cache. If the specific data don't exist, it will be got from database to return and finally put into the cache. The data stored in the cache is frequently accessed data and low efficiency query data. Data in the cache is regularly synchronized to the database by background threads, such as the history data of personnel counting which is stored

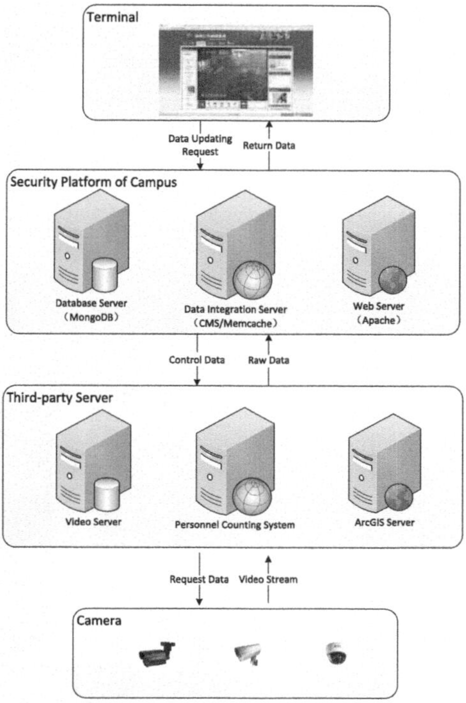

Fig. 27.6 System deployment

by buffer queue. Accessing data directly from the cache reduces the frequency of interaction with the database and improve the system performance.

27.5 Experiment

27.5.1 Installation and Deployment

As is shown in Fig. 27.6, in general, campus security platform is consisted of the CMS server, video server and the database of MongoDB. Based on the B/S architecture, the platform provides browser with access interface and calls the third-party system interface, such as license plate recognition system, the vehicle trajectory analysis system and personnel counting system, to support the required data.

27.5.2 User Interfaces

The display of statistic function, video surveillance and tracking system are as followed in Fig. 27.7.

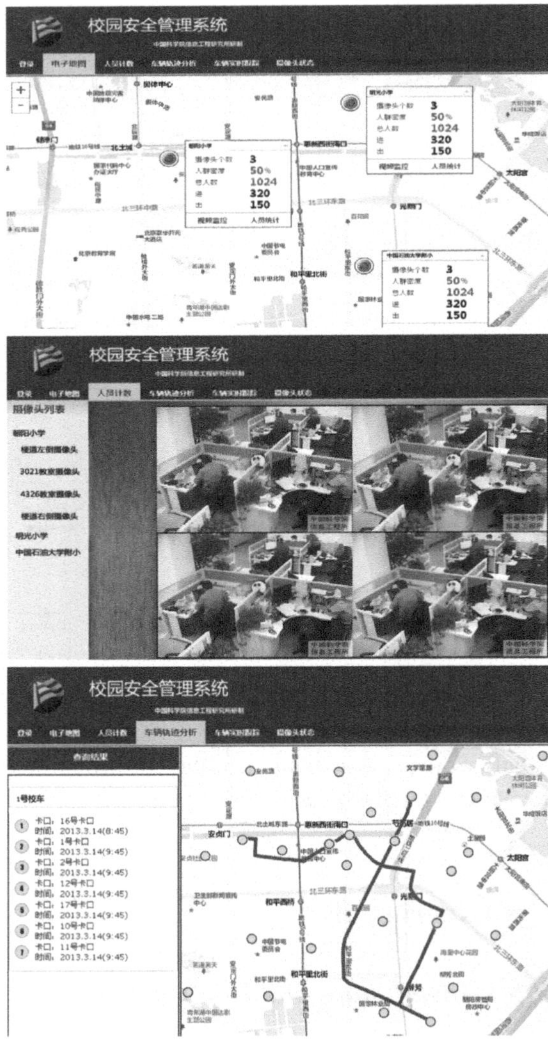

Fig. 27.7 User interfaces

Fig. 27.8 Performance evaluation

(a)

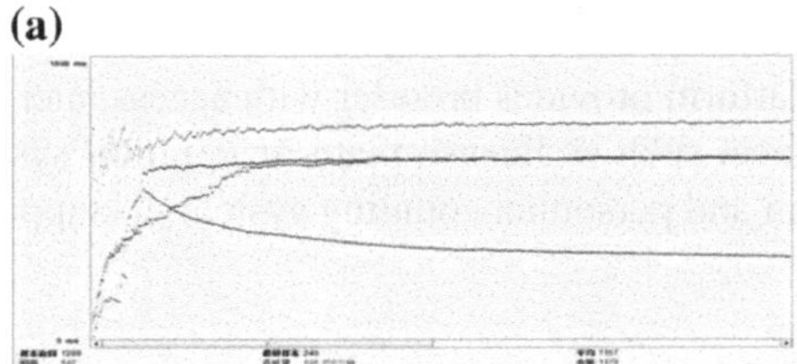

(b)

Label	# Samples	Average	Median	90% Line	Min	Max	Error %	Throughput	KB/sec
Login	400	902	888	1325	121	4671	0.00%	5.0/sec	52.9
map	400	1290	1312	1831	265	2421	0.00%	5.0/sec	54.1
Headcount	400	1306	1322	1957	163	2492	0.00%	5.0/sec	54.3
总体	1200	1167	1179	1882	121	4671	0.00%	14.9/sec	160.5

27.5.3 System Performance

At present, this platform has realized data storage and cluster analysis for 60 personnel counting environment, about 200 cameras and 20 vehicle bayonets. Caching strategy and High concurrent access are adopted to improve the performance. Using message queue mechanism ensures that the real-time alarm information can be pushed to the user interface.

The security platform is consisted of a web server, a database server, an ArcGIS server, a video streaming server and a license plate recognition server and adopts performance testing tool Jmeter to test the system. Under 200 concurrent threads to access the system and the analysis of 1,200 test sample, the result is shown in Fig. 27.8. Obviously, this security platform has good access performance for its response time of 1,167 ms and the average throughput of 896.359 kbc.

27.6 Conclusions and Further Research

Massive sensor data in traditional security systems is independent and there is no convergence and analysis of information. Based on the IOT technology to collect security-related multiple source information, this article adopts NoSQL database to store sensor data , uses data mining to make cluster analysis of the raw data and provides users with a visualization way. The article also has implemented a security platform and given the architecture and implementation's method of it. At present, the research based on multiple source security information convergence is still in the development stage and there are many issues to be studied in the field of data mining algorithms and interactive video.

As for further study, we hope to extend the existing platform so that more fields can be used, such as corporate security, residential security management, urban security, and so on. At the same time, we also hope that by increasing the versatility of the page, different mobile devices are adapted to this platform so that users can achieve security monitoring at any time.

Acknowledgments The work has been funded by Program of Innovation Research Foundation of Institute of Information Engineering, Chinese Academy of Sciences (Y3Z0071H02).

References

1. Trump KS (1998) Practical school security: basic guidelines for safe and secure schools. Eugene, OR, ERIC, pp 2–4
2. Nance JP (2013) School security considerations after newtown. Stan L Rev Online 65:103
3. Sun LM (2006) Wireless sensor networks. Tsinghua University Press, Beijing

4. Kupchik A, Catlaw TJ (2012) Discipline and participation: the long-term effects of suspension and school security on the political and civic engagement of youth1. http://www.civilrightsproject.ucla.edu
5. Finn JD, Servoss TJ (2013) Misbehavior, suspensions, and security measures in high school: Racial/ethnic and gender differences. Closing the school discipline gap: Research to practice conference, Washington, DC
6. Liang LH, Ai HZ, Xu GY, Zhang B (2012) The research summary of face recognition. Chin J Comput 28(5):782–791
7. Gan JY, Zhang YW (2004) A new approach for face recognition based on singular value features and neural networks. Acta Electron Sin 32(1):170–173
8. Cui L (2012) The research of the logistics monitoring system based on WebGIS platform. http://www.cdmd.cnki.com.cn
9. Yuqi WXGLB (2001) Developing techniques analysis and implementation of WebGIS [J]. Comput Eng Appl 5:031
10. Kraak MJ (2004) The role of the map in a Web-GIS environment. J Geogr Syst 6:83–93

Chapter 28
Distributed T-Distribution-Based Intrusion Detection in Wireless Sensor Networks

Pu Cheng, Minghua Zhu and Xianzhong Liu

Abstract Detecting malicious attackers is a critical problem for many sensor network applications. In this paper, a distributed t-distribution-based intrusion detection scheme was proposed. Considering the spatial correlation in the neighborhood activities, our intrusion detection algorithm established a robust model for multiple attributes of sensor nodes using t-distribution. The robust model with an approximate parameter algorithm was exploited to detect malicious attackers precisely. Experimental results show that our algorithm can achieve high detection accuracy and low false alarm rate even when a few sensor nodes are misbehaving, and perform quickly with a lower computational cost.

Keywords Wireless sensor networks · Intrusion detection · t-distribution · Approximate estimation

28.1 Introduction

Wireless sensor networks (WSNs) have been widely used in various applications including those related to personal, industrial, business, and military domains [1]. And security problem become a critical issue for many of these applications. Especially in some cases, when some sensor nodes are compromised or captured,

---

P. Cheng · M. Zhu (✉) · X. Liu
Software Engineering Institute, East China Normal University,
ShangHai 200062, China
e-mail: mhzhu@sei.ecnu.edu.cn

P. Cheng
Software School, Henan University, Kaifeng 450004, China
e-mail: chengpul@163.com

X. Liu
e-mail: xzliu@sei.ecnu.edu.cn

X. Wang et al. (eds.), *Advanced Technologies in Ad Hoc and Sensor Networks*,
Lecture Notes in Electrical Engineering 295, DOI: 10.1007/978-3-642-54174-2_28,

traditional cryptography and authentication techniques may fail to work. Nevertheless, the constrained capability of low-cost sensors and the harsh or even hostile deployment environment of a sensor network make this security problem very challenging [2]. Therefore, it is necessary to design a robust, efficient, and lightweight intrusion detection scheme to prevent the attack coming from malicious sensors.

There are two types of intrusion detection techniques: signature-based detection and anomaly-based detection. Signature-based detection techniques match the known attack profiles with suspicious behaviors while anomaly based detection techniques detect unusual deviations from pre-established normal profiles to identify the abnormal behaviors [3]. In this paper, we focus on anomaly based detection.

Many of anomaly detection schemes have been developed by exploring the correlation among neighboring sensors, and this correlation is the most commonly modeled as multivariate Gaussian distribution. However, Gaussian distribution is sensitive to outliers. To avoid this issue, these anomaly detection schemes have to introduce robust methods for Gaussian distribution, such as Orthogonalized GnanadesikanCK-ettenring (OGK) [4], MCD, m-estimate [5], etc. Owing to these robust methods, the schemes have high accuracy, and also have high false alarm rate. The problem becomes more serious in some cases of the presence of few outliers or after the previous identification of most of outliers, which is the most common situation in WSNs. Moreover, these robust methods also bring relatively high computational cost to the anomaly detection schemes. The multivariate t-distribution has a similar shape as the Gaussian distribution but with much heavier tails. The heavy tails of the t-distribution increase the robustness in parameter estimation, since the outliers in the observations naturally have little overall impact on the parameters. This is in stark contrast to the Gaussian. But most of anomaly detection algorithms do not consider t-distribution despite its natural robustness to outliers. The main reason is lack of good estimation techniques for the parameters of t-distribution. Because of their iterative nature, traditional EM algorithms for the parameters of the t-distribution are computationally expensive and not suitable for the context of sensor networks.

In this paper, we propose a novel t-distribution-based algorithm for anomaly detection in wireless sensor networks. Our intrusion detection algorithm takes into consideration of the spatial correlation in the neighborhood activities by a robust model based on t-distribution to detect malicious attacker. Experimental results show that our algorithm can achieve high detection accuracy and low false alarm rate even when a few sensor nodes are misbehaving, and be performed quickly with a lower computational cost.

The paper is organized as follows. In Sect. 28.2, some related work are given. In Sect. 28.3, multivariate t-distribution is reviewed. The t-distribution based intrusion detection algorithm is presented in Sect. 28.4. The experimental results of comparing the algorithm proposed in this paper with other algorithms are also presented in Sect. 28.5. Finally, our work of this paper is summarized in the last section.

28.2 Relate Work

Agah et al. presented a noncooperative game approach in which the key is to find the most vulnerable node in a sensor network and protect it [6]. Silva et al. defined multiple rules that can be used to determine if a failure has happened and to raise an intrusion alarm if the number of failures exceeds a predefined threshold [7]. Su et al. developed eHIP—energy efficient hybrid intrusion prohibition system to improve the security of cluster-based sensor networks [8]. Such a system consists of authentication-based intrusion prevention (AIP) subsystem and collaboration-based intrusion detection (CID) subsystem to provide heterogeneous mechanisms that can meet the different demands of security levels in cluster-based wireless sensor networks to improve energy efficiency. Wang et al. analyzed the problem of intrusion detection in a Gaussian-distributed WSN theoretically, and provided an appropriate random sensor deployment strategy to determine critical parameters for intrusion detection [9]. Krontiris et al. formalized the problem of intrusion detection and identified necessary and sufficient conditions for its solvability. Based on these conditions, they developed a generic algorithm for intrusion detection [10]. Liu et al. proposed an insider attacker detection scheme, and which takes simultaneously into consideration of multiple attributes of sensor node to detect malicious attackers without the requirement for prior knowledge about normal or malicious sensor activities [11]. It has high accuracy and low false alarm rate when some sensor nodes are misbehaving. Li et al. improved the work in [3] using a group-based intrusion detection scheme. Both of these schemes employ OGK estimator to obtain the parameters of a Gaussian distribution. We will compare our scheme with the insider attack detection scheme in Sect. 28.5.

28.3 Motivation and T-distribution

The multivariate t-distribution has been widely used in many areas. Our major contribution is exploring the robust statistical property of the multivariate t-distribution with its approximate parameter estimation algorithm, and adopting it into the scenario of distributive intrusion detection.

28.3.1 Multivariate T-distribution

Given a collection of data samples $\mathbf{X} = \{x_1, x_2, \cdots, x_n\}$ with each x_i representing a q-dimensional random vector, assume that $\mathbf{x}$ following a q-variate t-distribution with v degrees of freedom as

$$p(\mathbf{x}, \Sigma, v) = \frac{\Gamma(v/2 + q/2)\,|\Sigma|^{-1/2}}{\Gamma(v/2 + q/2)(v + \pi)^{-1/2}} \left(1 + \frac{(x - \mu)^T \Sigma^{-1} (x - \mu)}{v}\right)^{-(v+q)/2} \tag{28.1}$$

where μ and Σ are the mean and covariate matrix of $\mathbf{x}$ and $\Gamma(\bullet)$ denotes the gamma function. The degree of freedom v control the thickness of the distribution tails and thus regulates the degree of robustness to outliers. When v goes to infinity, the t-distributions approach Gaussian distributions. Usually, an EM-based algorithm can be used iteratively to find the maximum likelihood solution for the parameters of the t-distribution. Numerous EM-based iterative algorithms have been developed to compute the maximum likelihood estimates for the parameters of the t-distribution. However, because of their iterative nature, these algorithms are computationally expensive.

28.3.2 Parameter Estimation for T-distribution

Due to their iterative nature, the EM algorithms are extremely costly and not suitable for limited resource available in WSNs. So we adopt the approximate parameter estimation scheme for t-distribution [12].

Firstly, we consider the special case for $\mathbf{x} \sim t(\mu, \Sigma, v)$ and $\mu = 0$, $\Sigma = \alpha(I)$, α is constant. In this special case the probability density function (pdf) of the norm of α is given by

$$f(||\mathbf{x}||) = \frac{2\,||\mathbf{x}||^{-1/2}}{B(v/2+q/2)(\alpha v)^{q/2}}\left(1+\frac{1}{\alpha v}\,||\mathbf{x}||^2\right)^{-(v+q)/2} \tag{28.2}$$

where $B(x, y) = \Gamma(x)\Gamma(y)\Gamma^{-1}(x+y)$ is the beta function.

Given sample data $\mathbf{X} = [x_1, x_2, \cdots, x_n]$, the following equations are used to estimate v and α.

$$E\{\log||\mathbf{x}||^2\} = \log\alpha + \log v + \psi_0(q/2) - \psi_0(v/2) \tag{28.3}$$

$$Var\{\log||\mathbf{x}||^2\} = \psi_1(v/2) - \psi_1(q/2) \tag{28.4}$$

where $\psi_0(x)$ is the digamma function and $\psi_1(x)$ is the trigamma function.

Let $z_i = \log||\mathbf{x}||^2 = \log\mathbf{x}_i^T\mathbf{x}_i$, and $\hat{v}$ be the estimate of v. we can obtain

$$\psi_1(\frac{\hat{v}}{2}) = \frac{1}{n}\sum_{i=1}^{n}(z_i - \bar{z})^2 - \psi_1(\frac{q}{2}) \tag{28.5}$$

where $\bar{z} = \frac{1}{n}\sum_{i=1}^{n} z_i$. Since Eq. (28.5) cannot be solved directly, the approximation is used

$$\psi_1(x) \approx \frac{x+1}{x^2} \tag{28.6}$$

We can compute the estimate

$$\bar{v} = \frac{1 + \sqrt{1 + 4b}}{b} \tag{28.7}$$

with

$$b = \frac{1}{n}\sum_{i=1}^{n}(z_i - \bar{z})^2 - \psi_1(\frac{q}{2}) \tag{28.8}$$

Finally, we use Eq. (28.3) to compute the estimate α for $\hat{\alpha}$

$$\hat{\alpha} = \exp\left\{\bar{z} - \log\hat{v} + \psi_0(\frac{\hat{v}}{2}) - \psi_0(\frac{q}{2})\right\} \tag{28.9}$$

For the general case of $\mathbf{x} \sim t(\mu, \Sigma, v)$, the estimate of μ is denoted as $\hat{\mu}$, which can be obtained by considering each dimension of the data separately and computing the sample median of the data. Due to our purpose of a computationally efficient approximation, an approximation is assumed that Σ for some α. So the previous discussed equations can be used to estimate v and α. This can be done by first computing $z_i = \log || x_i - \hat{\mu} ||^2$ and then directly applying Eqs. (28.7) and (28.9). In practice, the estimate $\hat{v}$ is a good approximation to v regardless of the structure of Σ. The slight positive bias may be due to the error in the approximation for the trigamma function given in Eq. (28.6). The scaling estimate $\hat{\alpha}$ also provides a good estimate for the mean of the diagonal entries of Σ. Hence all that remains is to estimate the relative scaling of the elements of Σ.

To estimate the relative scaling of the elements of Σ, we use the auxiliary matrix

$$\bar{\Sigma} = \frac{1}{n}\sum_{i=1}^{n}\frac{(x_i - \hat{\mu})(x_i - \hat{\mu})^T}{|| x_i - \hat{\mu} ||^{\beta}} \tag{28.10}$$

where $\bar{\Sigma}$ is similar to the sample covariance except that each sample is first scaled by the norm raised to a constant power β. Usually, β can be given by

$$\beta = \frac{2\log_2 q}{\hat{v}^2 + \log_2 q} \tag{28.11}$$

Note that the scaling term in the denominator of Eq. (28.10) is necessary in order to give a good approximation when v is small. We can now apply the estimated mean of the diagonal entries $\hat{\alpha}$ to obtain an estimate for Σ

$$\hat{\Sigma} = \frac{\hat{\alpha} q}{tr(\bar{\Sigma})}\bar{\Sigma} \tag{28.12}$$

Table 28.1 The collected information

Collected information	Detected attack behavior
Sensor sensed data	Fabricate information attack
Packet sending rate	Energy exhausting attack
Packet dropping rate	Select forward attack, black hole attack
Packet mismatch rate	Message alter attack
Packet receiving rate	Sink hole attack
Packet sending power	Hello attack, worm hole attack

28.4 Anomaly Detection Algorithm

In this section, we provide a solution for detecting malicious attackers whose behaviors are "abnormal" with respect to normal sensors using t-distribution based model with the approximate parameter estimation.

We assume that sensors are randomly deployed in the spatial domain and there are no malicious nodes at the beginning of the sensor network deployment. Every sensor monitors the activities of the sensors within its neighborhood. To detect faulty or compromised sensors, the associated neighborhood sensor set $N(s)$ of each sensor s needs to be first constructed as follows:

$N(s) = s_n$ subject to : $dist(s, s_n) \leqslant D$

where $dist(s, s_n)$ represents either the Euclidean distance or one-hop, or multi-hop distance between sensor s and s_n, and D is the distance constraint. At every time interval Δ_t, each sensor node s collect the monitoring information about its neighbor s_i in $D(s)$. The monitoring information includes sensed data, packet sending rate, packet dropping rate, etc, which is shown in Table 28.1. The collected information about s_i is expressed as a q-component attribute vector $x_i = \{x_{1i}, x_{2i}, \cdots, x_{qi}\}$ with each component describing s_i s activity in one aspect. So sensor s obtains a set of $\mathbf{x}$ of attribute vectors, where $\mathbf{x} = \{x_1, x_2, \cdots, x_n\}$ and n is the number of monitoring sensor nodes in $D(s)$.

Considering the presence of outliers or noises in the dataset, we assume that $\mathbf{x}$ form a sample of a multivariate t-distribution instead of a multivariate Gaussian distribution. And $\mathbf{x}$ is distributed as $t(\mu, \Sigma, v)$, following a q-variate t-distribution with mean vector μ, variance-covariance matrix Σ, and degrees of freedom v. For Gaussian model which $\mathbf{x}$ follows a Gaussian distribution, the Mahalanobis squared distance $(\mathbf{x} - \mu)^T \Sigma^{-1} (\mathbf{x} - \mu)$ is distributed as χ_q^2, where χ_q^2 is the chi-square distribution with q degree of freedom. However, for robust model where $\mathbf{x}$ follows a multivariate t-distribution, has an F-distribution with q and v degrees of freedom. Therefore, the probability that $\mathbf{x}$ satisfies $(\mathbf{x} - \mu)^T \Sigma^{-1} (\mathbf{x} - \mu)/q > F_\alpha(q, t)$ is α, where $F_\alpha(q, t)$ is the upper (100α)th percentile of a F distribution.

Let $\hat{t}$, $\hat{\mu}$ and $\hat{\Sigma}$ be the estimates of t, μ and Σ, respectively. Then the probability that $\mathbf{x}$ satisfies $(\mathbf{x} - \hat{\mu})^T \hat{\Sigma}^{-1} (\mathbf{x} - \hat{\mu})/q > F_\alpha(q, \hat{t})$ is expected to be roughly α. Let $d(x_i) = ((x_i - \hat{\mu})^T \hat{\Sigma}^{-1} (x_i - \hat{\mu}))_{0.5}$. Sensor s_i should be treated as an outlier if

$d^2(x_i)/q$ is unusually large. In our scheme, sensor s declares s_i as an abnormal node if $d^2(x_i)/q > \theta_0$. The estimation of $\hat{t}$, $\hat{\mu}$ and $\hat{\Sigma}$ and determination of threshold θ_0 will be discussed below in details.

To outlier detection, the robust estimation for the parameters $\hat{t}$, $\hat{\mu}$ and $\hat{\Sigma}$, is needed. Rather than using computationally expensive EM method, we adopt a faster approximate method in Sect. 28.3.1. Correspondingly, we obtain $\hat{t}$, $\hat{\mu}$ and $\hat{\Sigma}$ in light of the approximate parameter estimation algorithm in Sect. 28.3.1. However, it is noted that we must be careful to ensure that $|| x_i - \mu ||^2 \neq 0$, when n is very small. One way to do this is to store the data in the last k time interval (from time interval Δ_{t-k+1} to Δ_t) and use these to compute the estimation. Due to the temporal correlation in sensor networks, this way also help to get more precise intrusion detection results. For convenience, we just consider the case of only one time interval (i.e., the current time interval Δ_t).

Threshold θ_0 plays an important role in checking whether or not sensors are outlying sensors. According to the previous discussions, θ_0 seem to be $qF_\alpha(q, \hat{t})$ for t-distribution. However, as the result of the heavy-tail characteristic of the t-distribution and bias, the threshold is observed in extensive preliminary studies (not reported) to be larger than is required and thus will fail to identify potential outliers [13]. To avoid this issue, one option is to simply select k nodes with the largest k Mahalanobis distance from $N(s)$, where $k = n \times p_0$ with p_0 denoting the sensor outlying probability. The estimation of p_0 can be studied using empirical data, which is often difficult to obtain. Another selection method focuses on choose a function as the adaptable threshold. At least a proper function should satisfy the following conditions: (1) Values of the function depend on the parameter v and q, similar to $qF_\alpha(q, t)$. (2) Values of the function should be less than $qF_\alpha(q, t)$ for a small v and larger than $\chi_q^2(\alpha)$ for a large v. When v is too small, a slight bias will can dramatically change the values of $qF_\alpha(q, t)$. And that also is one of the reasons for failure of $qF_\alpha(q, t)$. When v is too large, t-distributions approach Gaussian distributions, and that means there are a few outliers in sample set. So a larger threshold will decrease the false alarms in the outlier detection. We have experimental validated that a good choice for θ_0 can be given by

$$\theta_0 = \log_2 q + 2 + \frac{2 \log_2 q}{\hat{v}^2 + \log_2 q} \tag{28.13}$$

This approach will be adopted in our simulation studies. After calculating the Mahalanobis distance for each neighbor in $N(S)$ and comparing with θ_0, sensor s gets a set $D(s)$ of suspicious neighbors. These sets can be combined through the majority vote to reach a result with a higher accuracy. Each sensor s broadcasts a message including $D(s)$ to its neighborhood $N(s)$. At the same time, sensor s will receive the massages from others and count the number of outlying neighbors. A neighbor s_i is determined to be a malicious attacker, if the number of the messages which report s_i is abnormal is less than half of all received messages. Then the sensor s will report to the base station, which should isolate the attacker s_i if multiple reports about s_i have been received.

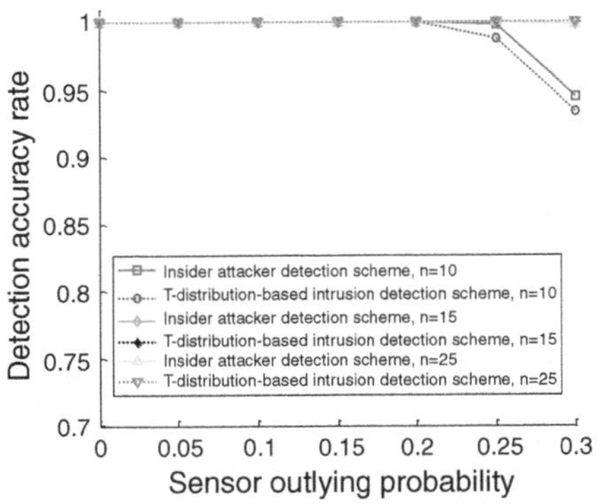

Fig. 28.1 Comparison of detection accurate rate under different average number of neighbors and sensor outlying probability: $q = 3$

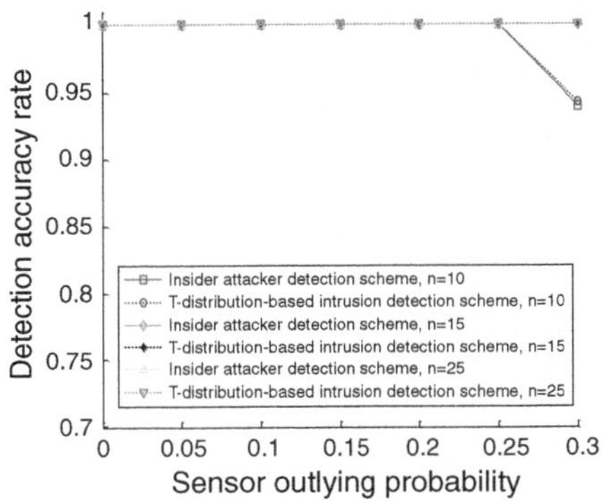

Fig. 28.2 Comparison of detection accurate rate under different average number of neighbors and sensor outlying probability: $q = 10$

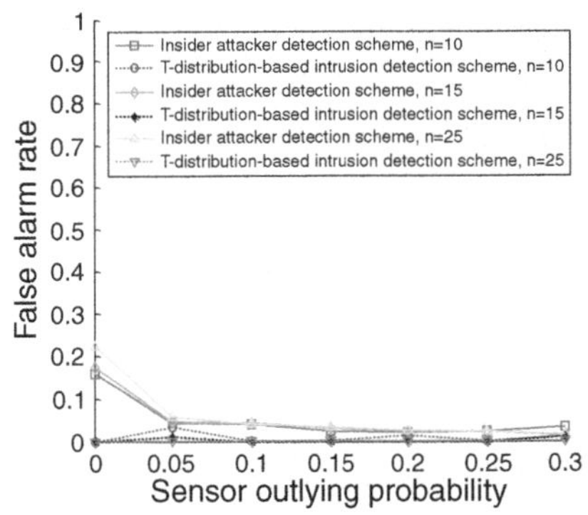

Fig. 28.3 Comparison of false alarm rate under different average number of neighbors and sensor outlying probability: $q = 10$

28.5 Experiment

In this section, we comparatively evaluate our algorithm and the insider attacker detection algorithm, and report our evaluation results on the task of malicious attacker detection.

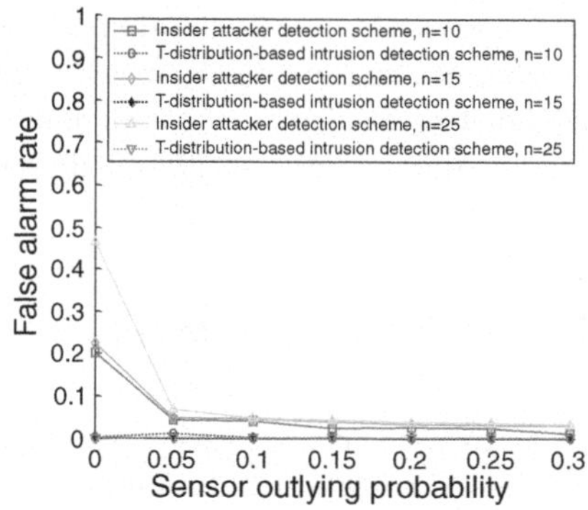

Fig. 28.4 Comparison of false alarm rate under different average number of neighbors and sensor outlying probability: $q = 10$

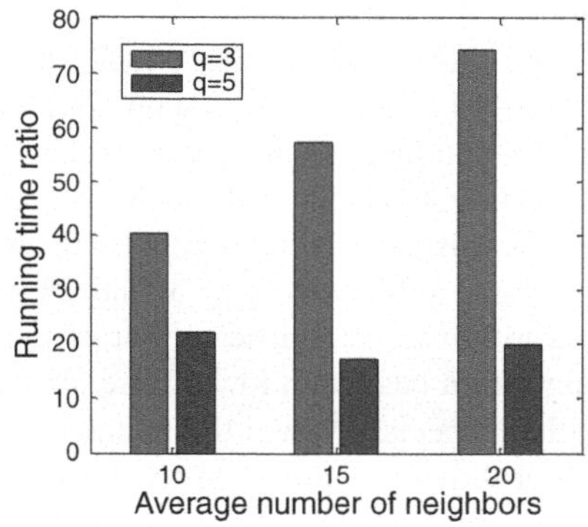

Fig. 28.5 Running time ratio under different average number of neighbors and the number of attributes

28.5.1 Experiment Setup

We carry out several experiments on synthetic data using MATLAB. In these experiments, we consider 900 sensors are randomly deployed in a 30×30 square region. Let q be the number of monitoring attributes of each sensor. The attribute values of normal sensors are drawn from $N_q(\mu_1, \Sigma_1)$. We set $\mu_1 = (\mu_{11}, \mu_{12}, \ldots, \mu_{1q})$, $\Sigma_1 = \rho_i ones(q) - (\rho_i - 1)eye(q)$, where $ones(q)$ is an $q \times q$ matrix of 1s, $eye(q)$ is the $q \times q$ identity matrix, $\rho_1 = 0.9$, $\rho_2 = -0.5$ and $\rho_3 = 0$. ρ_i indicates how much correlation among attributes, $i \in \{1, 2, 3\}$. This presents a coarse summary of all the possibilities in the real world scenarios [11]. We randomly add a noise e following the normal distribution $N_q(\mu_2, \Sigma_2)$ to the normal data in order to simulate the abnormal sensors, where $\mu_2 = 0.5\mu_1$ and $\Sigma_2 = \rho_j ones(q) - (\rho_j - 1)eye(q)$, $j \in \{1, 2, 3\}$. So the attribute vector of the outlying sensor s_i is simulated as.

$$x_i = x_i + e \tag{28.14}$$

To evaluate the performance of our algorithm, we use three metrics: detection accuracy rate (DA rate), and false alarm rate (FA Rate), and Running time ratio

(RT Ratio). DA rate is the ratio of the number of attackers detected to the total number of attackers. FA Rate is the ratio of the number of normal sensors that are regarded as attackers to the total number of normal sensors. RT Ratio is the ratio of running time of our algorithm to the running time of other comparative algorithm.

28.5.2 Experiment Result

In our experiments, we vary separately sensor outlying probability p and the average number of neighbors n. In both tests, we evaluate DA rate, FA Rate and RT Ratio for our algorithm, and also evaluate these metrics of the insider attacker detection scheme for comparison.

As shown in Figs. 28.1, 28.2, 28.3, 28.4 and 28.5, our algorithm can effectively identify insider attackers with a high detection accuracy and a low false alarm and faster computation than the insider attacker detection scheme in most cases. Our detection algorithm can reach a high detection accuracy (>92 %) when as many as 30 % sensors are outlying, in both cases of $q = 5$ and $q = 10$.

Figures 28.1, 28.2, 28.3, and 28.4 show that the detection accuracy rate of our algorithm is near close to that of the insider attacker detection algorithm but our algorithm has much lower false alarm rate than the insider attacker detection algorithm especially when the number of outlying sensors are very few. That help our algorithm to be more realistic for practical application because in the most cases of there are a few anomaly nodes in sensor networks especially after the outliers being identified and isolated.

And we also observe that for our algorithm the increase in the number of outlying sensors lead to few increase in the false alarm rate and few decrease in detection accuracy rate. Such a nice property results from the robust model based on t-distribution and the proper choice for the threshold.

Figure 28.5 shows that our algorithm is $10 \sim 70$ times faster than the insider attacker detection algorithm. Performing fast with a low computational cost which is another important feature of our algorithm. This feature can reduce the power consumption for intrusion detection in sensor networks, and also low the detection time.

28.6 Conclusions

In this paper, we proposed the t-distribution-based intrusion detection scheme for wireless sensor networks to detect malicious attacker. Our experiment results show that our scheme can effectively identify malicious attacker with a high detection accuracy and a low false alarm. At the same time, it can also perform quickly with a lower computational cost.

In addition, our experiment is performed on synthetic data which ignores a variety of noises and conflicts. For our algorithm, Deployment in real sensor networks and validation on realistic data is the future work.

Acknowledgments This paper was supported by National Science and Technology Major Project of the Ministry of Science and Technology of China. (Grant No. $2010ZX03006-001-01$), and National Program on Key Basic Research Project of China. (Grant No. $2011CB302902$).

References

1. Akyildiz IF, Su W, Sankarasubramaniam Y et al (2002) A survey on sensor networks. IEEE Commun Mag 40(8):102–114
2. Liu F, Cheng X, An F (2006) On the performance of in-situ key establishment schemes for wireless sensor networks. In: IEEE GLOBECOM. IEEE Press, San Francisco, pp 1–5
3. Li GR, He JS, Fu YF (2008) Group-based intrusion detection system in wireless sensor networks. Comput Commun 31(18):4324–4332
4. Yohai YJ, Zamar R (1988) High breakdown-point estimates of regression by means of the minimization of an efficient scale. J Am Stat Assoc 86(402):403–413
5. Maronna RA, Martin RD, Yohai VJ (2006) Robust statistics: theory and methods. Wiley Publisher, New York
6. Agah A, Das S, Basu K, Asadi M (2004) Intrusion detection in sensor networks: a non-cooperative game approach. In: The 3rd IEEE international symposium on network computing and applications, pp 343C–346
7. Silva AD, Martin M, Rocha B et al (2005) Decentralized intrusion detection in wireless sensor networks. In: The first ACM international workshop on quality of service and security in wireless and mobile networks, pp 16C–23
8. Su W, Chang K, Kuo Y (2007) eHIP: an energy-efficient hybrid intrusion prohibition system for cluster-based wireless sensor networks. Comput Networks 51(4):1151–C1168
9. Wang Y, Fu WH, Agrawal DP (2013) Gaussian versus uniform distribution for intrusion detection in wireless sensor networks. IEEE Trans Parallel Distrib Syst 24(2):324–355
10. Krontiris I, Benenson Z, Giannetsos T et al (2009) Cooperative intrusion detection in wireless sensor networks. In: The 6th European conference on wireless sensor networks. Springer, Cork, pp 263–278
11. Liu F, Cheng XZ, Chen D (2007) Insider attacker detection in wireless sensor networks. In: The 26th IEEE international conference on computer communications. IEEE Press, Anchorage, pp 937C–1945
12. Aeschliman C, Park J, Kak AC (2010) A novel parameter estimation algorithm for the multivariate t-distribution and its application to computer vision. In: The 11th European conference on computer vision. Springer, Crete, pp 594–607
13. Chen T, Martin E, Montague G (2009) Robust probabilistic PCA with missing data and contribution analysis for outlier detection. Comput Stat Data Anal 53(10):3706–3716

Chapter 29
Weighted Localization for Underwater Sensor Networks

Shaobin Cai, Guangzhi Zhang and Shilong Liu

Abstract ELSN (Efficient Localization for large-scale underwater Sensor Networks) gets the node location by solving equations. However, sometimes the unique solution (no solution or multiple solutions) cannot be obtained from equations because of measurement error. In this case, the node cannot be located. On the basis of ELSN algorithm, a new localization algorithm is proposed. It transforms the process of solving equations into looking for the best point of intersection of the three circles in the plane, and regards the point as the potential location of the node. First, in this algorithm , the node is projected into a two-dimensional plane to reduce the computational complexity of the algorithm. Second, every three reference nodes are selected as a combination at random. That is to say, each triplet of nodes which represent a triplet of equations forms three tangent or intersecting circles. Based on the positional relationship of three circles, an optimal point of intersection is served as a potential location of the target node, and the residuals of each triplet are served as a potential weight. Finally, the weighted results of all potential locations are considered as the final position of the node.

Keywords Localization · Potential location · Reference node · Residuals · Weight

As a necessary means of marine monitoring, acoustic sensor networks have a wide spectrum of aquatic applications such as real-time monitoring of the target waters, exploration of marine resources, marine environment monitoring, and auxiliary navigation. In order to make sense, the nodes information needs to be combined with

---

S. Cai (✉) · G. Zhang · S. Liu
College of Computer Science and Technology, Harbin Engineering University, Harbin 150001, China
e-mail: caishaobin@hrbeu.edu.cn

X. Wang et al. (eds.), *Advanced Technologies in Ad Hoc and Sensor Networks*, Lecture Notes in Electrical Engineering 295, DOI: 10.1007/978-3-642-54174-2_29,

its location. So sensor networks node localization technology is one of the key technologies. There are many studies about the wireless sensor network node localization [1–6].

The special features of acoustic channels (high error rate, low bandwidth, and long propagation delay) cause many constraints on the localization schemes for underwater sensor networks. So, how to locate the node in underwater sensor networks is a great challenge.

In [7] is proposed a localization algorithm, which combines three-dimensional Euclidean distance estimate method [8] and recursion localization estimate method together [9]. But there are still unsolved issues because of measurement error. Sometimes it may not be able to solve the equations directly which consist of Euclidean formulas to get a unique solution [10]. So some nodes that receive a lot of information cannot be located.

Integrating the high weight idea to ELSN algorithm, we propose a scheme called Weighted Localization for Underwater Sensor Networks (WUWSN). It is not to determine the coordinates of the target node by directly solving equations, but to transform the solving equations into finding the best plane intersection point of the circle as a potential position of the node, which takes full advantage of the collected information to locate the node.

29.1 ELSN Algorithm

ELSN algorithm combines three-dimensional Euclidean distance estimate method and the recursive localization estimate method together. There are several types of nodes, such as reference nodes, non-localized nodes, and so on. Reference nodes contain the anchor node and the non-localized node whose confidence value is higher than the stipulated threshold.

In the initialization phase, all the anchor nodes label themselves as reference nodes and set their confidence values to 1. If the confidence value is higher than the threshold value, the non-localized node is added to the reference node set after it is located. There are two types of messages: localization messages and beacon messages. Localization messages are used for information exchange between the non-localized node and the reference node, while beacon messages are designed for distance estimate. During the localization process, each node (including reference nodes and non-localized nodes) broadcasts beacon messages periodically which contain its id. All the neighboring nodes which receive this beacon message can estimate their distances to this node using techniques.

Each non-localized node maintains a counter n, of the localization messages it broadcasts. A threshold value K is set to limit the maximum number of localization messages which are sent by each node. Each non-localized node also keeps a counter m, of the reference nodes to which it knows the distances. Once the localization process starts, each non-localized node keeps checking m. There are two cases:

(1) When $m < 4$, the non-localized node broadcasts a localization message including the location of the reference node, its estimated distance to the reference node and messages received possibly from other non-localized nodes. Then the node which has received messages can estimate the distances from itself to more non-neighboring reference nodes with the three-dimensional Euclidean distance estimate method. The method above is commonly referred to multi-hop node distance estimate to more non-neighboring reference nodes. m will updated according to the known reference nodes, then the node will return to check m.

(2) When $m \geq 4$, the non-localized node selects four non-collinear reference nodes which have high confidence value. The beacon message includes the coordinates, id number, and some other information about the reference node. If the obtained confidence value is higher than the predefined threshold, the confidence value will be calculated after the estimate procedure of the position coordinates. So this located node will be added to the reference node set. Then, it broadcasts his location messages including its coordinates, id, and confidence value. It will go back to (1) if the confidence value is higher than the predefined threshold [10].

29.2 WUWSN

Unlike ELSN, WUWSN does not always check whether $m \geq 4$ triggers the localization procedure of the target node. Instead, it starts Euclidean distance estimate to calculate its own coordinates directly. In this period, the non-localized node generally has collected enough information. So, equations can be formed by Euclidean distance estimate method. WUWSN is not to determine the coordinates of the non-localized node by solving equations directly, but to transform the solving equations to find the best plane intersection point of the circle as a potential position of the node. The basic steps are as follows. First, the three-dimensional problem is transformed into a two-dimensional localization problem. Equations are listed using Euclidean distance estimate method and the unknown elements are eliminated in order to leave the coordinates of target nodes only. Then potential location of the node is sought among circles' intersection points. Finally, the weighted value of potential locations is regarded as the final result of the node position.

29.2.1 Euclidean Distance Estimate

The basic idea of Euclidean distance estimate is to estimate the distance between two non-neighboring nodes based on the known one-hop distance. Thus, the non-localized node can estimate the distance between itself and the reference node which might be multiple hops away from it.

This method is illustrated in an example shown in Fig. 29.1. A non-localized node E wants to estimate its distance to anchor node A, and A is a two-hop neighbor of E.

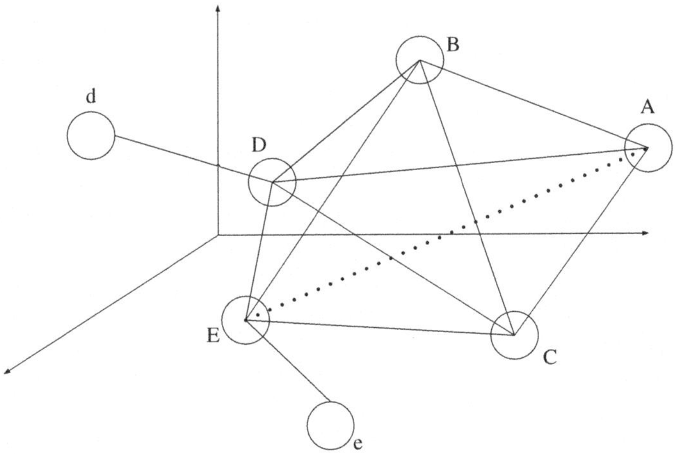

Fig. 29.1 Three-Dimensional Euclidean Distance Estimate

Then node E needs to know the estimated distance to three one-hop neighbor nodes at least. The nodes A, B, C, D are not co-plane while any three nodes out of the nodes A, B, C, D, E are not co-line. Node E should have the length information of EB, BA, EC, CA, ED, DA, DB, DC, BC. Three-dimensional Euclidean distance estimate is described as follows. First, node E uses edge BA, CA, BC to construct the basic localization plane while node E knows the length of edge DA, DB, DC. Then node E can estimate the coordinates of node D and node D may have two possible coordinates. Because the lengths of EB, EC, and ED are known, node E will have four possible positions correspondingly if node D has two possible positions. Node E can select the possible values of node E after it receives more location information from other reference nodes [10].

This process above can be also illustrated in solving equations equivalently. Coordinates of A, B, and C are known, and coordinates of D and E are unknown. Nodes coordinates of D and E are denoted by (x_D, y_D) and (x_E, y_E). Nine equations can be listed based on the known edge distances with the unknown elements (x_D, y_D) and (x_E, y_E). Then, we can get the equations with the only unknown elements (x_E, y_E) after eliminating unknown elements (x_D, y_D). There will be four possible values corresponding to roots of equations because of the non-sufficient information.

In this case, more information can be gathered to solve equations that have a unique solution. According to the algorithm, the simplest method is to collect the information of neighbor nodes. If node D knows the distance to the node d and node E knows the distance to the reference node e, the equations will have a unique solution. In short, the unknown elements of the equations are the only coordinates of the target non-localized node.

Now we will consider the case that the information just makes equations to have a unique solution. Because the presence of underwater measurement error is inevitable, the equations may have no solution. Even if the node has collected enough information, it cannot realize its own localization because there are some measurement errors.

Under normal circumstances, the collected information of the node is more than the information which leads to a unique solution. That is to say, the equations is an over-determined system. In fact, measurement errors are inevitable, so the result may deviate from the true value.

29.2.2 Potential Location

A three-dimensional localization problem is transformed into a two-dimensional localization problem. Each node is equipped with a pressure sensor which can measure the underwater depth accurately. The reference node is projected to the plane where the non-localized node exits. With the help of Euclidean distance estimate method, equations can be formed and coordinates of non-localized nodes are also contained in the unknown elements of the equations. Then, the unknown elements like coordinates of non-target nodes are eliminated just to retain the coordinate of target node. When the equations just have three entries, there is only a unique solution. The collected information by the node may be more than the information leading to a unique solution. That is to say, the equations are an over-determined system. So, the entries of the equations about a target node may be more than three. That is to say, the entries out of equations about a target node are more than three after unknown elements are eliminated. And any triples out of equation system are equivalent to the equation system itself.

$$\begin{cases} \sqrt{(x - x_1')^2 + (y - y_1')} = d_{r_1}' \\ \sqrt{(x - x_2')^2 + (y - y_2')} = d_{r_2}' \\ \quad\cdot \\ \quad\cdot \\ \sqrt{(x - x_N')^2 + (y - y_N')} = d_{r_N}' \end{cases} \tag{29.1}$$

In Eq. (29.1), (x, y) is the coordinates of the target non-localized node, and (x_i', y_i') is the projected coordinate of the reference node. N is the number of the reference nodes whose messages are received by the target node and $N \geq 3$. d_{r_i}' is the distance from non-localized node to the i'th reference node.

Each triple (C_N^3) of N equations about one non-localized node corresponds to three reference nodes. Based on the positional relationship of the three circles in the plane, an optimal intersection point is selected to be as a potential location of the non-localized node, and the following cases are some typical examples.

We regard the projected points A_1', A_2', A_3' as the center, the distances from A_1', A_2', A_3' to the non-localized node(red dots in the figure) as radius respectively. Then three circles are formed. Figure 29.1 shows the position relationship.

(1) Figure 29.2a shows the case that three circles intersect with one another. As can be seen from the figure, we regard A_1', A_2' as the center respectively, the two circles

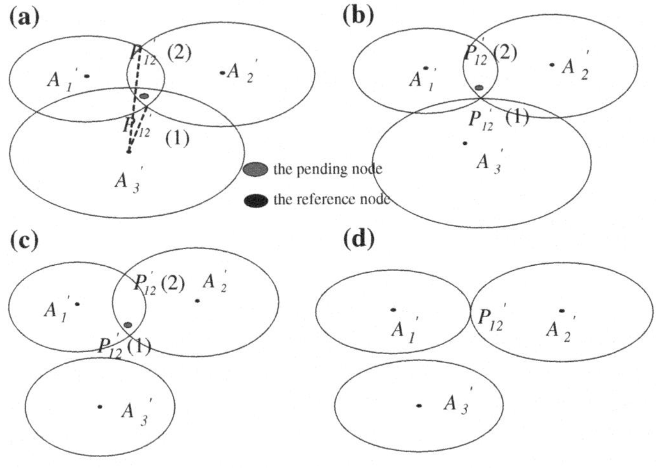

Fig. 29.2 Intersecting to select potential locations. **a** every two *circles* are intersecting, **b** a unique solution, **c** no intersection of two *circles*, **d** two *circles* are tangent

intersect at two points $P'_{12}(2)$, $P'_{12}(1)$. It may appear that two intersection points are both within the circle A'_3 ,or just one intersection point is within it. It may be possible that neither of two intersection points is within the circle A'_3. If $P'_{12}(1)$, $P'_{12}(2)$ meet Eq. (29.2), $P'_{12}(1)$ is selected to be as the potential position, or $P'_{12}(2)$ will be selected. N refers to the number of the reference nodes whose messages are received by the non-localized node.

$$\sum_{i=1, i \neq j,k}^{N} \left| \left\| A'_i - P_{jk}(1) \right\| - d'_{r_i} \right| < \sum_{i=1, i \neq j,k}^{N} \left| \left\| A'_i - P_{jk}(2) \right\| - d'_{r_i} \right| \qquad (29.2)$$

(2) Figure 29.2b shows the case that the three circles have a unique intersection point. It is well-reasoned to put the unique intersection point as the estimated position of the non-localized node, and there is no need to participate in the subsequent weight operation.

(3) Figure 29.2c shows that the two circles do not intersect. In the figure, because the sum of d'_{r1} and d'_{r3} is bigger than segment $A'_1A'_3$, the circles A'_1 and A'_3 have no intersection point. Then $A'_1A'_3$ is divided according to the ratio of $d'_{r_1} : d'_{r_3}$, then the point of demarcation that will participate in the subsequent weight calculation is right the potential location. If A'_1 is within A'_3 entirely, segment $A'_1A'_3$ will be extended to intersect with circles A'_3 *and* A'_1 ,and the ring $A'_3A'_1$ is divided according to the ratio of $d'_{r_1} : d'_{r_3}$. The intersection point determined by Eq. (29.2) will be the potential position of the node and it will be participant in the subsequent weight operation.

(4) Figure 29.2d shows the two circles are tangent. Circles A'_1 *and* A'_2 tangent at Point P'_{12}, so P'_{12} is the potential location determined by the two circles. The potential location is denoted by $P_{jk}(\hat{x}_{jk}, \hat{y}_{jk})$. Subscript jk is used to note which two circles are intersecting and $j \neq k$.

The location of a node can be simply described as the average estimate of the potential locations. However, in order to increase localization accuracy further, the average estimate is replaced by weights related to residuals in this paper.

If $\hat{x}$ is the solution of the equation $f(x) = b$, the residual is defined as $b - f(\hat{x})$. The residual indicates the error between $f(\hat{x})$ and true-value b and the error is $\hat{x} - x$. Not knowing x needed in calculating error, it is a natural choice to calculate the residual instead of the error. The relationship between N residuals and $P_{jk}(\hat{x}_{jk}, \hat{y}_{jk})$ is described as Eq. (29.3).

$$\begin{cases} \hat{\varepsilon}_{jk}(1) = \mid \sqrt{(\hat{x}_{jk} - x_1')^2 + (\hat{y}_{jk} - y_1')^2} - d_{r_1}' \mid \\ \hat{\varepsilon}_{jk}(2) = \mid \sqrt{(\hat{x}_{jk} - x_2')^2 + (\hat{y}_{jk} - y_2')^2} - d_{r_2}' \mid \\ \quad\vdots \\ \hat{\varepsilon}_{jk}(N) = \mid \sqrt{(\hat{x}_{jk} - x_N')^2 + (\hat{y}_{jk} - y_N')^2} - d_{r_N}' \mid \end{cases} \tag{29.3}$$

The weight of potential position $P_{jk}(\hat{x}_{jk}, \hat{y}_{jk})$ is described as Eq. (29.4). The accuracy is inversely proportional to the sum of residuals and the reciprocal of all the residuals represent the weight factor.

$$W_{jk} = \frac{1}{\sum_{i=1}^{N} \hat{\varepsilon}_{jk}(i)} \tag{29.4}$$

Finally, from the perspective of the weight, the non-localized sensor node is estimated as follows:

$$(x, y) = \frac{\sum_{j=1}^{N-1} \sum_{k=j+1}^{N} W_{jk} * P_{jk}(\hat{x}_{jk}, \hat{y}_{jk})}{\sum_{j=1}^{N-1} \sum_{k=j+1}^{N} W_{jk}} \tag{29.5}$$

If the number of messages received by the non-localized node is less than three, the method above is not applicable, so other methods should be accepted. If the node gains only two beacon messages, the method is shown as follows. Drawing two circles with A_1', A_2' as the center and d_{r_1}', d_{r_2}' as the radius, respectively, the two circles intersect at two points $P_{12}'(2)$, $P_{12}'(1)$. The intersection point $P_{12}'(2)$, $P_{12}'(1)$ will send messages within their own communication radius and the reference node that receives this message will reply. In this case, between $P_{12}'(2)$, $P_{12}'(1)$, the one that receives more feedback messages will be selected as the estimate location. If the number of messages is less than two, Euclidean distance estimate will be applied directly.

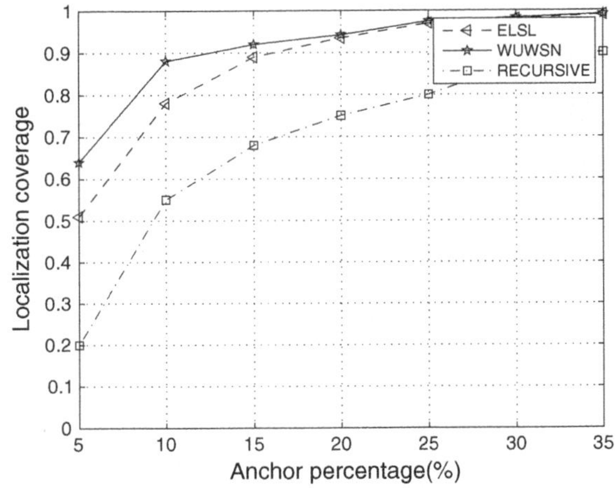

Fig. 29.3 The comparison chart of ELSN, WUWSN, and RECUSIVE in the localization coverage

29.3 Simulation Analysis

MATLAB is used to simulate WUWSNELSN and RECURSIVE [9]. Parameters are set as follows: (1) A 1000 * 1000 * 1000 m three-dimensional region which can drift within 20 * 20 km area with the water current. (2) The confidence threshold is 0.98. (3)Anchor nodes proportion is from 5 to 35 % while the increment interval is 5 %. (4) the radius is 200 m, the transmitting interval of beacon messages is 100 s. The simulation environment is described as follows: the water acoustic channel model uses Rayleigh fading channels model. ToA method is used to measure the distance, and the movement model is Lagrange movement oceanographic models [11, 12]. Each simulation experiment is operated 50times and the 90 % of the centralized results interval is set as the experiment result.

29.3.1 Localization Coverage

Figure 29.3 shows the comparison in localization coverage among ELSN, WUWSN, and RECUSIVE when the percentage of anchor nodes varies from 5 to 35 %. The localization coverage of ELSN is higher than RECUSIVE because Euclidean distance estimate method is used to calculate multi-hop distance in ELSN. When the ratio of anchor nodes is small, the localization coverage ratio of ELSN is lower than that of WUWSN, and otherwise the opposite. In the previous case, because of weighted average method, each time WUWSN will be likely to choose the maximum of potential reference nodes affording the location information, so most nodes can receive the appreciate location messages. But sometimes the non-localization node cannot get the unique solution of equations. Therefore, the node will not be able to complete the localization procedure. So the localization coverage of WUWSN is

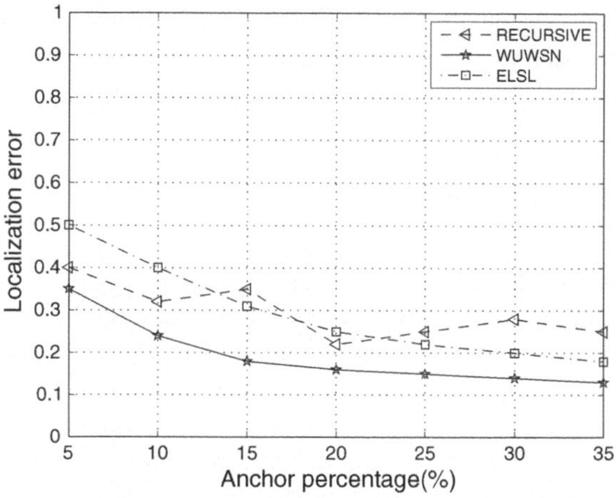

Fig. 29.4 The comparison chart of ELSN, WUWSN, and RECUSIVE in the average localization error

higher than that of ELSN. In the latter case, because the information afforded by the anchor nodes is already wealthy, the reference nodes which are added recursively from ordinary nodes have became non-key factors in this case. So ELSN is close to WUWSN in the localization coverage.

29.3.2 Localization Error

Figure 29.4 shows the comparison in localization error among ELSN, WUWSN, and RECUSIVE when the percentage of anchor nodes varies from 5 to 35 %. Because of weighted average method, each time WUWSN will likely be able to choose the maximum amount of potential reference nodes affording location information, so localization error of WUWSN is the least. When the anchor node ratio reaches 20 %, the three localization error curves begin to reduce, but not so sharply.

29.3.3 Communication Cost

Now we begin to consider the comparison in the communication cost among ELSN, WUWSN, and RECUSIVE when the percentage of anchor nodes varies from 5 to 35 %. When the proportion of anchor nodes is low, communication cost of RECUSIVE is the lowest. This is because only a part of nodes send messages in RECUSIVE. But in WUWSN and RECUSIVE, anchor nodes will send beacon messages while some non-localized nodes which do not receive enough beacon messages will also send localization messages. When the proportion of anchor nodes becomes higher, the localization errors play an important factor in communication cost, so the communication costs of ELSN and WUWSN with lower localization errors become low.

Because the three-dimensional space is transformed into two-dimensional space, less messages are needed to be sent. As less messages are needed to be sent, WUWSN is lower than ELSN in communication cost.

29.4 Conclusions

WUWSN transforms solving Euclidean equations into finding the best plane intersection point of circles and regards weighted sum of all the potential position as the final position. Compared with ELSN, WUWSN has an obvious advantage that a reasonable estimate location could be found even if there is no enough effective information. Another advantage is that residuals make location estimate more accurate. The communication traffic becomes less because three-dimensional space is transformed into two-dimensional space. The simulation result shows that compared with ELSN, WUWSN is lower in both the average localization error and the average communication cost, but is higher in the localization coverage.

Acknowledgments The work is supported by the National Science foundation of China (41176082, 61073182, 40827003, 61073183), the Fundamental Research Funds for the Central Universities (HEUCF1006), and Young backbone teacher project of Heilongjiang province (1155G15).

References

1. Wang F, Shi L, Ren F (2005) Self-localization systems and algorithms for wireless sensor networks. J Softw 15(6):857–868 (in Chinese)
2. Shanshan W, Jianping Y, Zhiping C, Guomin Z (2008) A RSSI-based self-localization algorithm for wireless sensor networks [J]. J Comput Res Devel 45(1):385–388
3. Cui X, Liu I, Fan X (2009) A distributed anchor free localization algorithm in sensor networks. J Comput Res Dev 46(3):425–433 (in Chinese)
4. Langendoen K, Reijers N (2003) Distributed localization in wireless sensor networks: a quantitative comparison. Comput Netw 42(4):499–518
5. Niculescu D, Nath B (2004) Position and orientation in ad hoc networks. Ad Hoc Netw 2(1):133–151
6. Shaobin Cai, Xi Li, Ying Tian et al (2010) Alternating combination trilateration based on circle-selection. J Comput Res Dev 46(2):238–244 (in Chinese)
7. Zhou Z, Peng Z, Cui JH et al (2011) Scalable localization with mobility prediction for underwater sensor networks. IEEE Trans Mob Comput 10(3):335–348
8. Niculescu D, Nathi B (2001) Ad-hoc localization system (APS). In: Proceeding of IEEE GLOBECOM01. Piscataway, pp 2926–2931
9. Albowitz J, Chen A, Zhang L (2001) Recursive position estimation in sensor networks. In: Proceedings of the IEEE international conference on network protocols (ICNP) 01. Los Angeles, pp. 35–41
10. Zhou Z, Cui J, Zhou S (2010) Efficient localization for large-scale underwater sensor networks. Ad Hoc Netw 8(3):267–279
11. Caruso A, Paparella F, Vieira L et al (2008) Meandering current model and its application to underwater sensor networks. In: Proceeding of INFOCOM08. Phoenix, pp 221–225
12. Bower AS (1991) A simple kinematic mechanism for mixing fluid parcels across a meandering jet. J Phys Oceanogr 21(1):173–180

Chapter 30
The Summarize of Medium Access Control Protocol in RFID

Zhijie Han, Yin Xu and Ruchuan Wang

Abstract RFID (Radio Frequency Identification) is a technology which do not need to be established the direct contact but with the optical contact between the system and the objective, identify a specific target with a radio signal. One of the main advantages of RFID is multitarget identification; at the same time, but there is a problem which is called the tag collision. The ALOHA-based probabilistic algorithm had a tag hunger problem, Tag cannot be identified in a limited time, although the Binary search tree algorithm has solved this problem, but also with shortcomings of longer period of recognition, and the large amount of tag communication. This article organized various types of literature reported collision prevention mechanism used in RFID systems and documents provided the latest and more comprehensive MAC protocol herein compared with others and provided a new summary for RFID MAC protocol. A detailed description of different ways solving the reader and tag collision, so as to provide a more convenient and efficient research references for the readers of this article.

Keywords RFID · MAC · Reader · Tags · Collision · Anti-collision

Z. Han (✉) · Y. Xu · R. Wang
College of Computer and Information Engineering, Henan University,
Kaifeng 475001, Henan, China
e-mail: hanzhijie@126.com

Y. Xu
e-mail: conan0981@163.com

R. Wang
e-mail: wangrc@njupt.edu.cn

Z. Han · Y. Xu · R. Wang
College of Computer, Nanjing University of Posts and Telecommunications,
Nanjing 210003, Jiangsu, China

X. Wang et al. (eds.), *Advanced Technologies in Ad Hoc and Sensor Networks*,
Lecture Notes in Electrical Engineering 295, DOI: 10.1007/978-3-642-54174-2_30,

30.1 Introduction

As one of the focus of the current networking technology, radio frequency identification (RFID) technology can be said that one of its core infrastructure technology, RFID is a communication technology that does not require line of sight and it can withstand the harsh physical environment, identifies specific objectives through the wireless signal, and it also ensure low cost and high performance, while allowing multiple labels identified. The same as the other communication technologies, RFID systems also need the media access control (Medium Access Control, MAC) protocol to prevent different types of collisions, as these collisions wastes network resources and slow down the reading process. Since the limited processing and storage capacity of RFID tags, the designing of MAC protocol in RFID systems has also become one of the key links.

30.2 RFID Systems Theory

30.2.1 RFID System Introduction

A typical RFID system consists of the application host, RFID reader, and a range of tags. Tag is used to store certain information while the capacity is generally 32 bit–3,200 B. Tags can be divided into active and passive. Passive tags have no direct power supply, which produce energy by receiving electromagnetic waves emitted by the reader, in order to supply sufficient to meet the treatment and the energy require when communication begins. Passive RFID tags are the most common type of label on the market currently. Since there is no power, its processing and communication capabilities is limited, therefore, passive tags can only handle simple state work, no media monitoring capabilities. While the active tags have a direct power supply to meet the energy needs when processing tags, and active tags can be processed with some, such as temperature or pressure data. RFID tag reader is the convergence part, but also the next level of the application hosts. Figure 30.1 illustrates this master-slave structure.

30.2.2 RFID Collision and its Classification

When a RFID reader navigates to a tag, indicates that the tag is located in the readers recognizable area. In general, the recognition area is a physical area covered by electromagnetic waves generated by the reader. In this area, the reader may provide energy to the tag, receive tag signal, and make the decoding operation of the tag (This process is often referred to singulation). When singulation used in a recognition area of the reader, the reader can only identify one tag, while one tag can be only read

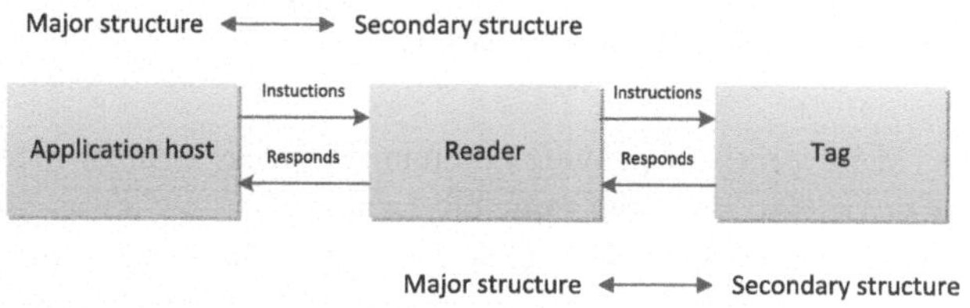

Fig. 30.1 RFID system master-slave structure

by one reader. A tag in recognition area of the reader and reader with overlapping identification area can access the wireless medium for data communication at the same time. However, this simultaneous wireless medium access will result in a data collision, and affect the performance of the RFID system. To maintain the operability of the system, effective media access control (MAC) mechanism is necessary. Like the other RF systems, the main purpose of MAC mechanism is that through an active or a passive way to regulate access of the media in order to reduce the effect of the collision. Initiative, the effective dissemination of sharing media access requests information to avoid collisions; passive mechanism is the response collision and trying to speed up the recovery process after a system crash. Traditional methods to avoid a collision, for example, carrier sense multiplexer (Carrier Sense Multiple Access, CSMA), it is not suitable for RFID systems, particularly when the energy of passive tag is limited, and the communicated situation when it is based on backscattering modulation. Such avoidance mechanism increased the cost of the entire tag, at the same time, reduced the individual potential reader identify areas. Therefore, RFID systems tend to use passive mechanisms to mitigate collisions. In RFID systems, the collision can be classified to the following categories based on the entity type.

(1) Multitag with a single reader collision: these collisions occurred in one reader's recognition area mostly, more than one tag tries to correspond the request of the reader. Multitag with a single reader collision causing great harm, especially in the environment using passive tags. Such collisions will lead to lower literacy rates, waste of resources, delay increases, and other consequences.

(2) Multireader with a single tag collision: these collisions occurred in a single tag is identified by multiple readers. In this case, multiple readers attempt to singulation a single tag, this will result in a label interval condition error. A result, the label will not be detected.

(3) The reader and the reader collision: Such collisions are caused by the conventional frequency interferences, multiple readers with region of mutual interference will be locking by the same frequency. Existing mechanisms like hopping, dynamic frequency allocation, and dynamic energy regulation can hide these collisions. This article organized various types of literature that have reported collision prevention mechanism used in RFID systems and documents provided the latest and more comprehensive MAC protocol [1, 2] herein compared with others.

30.2.3 The MAC Mechanisms in RFID

In RF systems, MAC protocol provides a channel access control mechanism that allows multiple devices to share the same physical media.

Among them, the most famous MAC technology is the MAC protocol based on CSMA, Multiple Access with Collision Avoidance (MACA), and the Aloha protocol. RFID system use half-duplex communication, the point to point connection between reader and tag, and the reader and tag communication is based on backscatter modulation. In backscatter modulation, the tag using the reflected signal, the serial data is sent through the antenna. Traditional technique does not use the MAC on the tags due to the characteristics of reflection communication, as these labels cannot monitor the media, monitoring the collision or listen to obstruction exists of the other channels. This means that all conditions for collision avoidance and resolution cannot be achieved on the tags. Thus, in RFID systems, the MAC protocol confined to solve data collision on the reader, this results in an anti-collision algorithm that no matter tags or readers collisions are applicable, which based on Aloha anti-collision solutions. RFID system operating conditions are: very high active period in a short time, while others have a lot of time in a relatively static inactive. For a reader, the number of tags within the recognition area is usually uncertain. This problem led to the need to design scalable needs and adaptive protocols. Generally, there are four kinds of methods that can be used to solve the multichannel wireless technology access problems. These methods are based on time and spatial (location), the frequency and the code system to regulate media access. Code Division Multiple Access, (CDMA) using spread spectrum technology to support high rate data multiplexed. In conventional spread spectrum technology, each user encodes the user data packet, use orthogonal code so that multiple users can simultaneously transmit data. However, the complex receiver design and high energy demand or calculation limited the use of CDMA technology in the RFID system. References [3] reported the solutions based on combination of TDMA and CDMA technology. But it has not conducted in-depth research for CDMA scheme used in the RFID system. Space Division Multiple Access (SDMA) [4, 5] technology spatially multiplexed channels. The principle of SDMA operation is that, there is only limited number of RFID tags at the same time and the same place, independent of total tags that may arise in the identification area. Thus, these RFID tags can be isolated in the space so that the interference caused by the other labels to a minimum. By adjusting the reader (energy Control [6–8]), the use of adaptive arrays (adaptive array), Multiple Input Multiple Output, (MIMO) antenna technology [9], and directional antenna technology of electronic control, which can be formed by many space-isolated areas. Frequency division multiple access (FDMA) technique uses a plurality of channels, each channel using a different carrier frequency for communication. In the RFID system, this can be achieved by possibilities of reader radio frequency and a plurality of tags for locking. Reader uses radio frequency to synchronize and publish identification command. Tag uses one of a number of available frequencies. The advantage of FDMA is the use of nondisturbing frequency in a multitag reader and multiple simultaneous communications. However, FDMA techniques is not widely used in the RFID system, because the FDMA technology which is not practical for the tag, because the reader requires a relatively high cost, and each receiver must also provide reception channel alone, which limits the frequency division multiple access technology applications in certain situations. Time division multiple access (TDMA) technology is to be able to take advantage of the channel between the potential participants and divide according to the time dimension. For example, in GSM, Bluetooth, and IEEE802.16 (Wimax), etc. All these major networks, time division multiple access technology is often used for improvement, and in most cases in a mixed manner. So far, In RFID systems, time division multiple access technology is the most important media access protocol. This is because compared to FDMA, CDMA, and FDMA technology, TDMA technology is simple, and relative to the passive tag, it has a lower processing cost, and is not complicated in the calculation, the process, and capital consumption.

30.3 Collision Avoidance Algorithm and its Summary

Tag collision mechanism frequently use tree-based algorithms or probabilistic frame Aloha program tag number-based logical division, turn it into a set of labels and can be more easily processed. Use traditional competing solutions such as collisions between the timing, interfere with learning, color scheme to deal with the Collisions between the readers or between the reader and the tag. In the RFID system, the most common source of collision is tag collision. Tag collision mainly occurs in an area of reader's recognition with multiple tags at the same time, while corresponding to the reader's request command. Since the passive tag has no carrier sense or communication ability, it is designed to be reader-driven way. Therefore, readers use acquisition technology to solve tag collision, this acquisition technology is known as collision. The collision solutions were classified into deterministic or probabilistic mechanism. In deterministic mechanism, the reader division read a tag set in a given time. Divisions are based on the previous competitive information obtained by the process cycle and try to reduce the next cycle of competitive information. According to their division method, uncertainty of collision mechanisms are classified according to the algorithm based on tree. Deterministic mechanism implements the division of the branches by using the tag's serial number (identification number) or random numbers. In given conditions, a deterministic method may take quite a longtime. However, this does not cause hunger of tag. When it is in hunger, tags cannot be read in a longtime; What's more, the label may not be read at all. In probabilistic mechanism, the reader communicates through the frame with a specific length, and tag send a specific time slot in the frame randomly. The size of the frame is according to the previous interactive identification Cycle (Interrogation Cycle) process of information, and according to the density and distribution to improve the adaptability of the tag. Really process has been repeated, until all tags have been read. Because of lower consumption, so the probabilistic method of execution process is very fast, but will be affected by the label hunger. Tables 30.1 and 30.2 lists the performance comparison between the common collision algorithms.

Table 30.1 Deterministic anti-collision algorithm comparison

Algorithm	Speed	Cost	Stateful	Mobility	Scalability	Price
Energy-aware [10]	√	√	–	↓	↓	√
Hybrid [11]	√	√	↓	√	–	↓
Memoryless [12]	√	↓	√	–	–	√
Prefix-randomized [13]	–	√	↓	–	–	–
Counter and stack [14]	√	–	↓	√	√	↓

Table 30.2 Probabilistic anti-collision algorithm comparison

Algorithm	Speed	Cost	Stateful	Scalability	Price
Static [8]	↓	√	–	↓	–
Reservation [15]	√	↓	↓	–	↓
Estimation [16]	√	↓	↓	–	–
Counter and stack [17]	√	–	↓	√	↓

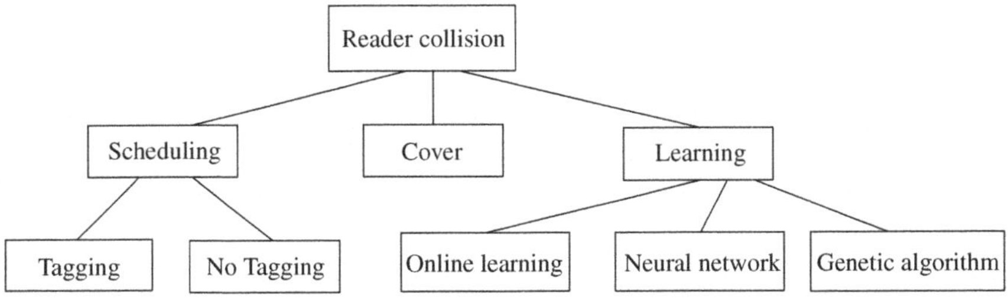

Fig. 30.2 Pattern classification of reader collision

The reader with overlapping identify that regional may cause interference when tag identification. As other RF system, running a reader may also affect other reader even to identify region overlap. This case will lead to reader collision [18, 19], more use may produce reader collision frequency, RFID tag type limits the function of the equipment. These devices cannot distinguish multiple readers, or can they be mass production and frequency to communicate. Figure 30.2 shows the classification of reader collision. Resolve collisions based on a variety of mechanisms such as scheduling, covering, and learning.

30.3.1 RFID Collision and its Classification

The scheme based on scheduling scheduled the frequency of the reader and the associated slots. Scheduling can be centralized or distributed, and supports frequency allocation both static and dynamic. Meanwhile, the scheduler needs the supports

Table 30.3 Comparison of the reader collision algorithm

Method	Algorithm	Centralized control	Distributed control	Fixed channel	Dynamic channel
DCS [20]	Scheduling	–	√	√	√
VDCS [21]	Scheduling	–	√	–	√
HiQ [22]	Learning	√	√	–	√
Neural network	Learning	–	√	√	–
Anneal arithmetic [25]	Learning	√	–	√	–
Genetic algorithm	Learning	√	–	√	–
Redundant [26]	Cover	–	√	–	–
CC [24]	Cover	√	–	√	–

of tags, namely tags need additional storage space to store additional data. However, scheduling does not need tag support under normal circumstances. Such as color waveform algorithm [20, 21] is this category. On the other hand, learning-based approach is based on hierarchical online learning, genetic algorithms, and neural network. Learning methods minimize collisions by learning collision mode and allocation frequencies of learning-based mode. For example, HiQ algorithm [22] is based on such methods. Literatures have been reported such alternate methods like use beacon channel [23], center coordinator [24], as well as covering issues to handling reader collision. Table 30.3 lists the performance comparison between common reader collision algorithms.

30.4 Conclusion

This article is intended to provide a new summary for RFID MAC protocol. A detailed description of different ways solving the reader and tag collision, so as to provide a more convenient and efficient research references for the readers of this article.

Acknowledgments The subject is sponsored by the National Natural Science Foundation of P. R. China (No.611700656137301761171053613731361103195) the Natural Science Foundation of Jiangsu Province (BK2012833) Scientific and Technological Support Project of Jiangsu ProvinceNo. BE2012183BE2012755 Natural Science Key Fund for Colleges and Universities in Jiangsu Province (11KJA52000112KJA520002) Project sponsored by Jiangsu provincial research scheme of natural science for higher education institutions (11KJB520016) Scientific Research and Industry Promotion Project for Higher Education Institutions (JHB2012-7) Science and Technology Innovation Fund for higher education institutions of Jiangsu Province (CXZZ12-0479) Doctoral Fund of Ministry of Education of China (20113223110002).

References

1. Shih D, Sun PL, Yen DC, Huang SM (2006) Taxonomy and survey of RFID anti-collision protocols. Comput Commun 29:2150–2166
2. Tang Z, He Y (2007) Research of multi-access and anti-collision protocols in RFID systems, 2007 IEEE international workshop on anti-counterfeiting, security, identification. Xiamen, pp 377–380
3. Mutti C, Floerkemeier C (2008) CDMA-based RFID systems in dense scenarios: Concepts and challenges, 2008 IEEE international conference on RFID. Los Vegas, pp 215–222
4. Vandenameele P (2001) Space division multiple access for wireless local area networks. Kluwer Academic Publishers, Norwell
5. Finkenzeller K (2003) RFID handbook: Fundamentals and applications in contactless smart cards and identification. Wiley, England
6. Ali K, Hassanein H, Taha AM (2007) RFID anti-collision protocols for dense passive tag environments, In: Proceedings of the 32nd IEEE conference on local computer networks. Ireland, pp 819–824
7. Kim J, Lee W, Kim E, Kim D, Suh K (2007) Optimized transmission power control of interrogators for collision arbitration in UHF RFID systems. IEEE Commun Lett 11:22–24
8. Alsalih W, Ali K, Hassanein H (2008) Optimal distance-based clustering for tag anti-collision in RFID systems, 33rd IEEE conference on local computer networks. Montreal, pp 266–273
9. Lee J, Kwon T, Choi Y, Das SK, Kim K (2004) Analysis of RFID anti-collision algorithms using smart antennas. In: Proceedings of the 2nd international conference on embedded networked sensor systems SenSys 04. Baltimore, pp 265–266
10. Zhou F, Jin D, Huang C, Hao M (2003) Optimize the power consumption of passive electronic tags for anti-collision schemes. In: Proceedings of 5th international conference on ASIC, vol. 2. Beijing, pp 1213–1217
11. Ryu J, Lee H, Seok Y, Kwon T, Choi Y (2007) A hybrid query tree protocol for tag collision arbitration in RFID systems, ICC 07: IEEE international conference on communications. Glasgow, pp 5981–5986
12. Myung J, Lee W, Srivastava J, Shih TK (2007) Tag-splitting: Adaptive collision arbitration protocols for RFID tag identification. IEEE Trans Parallel Distrib Syst 18:763–775
13. Zhang N, Vojcic B (2005) Binary search algorithms with interference cancellation RFID systems, MILCOM 2005: IEEE military communication conference, vol. 2. Atlantic City, pp 950–955
14. Chen W, Horng S, Fan P (2007) An enhanced anti-collision algorithm in RFID based on counter and stack, ICSNC 07: Proceedings of the second international conference on systems and networks communications. Cap Esteral, French Riviera, pp 21–24
15. Wong CP (2007) Grouping based bit-slot ALOHA protocol for tag anti-collision in RFID systems. IEEE Commun Lett 11:946–948
16. Cha J, Kim J (2006) Dynamic framed slotted ALOHA algorithms using fast tag estimation method for RFID system, CCNC2006: 3rd IEEE consumer communications and networking conference, vol. 2. Los Vegas, pp 768–772
17. Khandelwal G, Yener A, Lee K, Serbetli S (2006) ASAP: A MAC protocol for dense and time constrained RFID systems, ICC 06: IEEE international conference on communications, vol. 9. Istanbul, pp 4028–4033
18. Engels DW, Sarma SE (2002) The reader collision problem, 2002 IEEE international conference on systems, man and cybernetics, vol. 3, pp. 641–646, Hammamet, pp 641–646
19. Leong KS, Ng ML, Cole PH (2005) The reader collision problem in RFID systems, MAPE 2005:IEEE international symposium on microwave, antenna, propagation and EMC technologies for wireless communications, vol. 1. Beijing, pp 658–661
20. Waldrop J, Engels DW, Sarma SE (2003) Colorwave: An anti-collision algorithm for the reader collision problem, ICC 03: IEEE international conference on communications, vol. 2. Anchorage, pp 1206–1210

21. Waldrop J, Engels DW, Sarma SE (2003) Colorwave: A MAC for RFID reader networks, WCNC 2003: 2003 IEEE wireless communications and networking, vol. 3. New Orleans, pp 1701–1704
22. Ho J, Engels DW, Sarma SE (2006) HiQ: A hierarchical Q-learning algorithm to solve the reader collision problem, SAINT workshops 2006: International symposium on applications and the internet workshops. Phoenix, pp 88–91
23. Birari SM, Iyer S (2005) Mitigating the reader collision problem in RFID networks with mobile readers, 13th IEEE international conference on networks. Kuala Lumpur, pp 463–468
24. Wang D, Wang J, Zhao Y (2006) A novel solution to the reader collision problem in RFID system, WiCOM 2006: International conference on wireless communications, networking and mobile computing. Wuhan, pp 1–4
25. Lin C, Lin F (2007) A simulated annealing algorithm for RFID reader networks, WCNC2007: IEEE wireless communications and networking conference. Hong Kong, pp 1669–1672
26. Carbunar B, Ramanathan MK, Koyuturk M, Hoffmann C, Grama A (2005) Redundant reader elimination in RFID systems, IEEE SECON 2005: 2005 Second annual IEEE communications society conference on sensor and Ad Hoc communications and networks. Santa Clara, pp 176–184

Chapter 31
Wireless Sensor Network Data Storage Optimization Strategy

Weibei Fan, Zhijie Han and Zhengwei Guo

Abstract Wireless sensor network (WSN) is developing to be a new data-centric computer technology fields, which combined with wireless communication technology, computer network technology, sensor technology, and micro-electromechanical technology. WSN is a data-centric network, the application layer users most concern is that how to transmit and store the data of perception and monitor each nodes effectively rather than the situation of deployment scenarios and the underlying hardware. So, the data storage is an important area of research in wireless sensor network. This paper studies the status of the data storage strategy, introduced several data storage ideas, analyzes their advantages and disadvantages, and the future development are put forward.

Keywords WSN · Data storage · Research focus · Development trend

31.1 Introduction

The emergence of WSN is promoting the social progress and development of science and technology, which has become the research focus in the domestic and international numerous scholars. WSN has its own characteristics: (A) Node Energy Limited. By the limitation of price, power consumption, and the volume, its computing power and memory space is significantly weaker than the ordinary computer;

W. Fan (✉) · Z. Han · Z. Guo
Department of Computer Science and Engineering, Henan University,
Kaifeng 475001, Henan, China
e-mail: fanweibei@163.com

Z. Han
e-mail: hanzhijie@126.com

Z. Guo
e-mail: gzw@henu.edu.cn

X. Wang et al. (eds.), *Advanced Technologies in Ad Hoc and Sensor Networks*,
Lecture Notes in Electrical Engineering 295, DOI: 10.1007/978-3-642-54174-2_31,

(B) Power supply capacity is limited. Each nodes supplied by the battery, and the battery capacity is small; (C) Self-organization. The layout and expansion of the network does not depend on any pre-network facilities; (D) No center. WSN is a peer-to-peer network, all nodes have equal status; (E) The dynamic topology. Nodes can be moving anywhere; (F) Node number is numerous, densely distributed. There are usually hundreds of thousands of sensor nodes in a monitoring area. Therefore, it is necessary to research a suitable data storage strategy for the special environment of WSN. At present, according to the data characteristics of sensor networks, there are three kinds of storage. There are external storage, local storage, and data-centric storage. This paper introduced the researching status of the storage of WSN at home and abroad; After analyzing the characteristics of this storage strategy, then summarized some advanced storage strategy by other authors; Finally given some conclusions, and prospected about the future research direction in this field.

31.2 Classification of Data Storage Strategy

31.2.1 External Storage Strategy

In External storage strategy, all sensor nodes of WSNs sent data to the sink nodes collected by monitoring for centralized storage [1]. In the external storage, the nodes will be transmitting data to base station and save after collecting, [12] it is not need to send data query by base station. In order to deal with numerous data of nodes, base station can take some specifically for the development of WSN data management software for data processing and storage, such as wide application of so-called COUGAR [2] and TinyDB [3] data management system, etc. Sink node is connected to the sensor network and external network, generally has strong ability of data storage, processing and communication, the centralized storage of data to the node, data aggregation, and query in the late will be very convenient and flexible.

Due to all the data perceived by the nodes will be sent to base station, a query is also needed to base station if the nodes need to query data, thus the base station undertakes the task of data storage and query. Because of the busy communication task, especially the nodes layout density and the number is huge, base station would be a bottleneck with a bigger time delay. At the same time, the consumption of nodes close to the base station will be faster [4], affect the survival time, and quality of service system.directed diffusion algorithm [5] first before querying, then choose an optimal route path between the sink nodes and data nodes. By using this path for subsequent data acquisition directly, thus reducing the blindness of data forwarding, improving the efficient of querying and saving energy consumption of the relevant nodes. However, this method is to establish subsequent queries efficiently optimal path has been built in the early on this premise, and its preliminary work energy consumption is relatively large, so in general this method is suitable for the purpose of continuous queries. An improvement measure has been put forward for directed diffusion routing algorithm [6], the increasing node energy consumption, causing the network congestion problem in the previous query request flooding. Combined with query request to make the nodes that could not complete the task not to participate in query request diffusion as far as possible. These improvements take measures to reduce the energy consumption of nodes in different ways, but the basic design thought is not change. It still take this design pattern of the first query command flooding, and then the query result comes back. The disadvantages still exist, such as flood routing mechanism of repeated data forwarding, node resources, and blind using, etc.

31.2.2 Local Storage Strategy

Local storage is a kind of network data storage strategy, its data is stored in the acquisition node. When nodes receive data queries, then send data to the consumption nodes. Local storage strategy, all sensor nodes in monitoring events or data directly to the data stored in the local at the same time, to store the data phase of energy consumption caused by communication down to zero. It brought difficulties to the data query of late stage. First of all, it is hard to find data and the query process is complicated. Since there is no global knowledge of data distribution, all nodes in WSNs must participate in query request and result data forwarding while executing queries, which has caused great waste of energy of nodes.

Aimed at this situation, some improved measures have been put forward by related researchers. Adopting directed diffusion algorithm [5] first before querying, then choose an optimal route path between the sink nodes and data nodes. By using this path for subsequent data acquisition directly, thus reducing the blindness of data forwarding, improving the efficient of querying and saving energy consumption of the relevant nodes. However, this method is to establish subsequent queries efficiently and optimal path has been built in the early on this premise, and its preliminary work energy consumption is relatively large, so in general this method is suitable for the purpose of continuous queries. An improvement measure has been put forward for directed diffusion routing algorithm [6], the increasing node energy consumption, causing the network congestion problem in the previous query request flooding. Combined with query request to make the nodes that could not complete the task to participate in query request diffusion as far as possible. These improvements take measures to reduce the energy consumption of nodes in different ways, but the basic design thought is not change. It still take this design pattern of the first query command flooding, and then the query result comes back. The disadvantages still exist, such as flood routing mechanism of repeated data forwarding, node resources, and blind using, etc.

31.2.3 Data-Centric Storage

Data center storage is also known as distributed storage. Data center storage strategy is proposed first by Shenker in 2003. In this method, all the test events or sensory data corresponds to a specific name, which stored in the corresponding nodes in WSNs according to the name. For data query, according to event attributes, can also be convenient search to the data. So, the data storage for the center provides for data storage and query of the mediation mechanism, this mechanism is mainly done by describing the attributes of the data itself. Mediation mechanism commonly used are: hash map and data index.

In many applications, WSN will collect large amounts of data from the monitoring environment. If the data is distributed unevenly, events concentrated in certain areas of the network. It will lose data when a node is saturated or energy storage capacity is exhausted. So, it is necessary to play store data task distribution to other nodes, adopts the distributed storage, reduce the loss of data. Due to the use of data Intermediary mechanism, so you can easily find the data storage nodes, to provide efficient data queries, while reducing power consumption, reducing latency. Reducing the time and spatial redundancy of data has important significance, which could reduce the network traffic load and save energy. The data center stored the same event data to the same storage node, can be more efficient for data fusion. Compared to the other two types of storage, data center storage provides a data-based attribute information intermediary mechanism, making the balance of data storage and query cost. Although requires additional consideration when building the establishment of mechanisms for the intermediary, but through the protocol and algorithm design can make it to a minimum.

31.3 Based on Three Kinds of Storage Strategy of the Improved Algorithm

The data after the quantitative data sets need to send to the base station or users, generally adopt the strategy of external data store (Fig. 31.1).

WSN nodes are divided into two categories: ordinary nodes and storage nodes. Such as A only collect data, B and D forwards the received data; C is the storage node, in addition to collecting sensing data and forwarding data, storing data also. Sink node is a special kind of storage node. External storage was applied in the traditional data acquisition system widely. Low-Energy Adaptive Clustering Hierarchy [7] was put forward by Heinzelman, which used to collect data from the sensor network. LEACH algorithm selected part of the cluster randomly in the initial stage from the first node. The selected cluster head send nodes cluster structure to other neighbors. The advantage of clustering structure could reduce channel conflicts, save energy, and facilitate the hierarchy for data processing, such as data fusion and data compression. The problem of LEACH protocol is that when the network size is relatively large,

Fig. 31.1 Data transmission in WSN

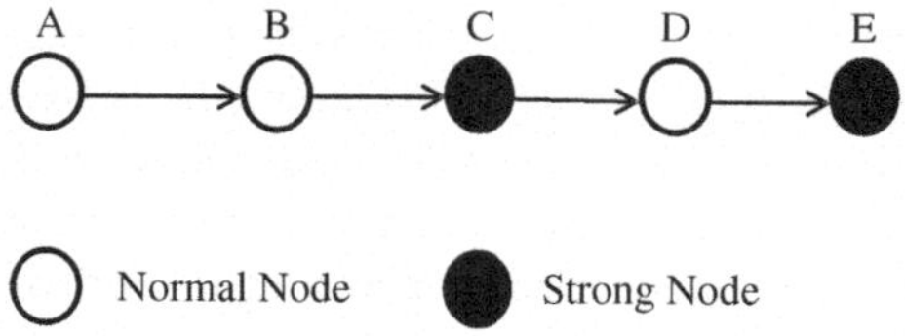

cluster head far away from the sink inches, it is hard to guarantee the communication link. Power-Efficient Gathering in Sensor Information Systems [8] (PEGASIS) has a improvement to LEACH protocol. In PEGASIS protocol, a structure of a chain data collection was built in the network. It is different to LEACH protocol, each nodes have the same transmission range and exchange data only with its neighbors in PEGASIS. Data gathered in the direction of the chain structure according to the fixed, and achieve the purpose of sending data to base station ultimately. Directed Diffusion Protocol [9] was raised in Intanagonwiwat, Nodes collected data and stored in the local node. when the convergence point needs to query data, nodes set weights and establish a gradient according to the environment, establish multiple sink nodes to connect to the data store path for data transmission. GEAR (Geographical and Energy Aware Routing) protocol [10] is the improvement of Directed Diffusion, which is to make the make the query message propagation area reduced to a query destination range. The agreement exchanged with the hello message between nodes. So that each node could get the position of neighbor nodes and the information of remaining energy. This could optimize the transmission path of data and reduce the energy consumption of the query process. Document promoting an adaptive ring index structure named Adaptive Ring-based Index [11] data storage. Some location information in the storage node was sent to index nodes, as a query data interface of sink node. This storage strategy requires each node knows the boundaries of whole monitoring area, but it is difficult to predefine the boundaries of WSN normally. Meanwhile, ARI did not work well with hot issues. The main idea of Two-Tier Data Dissemination [12] is: each data source node built a grid structure in the entire of WSN, and stored the observation data in this structure. The query could get the needed message only locally flood in a grid, thus save energy avoiding by the scope of entire WSN.

The above discussion by some foreign scholars make research work in terms of data storage strategy. Big work about the research of WSN has been made by Domestic scientific research institutions. By using of data-centric storage strategy, Li Zhi-gang proposed a data storage oscillation trajectory discovery mechanism [13]. The program requires that consumer's nodes and producer's nodes are stored in the corresponding oscillation path. According to local information and preset angle reflection, nodes data routing and forwarding. In order to ensure data query success rate, all oscillation should be the property of two intersection. Based mobile base stations for WSN data collection methods [14] is used to solve the traditional question of external storage hot spots appear around the base station.Hot-Spot-aware data storage strategy [15] is improved by Geographic Hash Table [16], By contrast,

SASS is different as follows: first depending on the space usage of storage nodes to extend storage, most existing methods are assigned in the same memory space for all events. After extending storage, the event can be forwarded to the new storage node directly instead of transmitting through the main storage. This could reduce the communication overhead of main memory. Gao Chao proposed Approximate Location Greedy Perimeter Stateless Routing [17] algorithm, the basic idea is that when the packets arrive fully nodes near the purpose, may terminate the perimeter transfer process. In particular, if an event or query packets to a node, the node from the target position is lower than a threshold, the node only by examining the adjacent nodes, will be able to determine the most closest node to the destination location.

In summary, the core idea of current data storage of WSN is: According to different application scenarios and design goals, saving the collected data in local nodes or base station directly, or storing in one or more nodes within the network according to a certain algorithm with distributed rules. This will provide an efficient, low-power query response.

This paper has studied a variety of existing storage strategies, through a comparative analysis found, that each of the above storage strategies are not fit for all applications, but only for a particular application. As the computing power and limited resources of WSN, the design algorithm should have a distributed computing and perform simple features.

31.4 Conclusion

In recent years, along with the surge in the volume of data, data analysis is more demanding. WSNs are producing data every minute, Mobile devices are everywhere, RFID, Internet service with hundreds of millions of users are generating a large number of interactive information at all times. PowerDrill [18], the Google new big data analysis techniques, which has a strong processing capacity in storing and querying big data. Carefully study the core of big data, and then have an optimization about data store of WSN. Generally speaking, to significantly reduce the data storage space in the nodes, which could reduce the energy consumption of the sensor nodes, ensure that storage load balancing, affect the query response latency directly. This could improve the overall performance of WSN and extend the life of the network. The further research of data storage strategy about WSN is the future trend.

Acknowledgments The subject is sponsored by the National Natural Science Foundation of P. R. China (No. 611700656137301761171053613731361103195) the Natural Science Foundation of Jiangsu Province (BK2012833) Scientific & Technological Support Project of Jiangsu Province No. BE2012183BE2012755 Natural Science Key Fund for Colleges and Universities in Jiangsu Province 11KJA52000112KJA520002) Project sponsored by Jiangsu provincial research scheme of natural science for higher education institutions (11KJB520016) Scientific Research & Industry Promotion Project for Higher Education Institutions JHB2012-7) Science & Technology Innovation Fund for higher education institutions of Jiangsu Province CXZZ12-0479) Doctoral Fund of Ministry of Education of China (20113223110002).

References

1. Greenstein B, Estrin D, Govindan R et al (2003) DIFS: a distributed index for features in sensor networks. In: IEEE international workshop on sensor networks protocols and applications, pp 163–173
2. Yao Y, Gehrke J (2002) The cougar approach to in-network query processing in sensor networks. Sigmod Rec 31(3):9–18
3. Madden SR, Franklin MJ, Hellerstein JM et al (2005) TinyDB: an acquisitional query processing system for sensor networks. ACM Trans Database Syst 30(1):122–173
4. Wu XB, Chen GH, Das SK (2008) Avoiding energy holes in wireless sensor networks with nonuniform node distribution. IEEE Trans Parallel Distrib Syst 19(5):710–720
5. Acu B, Reisig W (2006) Compensation in workflow nets. In: Proceedings of the 27th international conferenceon applications and theory of petri nets and other models of concurrency (LNCS), vol 4024. Springer, pp 65–83
6. Dong WL, Yu H, Zhang YB (2006) Testing BPEL-based webservice composition using high-level petri nets. In: Proceedings of the 10th IEEE international enterprise distributed object computing conference, HongKong
7. Heinzelman WR, Chandrakasan A, Balakrishnan H (2000) Energy-efficient communication protocol for wireless microsensor networks. In: The IEEE 33rd annual Hawaii international conference on system sciences (HICSS-33). Maui, pp 1–10
8. Lindsey S, Raghavendra CS (2002) PEGASIS: power-efficient gathering in sensor information systems. In: Proceedings of IEEE aerospace conference, pp 1125–1130
9. Intanagonwiwat C, Govindan R, Estrin D et al (2003) Directed diffusion for wireless sensor networking. IEEE/ACM Trans Netw 11(1):2–16
10. Yu Y, Govindan R, Estrin D (2001) Geographical and energy aware routing: a recursive data dissemination protocol for wireless sensor networks
11. Zhang WS, Cao GH, La Porta T (2007) Data dissemination with ring-based index for wireless sensor networks. IEEE Trans Mobile Comput 6(7):832–847
12. Greenstein B, Eslrin D, Govindan R et al (2003) DIFS: a distributed index for features in sensor networks. In: IEEE international workshop on sensor network protocols and applications, pp 163–173
13. Li Z-G, Xiao N, Zhu F-Y (2010) Large-scale wireless sensor networks based on the oscillation trajectory data storage and discovery mechanism. Comput Res Dev 14(11):1911–1918
14. Chen T, Guo D, Luo X (2011) Based mobile base stations for wireless sensor networks bookstore collection methods. J Natl Def Univ 7(02):49–53
15. Li Q-Q, Wu L, Wang Y (2013) Hot spot perception of wireless sensor network data storage strategy. Comput Sci 40(4):145–151
16. Ratnasamy S, Karp B, Shenker S et al (2003) Data-centric storage in sensornets with GHT, a geographic hash table. Mob Netw Appl 8(4):422–427
17. Gao C, Hu X-Y, Wang B-W, Gao H-L (2013) A low overhead routing scheme for data centric storage sensor networks. J Chinese Comput Syst 34(7)
18. Hall A, Bachmann O, Büssow R, Gănceanu S, Nunkesser M (2012) Processing a trillion cells per mouse click. Proc VLDB endowment 5(11):1436–1446

Chapter 32
Localization for a Hybrid Ocean Wireless Sensor Networks

Hanjiang Luo, Zhong Yang and Linfeng Huang

Abstract Ocean covers nearly 71 % of our planet's surface and deploying wireless sensor networks on the sea can perform monitoring tasks over the oceans. With some applications, we may deploy sensors both on the sea surface and underwater which forms a Hybrid Ocean Sensor Networks (HOSNs). In this paper, we discussed a localization scheme with the hybrid wireless sensor networks which combines the advantages of Wireless sensor networks (WSNs) and Underwater acoustic-sensor networks (UWA-SNs). We first introduced the structure of the hybrid wireless sensor networks and its applications, then we discussed a localization schemes MLA for moored nodes deployed underwater. We present the related work and finally discussed our future research of the hybrid ocean wireless sensor networks.

Keywords Wireless sensor networks · Underwater sensor networks · Monitoring sensor networks

32.1 Introduction

Deploying wireless sensor networks in the ocean, which called ocean sensor networks (OSNs), has recently been drawing much attention because of its potential applications ranging from ocean environment monitoring, structure monitoring, commercial exploitation, tactical surveillance to disaster prevention or other defense and security purposes, etc [1, 2].

OSNs are mainly formed by sensor nodes deployed on the sea surface or underwater depend on the specific applications. And with some applications, e.g., ocean monitoring and intrusion detection, a hybrid networks may need to deploy both sensor

H. Luo (✉) · Z. Yang · L. Huang
Department of Information Engineering, Zibo Vocational Institute,
Zhangdian liantong road, Zibo, Shandong Province, China
e-mail: luo.wireless@gmail.com

X. Wang et al. (eds.), *Advanced Technologies in Ad Hoc and Sensor Networks*,
Lecture Notes in Electrical Engineering 295, DOI: 10.1007/978-3-642-54174-2_32,

Fig. 32.1 Double-head node structure

nodes on the sea surface and underwater. Though the two types of sensor networks have different communication channels, i.e., wireless radio communication channel and underwater acoustic channel, such hybrid networks can combine the advantages of both types of networks in order to achieve three-dimensional monitoring tasks of the ocean [3].

In this paper, we proposed a hybrid ocean sensor networks which consist of double-head nodes, just as Fig. 32.1 shows, deployed on the sea surface and moored underwater acoustic sensor nodes deployed underwater. Such ocean hybrid sensor networks have many potential applications. For example, Surveillance security is a vital problem for harbor protection, border control, and the protection of commercial facilities, such as oil platforms and fisheries. So, we can deploy such hybrid sensor networks on the sea surface for harbor protection, as shown in Fig. 32.2, and we can also deploy such networks to protect the vast near-coast sea surface and busy harbor areas from intrusions of unauthorized marine vessels, such as trespassing boats and ships. The authors in paper [4] present an innovative solution for ship intrusion detection by deploying the wireless sensor networks with three-axis accelerometer sensors on the sea surface to detect ship-generated waves. However, the scheme

Fig. 32.2 Harbor protection with hybrid networks

requires a relatively dense network to achieve a high detection ratio especially with small boats. Combining with other sensors, such as acoustic sensors, can decrease the deployment density because sounds travel far underwater. The hybrid sensor nodes combine both sensors on the sea surface and underwater, thus, we can use such hybrid networks to protect harbor, just as Fig. 32.4 shows the hybrid networks can be deployed to detect boat intrusion.

For specific applications mentioned above, localization is an indispensable part of HOSNs. For example, energy-efficient geo-routing schemes require location information for making routing decision, aquatic monitoring, and underwater surveillance application demand high-precision localization [5–7]. In this paper, we mainly deal with the localization problems of HOSNs. We first introduced the structure of the hybrid ocean wireless sensor networks and its applications, then we discussed the localization schemes for moored node-deployed underwater.

The rest of our paper is organized as follows: Sect. 32.2 describes the hybrid ocean sensor networks localization characteristics and the potential applications of the hybrid networks. In Sect. 32.3, we discussed the localization algorithm MLA for the moored nodes. And the related works are presented in Sect. 32.4. The conclusions and future works are discussed in Sect. 32.5.

32.2 The Networks Model and Characteristics

In this section, we present the hybrid networks hardware, structure, characteristics, and its potential applications.

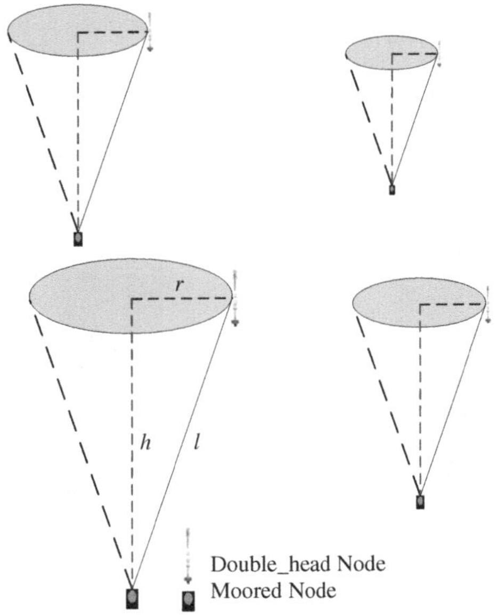

Fig. 32.3 The structure of the hybrid ocean sensor networks

32.2.1 The Hybrid Network Structure and Hardwares

The structure of the hybrid networks is shown in Fig. 32.3. The networks combine both the wireless sensor networks (WSNs) and the UWA-SNs together. These networks have two types of nodes: the double-head node and the moored node. The double-head nodes are shown in Fig. 32.1 and are mainly comprised by wireless sensor node, buoy, and underwater sensor node. When the double-head deployed on the sea surface, an anchor with a rope will tie to the node's buoy as shown in Fig. 32.3. The moored node is with the anchor, and when the moored node sink to the sea bed, and the double-head node floats on the sea surface with wireless sensor node above the sea surface, and the underwater water sensor node beneath the sea surface.

The wireless sensor node can choose node such as TelosB or IMote2. TelosB hosts a Texas Instruments MSP430 F1611 microcontroller running the Tiny Operating System (TinyOS) and it integrates with humidity, temperature, and light sensors on board. IMote2 composed of Processor Radio Board (IPR2400) and Basic Sensor Board (ITS400) which contain a three-axis ST Micro LIS3L02DQ accelerometer, an advanced temperature/humidity sensor, a light sensor, and a 4 channel A/D converter. The accelerometer can be used to measure waves with ocean monitoring applications. Because underwater sensor networks are still in its developing period, researchers can design their own underwater sensor node with specific application. The underwater sensor node may include protocol board, signal process board, power amplifier board, and transducer. The transducer is the main component of the underwater sensor node which transforms electronic signals into acoustic wave transmitting far away for

Fig. 32.4 Detection of boat with hybrid networks

underwater communication. One double-head node actually has two nodes and can be connected with each other using RS-232.

Due to ocean waves, winds, and the tide of the sea, when the double-head node deployed on the sea surface with the anchor on the sea bed, the node has a floating model as shown in Fig. 32.5. The floating model was first presented in [8], and it is a Restricted floating sensor (RFS) model. In this model, the sensor on the sea surface drifts randomly around the center of the anchor with waves or wind.

32.2.2 The Characteristics of the Hybrid Networks

The hybrid networks naturally combine the WSNs and UWA-SNs. However, the WSNs and UWA-SNs have different characteristics.

Just as the WSNs deployed on the land, using Telosb or other wireless nodes can firm ocean monitoring sensor networks to monitor the ocean tide, waves, temperature, salinity, etc. The WSNs consist of nodes with battery power supply or with solar-powered battery supply. Because it is hard and expensive to replace the sensor node's battery, using solar battery systems can achieve a perpetually powered system without a necessary periodical maintenance for battery replacement or recharging. Wireless nodes communicate with other nodes via their low-consumption radio modules. Compared with underwater acoustic communication, the radio module uses radio wave with specific radio frequency traveling at the speed of light (3×10^8 m/s). Thus, the propagation delay is very short. When deploying the WSNs on the sea surface, the networks can measure the tides, waves, winds, light, humidity, or other physical or chemical quantities of the ocean near the sea surface. WSNs also can send the collected data to the remote station via GPRS, satellite, or other radio communication methods. The networks can also use GPS for localization which is impossible underwater.

UWA-SNs are for scientific, industrial, and defense applications. However, UWA-SNs are very different from existing terrestrial sensor networks due to the properties of the underwater environment. First, UWA-SNs use acoustic signals to communicate, thus, the propagation delay is large due to the slow acoustic signal propagation speed (1.5×10^3 m/s). Second, the underwater acoustic communication channel

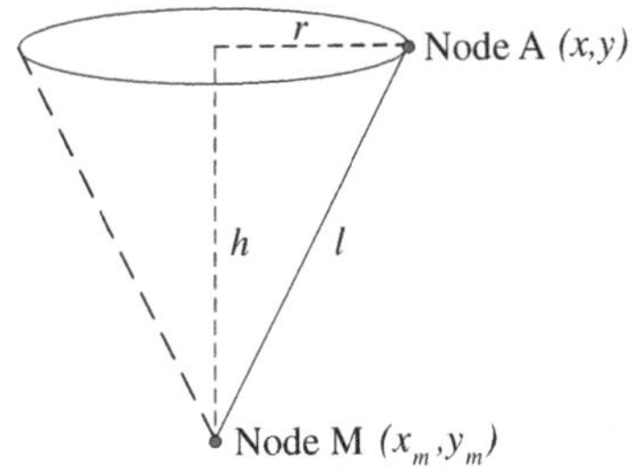

Fig. 32.5 The structure of the moored node and the double-head node

has limited bandwidth capacity because of the significant frequency and distance-dependent attenuation. Currently, the limit on available underwater bandwidth is roughly 40 km kbps. Third, the underwater sensors are battery powered and it is even harder to recharge or replace node batteries with underwater environments. Fourth, the energy consumption of acoustic modem in UWA-SNs is quiet different from those of typical radio transceivers, for the transmit power is often 100 times expensive than that of the receive mode. For example, the typical receive power of the WHOI micro-modem is about 80 mW, but the transmit power is 10 W [9].

32.2.3 The Potential Applications of the Hybrid Networks

Deploying the hybrid networks on the sea has many potential applications. Combined with the networks-deployed underwater, the networks can monitor the ocean, serve as a test bed for research, and for surveillance applications. As for ocean monitoring, sensor networks continuously observe the ocean to detect climate changes or ocean pollution which effects human and animal habitat. The hybrid networks can be used as a test bed for the research of both wireless sensor networks and underwater sensor networks because the hybrid networks combine both advantages of the networks. Next, we discuss the application of surveillance security of the hybrid networks in detail.

Surveillance security is a vital problem for harbor protection, border control, and the protection of commercial facilities, such as oil platforms and fisheries. And it is particularly challenging to protect the vast near-coast sea surface and busy harbor areas from intrusions of unauthorized marine vessels, such as trespassing boats and ships. The traditional methods of detecting ships are with radars or satellites which are very expensive. Except the high cost, the satellite image is easily affected by the cloud. And it is difficult to detect small boats or ships on the sea with marine radar due to the noise or clutters generated by the uneven sea surface. Wireless sensor networks are developed for terrestrial intrusion detection recently. These networks deploy sensors, such as magnetometers, thermal sensors, and acoustic sensors, in the monitored area to detect intruders. By deploying sensor networks on the sea in line, wireless sensor networks can protect harbor security with intrusion detection.

However, though such networks may work well on the land, it is challenging to deploy these sensors on the sea surface for ship detection. The main challenge is that when sensors are deployed on the sea surface, they are not static and tossed by ocean waves which make the sensors move around randomly [8]. These movements make most sensors, (e.g., magnetometers and thermal sensors), difficult to detect the intrusion. Due to the high cost and stability requirement, using the camera sensors is not a general solution in such scenarios. For the same reason mentioned earlier, the detection results can also be easily affected by the random movement of the cameras.

The authors in paper [4] present an innovative solution for ship intrusion detection by deploying the wireless sensor networks with three-axis accelerometer sensors on the sea surface to detect ship-generated waves. The scheme requires a relatively dense network, and combining with other sensors, such as acoustic sensors, can decrease the deployment density. Thus, we can use hybrid networks to protect harbor, just as Fig. 32.4 shows, the hybrid networks can be deployed to detect boat intrusion.

32.3 MLA: Moored Node Localization Algorithm Design

In this section, we present our localization scheme MLA for the moored node in the hybrid sensor networks. We assume that some double-head nodes equipped with GPS equipment, thus when they deployed in the ocean, they know their instant positions. And those nodes serve as seeds for localization of the hybrid networks. We also assume that the moored node underwater equipped with cheap pressure sensors, and the depth of node can be directly determined [10], thus knows its deployed depth h in the ocean.

As shown in Fig. 32.5, a double-head node deployed on the surface and the moored node linked with mooring rope with it form a big node in the networks. We assume that the mooring rope length is l and $l > h$. By node's physical MAC address binding, in each big node, the moored node and its surface underwater node know each other and can communicate with each other. Meanwhile, they can communicate with other underwater sensor nodes in their neighborhood.

Next, we first describe localization scheme for moored node underwater.

As shown in Fig. 32.5, equipped with GPS on the sea surface, node A can communicate with its moored node M. We assume that node M's position is (x_m, y_m). When node M receives the node A's instant position (x, y), it is easy to calculate the r by Right-angled triangle theorem:

$$r = \sqrt{l^2 - h^2} \tag{32.1}$$

where l is the length of the mooring rope, h is the depth of the moored node. Thus, the position of node M is on the circle centered with radius r.

As shown in Fig. 32.6, because of the rise and the fall of the tide regularly, the actual drifting area of the floating double-head node is a spherical crown. Tides are

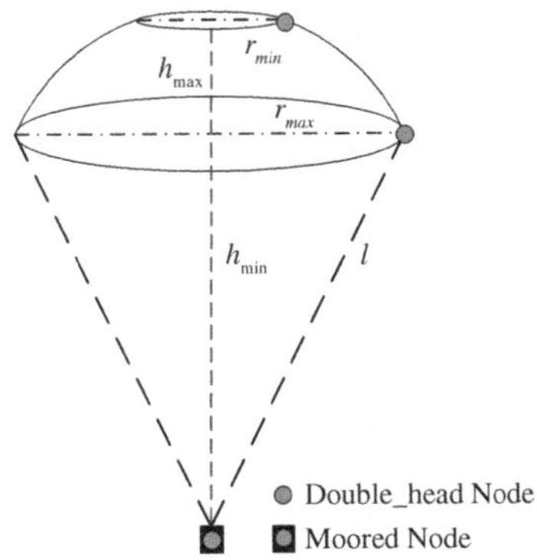

Fig. 32.6 The floating model of the double-head node

the rise and the fall of sea levels caused by the combined effects of the gravitational forces exerted by the Moon and the Sun and the rotation of the Earth. The high tides and low tides may happen one or two times each day and the times and amplitude of the tides at a locale are influenced by the alignment of the Sun and Moon, by the pattern of tides in the deep ocean and other factors. The fall between the high tide and the low tide may be as high as 10 m. As shown in Fig. 32.6, $h_{\max}$ presents the highest sea depth due to the highest tide level which leads to the minimum of radius of free drifting area of RFS, denoted as $r_{\min}$. Thus, at different time with different tide level, results with different drifting radius r. Using those positions of node A at different time reported to moored node M, node M can calculate its own position.

Let's suppose at different time points t_i (where $i = 1, 2, 3$), the positions of node A are (x_i, y_i), we can obtain:

$$\begin{cases} (x_m - x_1)^2 + (y_m - y_1)^2 = r_1^2 \\ (x_m - x_2)^2 + (y_m - y_2)^2 = r_2^2 \\ (x_m - x_3)^2 + (y_m - y_3)^2 = r_3^2 \end{cases} \tag{32.2}$$

where r_i is with the different tide which leads to different sea depth at different time.

Thus, when the moored node receives three positions with different h and different (x_i, y_i), the moored node can calculate its position (x_m, y_m) using Eq. (32.2). Once the moored node localized itself with its sea surface double-head node, these moored nodes can become seed nodes. With those seed nodes, the other moored node can localize its position with those formed underwater moored networks.

32.4 Related Work

Underwater acoustic sensor network localization can be broadly classified into two categories: range-based schemes and range-free schemes [11, 12]. Range-based schemes first measure or estimate distances or angles among nodes using Time of arrival (TOA), Time difference of arrival (TDOA), Angle of arrival (AOA), or Received signal strength indicator (RSSI) [13]. Then the schemes apply triangulation or multilateration to compute node's positions. Range-free schemes do not use range measurement to estimate distances between nodes, they use network topology or the position of surrounding anchor nodes to locate nodes, which can be generally classified into hop count-based and area-based schemes. The advantages of range-free schemes lie in their simplicity and the low communication costs.

Some papers deal with hybrid network localization. In paper [14], DNR was proposed with beacons sinking and rising in the water, the nodes can localize their position by passively listening to those beacons. DNR assumes that nodes are synchronized which increase the communication costs. In this paper, we proposed a localization scheme MLA for a hybrid ocean wireless sensor networks. The double-head node deployed on the sea surface sending different positions to its moored node underwater. Using these beacons, the moored node can localize itself.

32.5 Conclusions and Future Research

In this paper, we proposed a localization scheme MLA for our proposed hybrid ocean wireless sensor networks which combines the advantages of wireless sensor networks and underwater acoustic sensor networks. In MLA, the double-head node deployed on the sea surface sending different positions to its moored node underwater for the moored node to localize itself. In future, we will test the double-head node and to deploy the hybrid ocean networks to verify the localization scheme. We will also test the hybrid networks for ship intrusion with both sensors on the sea surface and acoustic sensor underwater.

Acknowledgments The research was supported by the Project of Shandong Province Higher Education Science and Technology Program No. J11LG88. We thank Professor Zhongwen Guo in Ocean University of China for discussions concerning this work, and the anonymous reviewers for their detailed comments.

References

1. Wang Y, Liu Y, Guo Z (2012) Three-dimensional ocean sensor networks: a survey. J Ocean Univ China 11(4):436–450
2. Heidemann J, Stojanovic M, Zorzi M (2012) Underwater sensor networks: applications, advances and challenges. Philos Trans R Soc A Math Phys Eng Sci 370(1958):158–175
3. Gao AG, Luo XL, Luo HJ (2013) A proposed hybrid wireless sensor networks deployed on the sea. Appl Mech Mater 347:990–994
4. Luo H, Wu K, Guo Z, Gu L, Ni LM (2012) Ship detection with wireless sensor networks. IEEE Trans Parallel Distrib Syst 23(7):1336–1343
5. Pompili D, Melodia T, Akyildiz I (2009) Three-dimensional and two-dimensional deployment analysis for underwater acoustic sensor networks. Ad Hoc Netw 7(4):778–790

6. Cui J, Kong J, Gerla M, Zhou S (2006) The challenges of building scalable mobile underwater wireless sensor networks for aquatic applications. IEEE Netw 20(3):12
7. Xie P, Cui J, Lao L (2006) VBF: vector-based forwarding protocol for underwater sensor networks. Lect Notes Comput Sci 3976:1216
8. Yang Z, Li M, Liu Y (2007) Sea depth measurement with restricted floating sensors. In: 28th IEEE international real-time systems symposium, RTSS 2007, IEEE, pp 469–478
9. Whoi: micro-modem overview. http://acomms.whio.edu/micromodem/
10. Akyildiz I, Pompili D, Melodia T (2004) Challenges for efficient communication in underwater acoustic sensor networks. ACM SIGBED Rev 1(2):3–8
11. Cheng W, Teymorian A, Ma L, Cheng X, Lu X, Lu Z (2008) Underwater localization in sparse 3d acoustic sensor networks. In: Proceedings of the 27th conference on computer communications, INFOCOM 2008, IEEE
12. Zhou Z, Cui J, Zhou S (2007) Localization for large-scale underwater sensor networks. Lect Notes Comput Sci 4479:108
13. Ko J, Shin J, Kwon S, Kim C (2008) Localization of sensor nodes in underwater acoustic sensor networks using two reference points. In: International conference on information networking, ICOIN, pp 1–5
14. Erol M, Vieira L, Gerla M (2007) Localization with dive'n'rise (dnr) beacons for underwater acoustic sensor networks. In: Proceedings of the 2nd ACM international workshop on underwater, networks, UUWNet'07, pp 97–100

Chapter 33
Graph Coloring-Based Clock Synchronization for Clustering Underwater Sensor Networks

Ying Guo

Abstract A graph coloring-based clock synchronization algorithm is proposed for clustering underwater sensor networks. It divides the nodes in deployment area into clusters, adjacent clusters that adapt different frequencies based on graph coloring principle to avoid interference. This method improves the efficiency of pairwise-based underwater clock synchronization algorithm. The simulation results show its feasibility.

Keywords Underwater sensor network · Clock synchronization · Graph coloring · Clustering

33.1 Introduction

Underwater Wireless Sensor Networks (UWSN) become more and more important in the applications of ocean monitoring, resource survey, and so on [1]. Clock synchronization is an essential service for UWSN [2]. Different from terrestrial sensor networks [3], underwater wireless sensor networks have limited bandwidth, limited transmission rate, high bit error rate, and long propagation delay [4]. Thus, terrestrial clock synchronization protocols cannot be used in underwater environment.

In the past few years, many clock synchronization algorithms for underwater wireless sensor network have been proposed. For example, Mobi-Sync [5], D-Sync [6],

This work was supported by the National Natural Science Foundation of China under Grant No. 61103196. Research Award Fund for Excellent Young Scientist of Shandong Province under Grant No. BS2012DX011. Qingdao Science and Technology Development Program under Grant No. 12-1-4-3-(16)-jch.

Y. Guo (✉)
Qingdao University of Science and Technology, Qingdao 266061, China
e-mail: guoying@qust.edu.cn

X. Wang et al. (eds.), *Advanced Technologies in Ad Hoc and Sensor Networks*, Lecture Notes in Electrical Engineering 295, DOI: 10.1007/978-3-642-54174-2_33,

TSHL [7], MU-Sync [8], Cloll-Sync [9], and so on. Some of them based on pairwise synchronization, such as Coll-Sync, which is a recently proposed collision- based clock synchronization scheme for mobile underwater sensor networks. It utilizes long propagation to detect collision in acoustic channel, and eliminates the bad effect of node mobility. However, it does not have efficiency in large-scale underwater sensor networks. In this paper, we solve this problem and propose a graph coloring-based clock synchronization algorithm for clustering underwater sensor networks.

33.2 Synchronization in Cluster

Nodes in deployment area are divided into clusters. Each cluster has a cluster head which has standard time. The cluster head plays the part of beacon node in the process of clock synchronization.

Every node has three states, which are synchronization state, silence state, and communication state. In synchronous state, the node operates synchronization algorithm to synchronize its clock to the beacon node. In silent state, the node does not send or receive any information. In communication state, the node communicates with other nodes as normal. The priority of these three states are: synchronization state > silence state > communication state.

In the pairwise synchronization process, the beacon node synchronizes other nodes according to their node ID successively. In order to avoid interference, nodes surrounding the cluster head and the node to be synchronized keep silent.

Beacon node determines the node to be synchronized and broadcasts its node ID. Other node decides its state by received information. If it is the node to be synchronized, it turns to synchronization state. If it is not the node to be synchronized, it turns to silence state. Then, the node to be synchronized transmits a synchronization request. All the nodes that receive this information turn to silence state.

As shown in Fig. 33.1, beacon node that intends to synchronize node 6 sends a synchronization request and contains the ID of node 6. Node 6 receives the information and turns to synchronous state as synchronous state has the highest priority. Nodes 5, 8, and 9 also receive the information; their ID is different from the ID broadcasted by beacon node, thus they turn to silence state. Then node 6 broadcasts synchronization request, nodes 3, 4, and 5 receive this information and turn to silence state. Other nodes that do not receive any information are still in communication state.

This method could avoid interference in the cluster, because the cluster only has one beacon node to synchronize another node. Underwater nodes are deployed sparsely and can only communicate with one hop neighbor that effectively overcomes the problem of hidden terminal. The algorithm of clock synchronization in cluster is shown as follows:

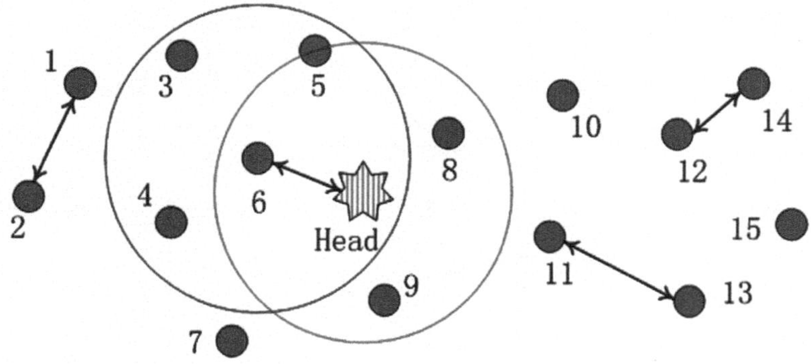

Fig. 33.1 Clock synchronization in cluster

```
if (Receive=sync ID)
   if (ID)
     Goto synchronization state;
   else
     Goto silence state;
   endif
endif
if (Receive=sync)
   Goto silence  state;
endif
```

33.3 Frequencies Distribution

Adjacent clusters adapt different frequencies to avoid interference. In order to make the number of frequencies as little as possible, we make use of graph coloring principle to distribute the frequencies.

Graph coloring is a special case in graph theory. Its basic idea is to assign colors to either edges or vertices of the graph subject to certain constraints. It contains edge coloring and vertex coloring, and both cases require the number of colors used to color the graph to be minimized. Edge coloring assigns colors to the graph edges, such that two edges incident on the same vertex receive different colors [10]. Vertex coloring assigns colors to nodes, such that two adjacent vertices have different colors [11].

We choose vertex coloring in our design because vertex coloring enables broadcast communications, which is needed to discover neighborhoods of a node and advertise synchronization information. Thus the graph coloring problem in our design deals with minimizing the number of colors to ensure that two adjacent clusters are colored different as different colors stand for different frequencies.

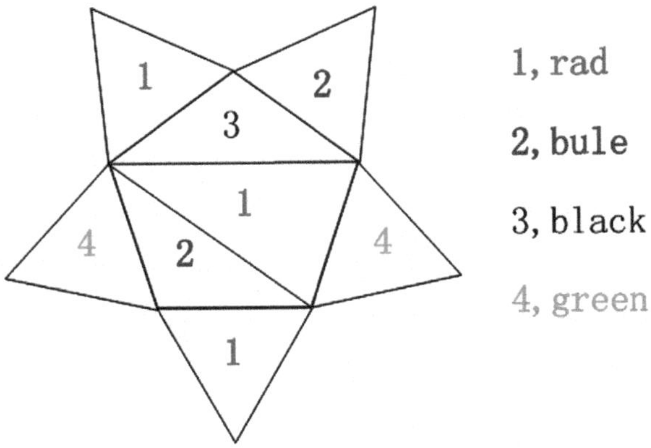

Fig. 33.2 Frequencies distribution

As shown in Fig. 33.2, adjacent clusters have different colors. According to the four-color principle, two-dimensional graph use at least four colors, that is to say four frequencies are allocated.

Many graph coloring mechanisms have been proposed. Backtracking method is one of the most used algorithms. The code of backtracking algorithm is shown as follows:

```
int n
int c[ ][ ]
int m            //All array subscripts start at 1.
for (i=1; i<=n; i++ ) //Initializing color[n] to 0.
    color[i]=0;
    k=1;
    while (k>=1)
       color[k]=color[k]+1;
       while (color[k]<=m)
       endwhille
       if Ok(k) break;
       else color[k]=color[k]+1;  //Search next color.
       endif
       if (color[k]<=m && k= =n)   //Output solution.
          for (i=1; i<=n; i++)
             cout<<color[i];
          endfor
          return;
       else if (color[k]<=m && k<n)
          k=k+1;  //Process next vertex.
       else
          color[k]=0;
          k=k-1;  //Backtracking.
      endif
   endwhile
endfor
```

Fig. 33.3 Error after synchronization

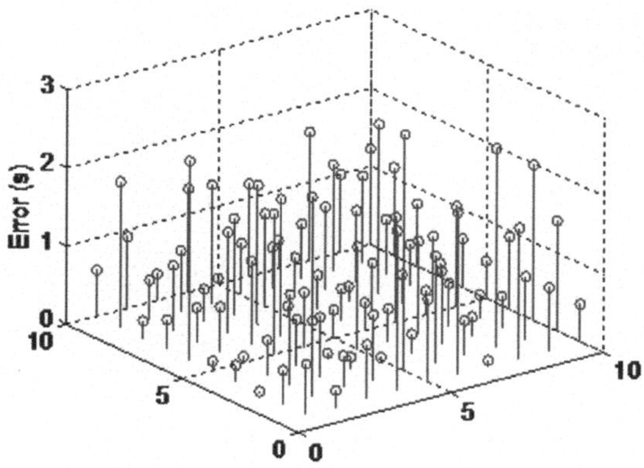

The algorithm used to decide whether the color of vertex K is conflict is as follows:

```
for (i=1; i<k; i++)
    if (c[k][i]= =1 && color[i]==color[k])
      return false;
    endif
  return true;
 endfor
```

33.4 Simulation Result

In this section, we make use our design to Cloll-Sync clock synchronization algorithm [9], and compare it with MU-Sync [8]. In the simulation, nodes to be synchronized have different internal clocks, and move with ocean current. The time interval of beacon node sends synchronization message packet is 1 s. The clock granularity is 1 μs, and the jitter at the receiving side is 15 μs. MU-Sync uses 25 reference information packets for linear regression computation. The timestamps used in the simulation come from MAC layer.

First, we discuss the performance of algorithm in cluster. We uniformly deployed 100 sensor nodes in accordance with 10*10 in a two-dimensional region, and the distance between two nodes is 1000 m. The beacon node is located in the corner of the deployment area. Beacon node synchronize other nodes according to node ID successively. Figure 33.3 shows the error of synchronization completed after 10^5 s. The average error of this method is 0.85 s, and the standard deviation 0.63 s.

Next, we research on the performance of frequencies distribution algorithm. We deployed 512 sensor nodes according to 8*8*8 in a three-dimensional region. Every cluster has 8 nodes, which includes a cluster head with standard time. Figure 33.4 is the result of frequency distribution based on graph coloring principle. There are 64 clusters and 8 frequencies are assigned.

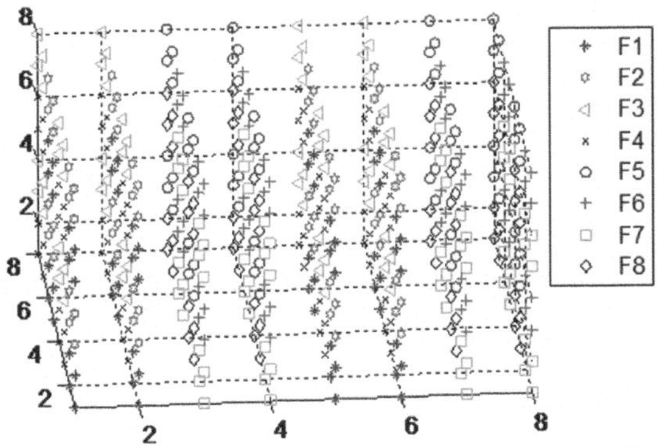

Fig. 33.4 Frequency distribution based on graph coloring principle

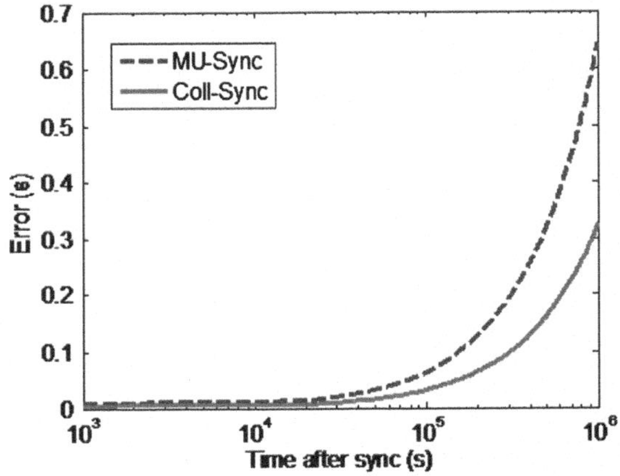

Fig. 33.5 Time deviation after synchronization

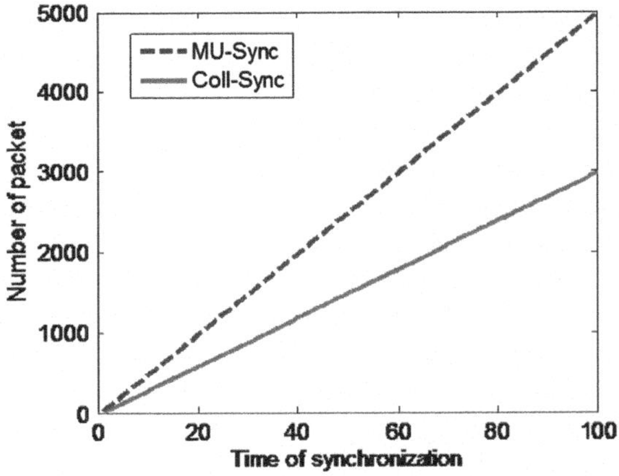

Fig. 33.6 Packet number

Figure 33.5 shows the local time deviation for a period after clock synchronization. As can be seen, the error of Coll-Sync less than MU-Sync. MU-Sync's local time deviation is large because it does not consider the propagation delay in the first time of linear regression estimation process.

Figure 33.6 shows the packet numbers of MU-Sync and Coll-Sync for 100 times of synchronization. Coll-Sync can achieve higher precision without linear regression calculation, thus the required number of packet is 74.9 % of MU-Sync.

The simulation results show that with the help of our design Coll-Sync can be implemented effectively with high precision and low energy consumption.

33.5 Conclusion

Pairwise clock synchronization algorithm, such as Coll-Sync, could not work efficiently in large-scale underwater sensor networks. We propose a coloring-based clock synchronization algorithm for clustering underwater sensor networks to solve this problem. It greatly expands the application scope of pairwise clock synchronization algorithm.

References

1. Yifan W, Zaichen Z, Guangguo B (2010) An overview on underwater acoustic sensor networks. In: 17th international conference on telecommunications, pp 779–783
2. Zhou Z, Cui J-H, Bagtzoglou A (2008) Scalable localization with mobility prediction for underwater sensor networks. In: IEEE INFOCOM'08, Mini-Conference, Phoenix, Arizona, USA
3. Wu YC, Chaudhari Q, Serpedin E (2011) Clock synchronization of wireless sensor networks. IEEE Signal Process Mag 8(1):124–138
4. Zhou Z, Peng Z, Cui J-H, Shi Z (2010) Efficient multipath communication for time-critical applications in underwater acoustic sensor networks. IEEE/ACM Trans Netw 19:28–41
5. Liu J, Zhou Z, Peng Z, Cui J-H (2010) Mobi-Sync: efficient time synchronization for mobile underwater sensor networks. IEEE GLOBECOM 2010(March):1–5
6. Lu F, Diba M, Curt S (2010) D-Sync: doppler-based time synchronization for mobile underwater sensor networks. In: Proceedings of the 5th ACM international workshop on underwater networks (WUWNET), Woods Hole
7. Syed AA, Heidemann J (2006) Time synchronization for high latency acoustic networks. In: 25th IEEE international conference on computer communications (INFOCOM), pp 1–12
8. Chirdchoo N, Soh WS, Chua KC (2008) MU-Sync: a time synchronization protocol for underwater mobile, networks, WUWNet'08, pp 35–42
9. Guo Y, Liu Y (2012) Collision based time synchronization for underwater sensor networks. Appl Mech Mater J 246–247:806–810
10. Gandham S, Dawande M, Prakash R (2005) Link scheduling in sensor networks: distributed edge colouring revisited. In: INFOCOM 2005, Miami, FL, USA
11. Hansen J, Kubale M, Kuszner L, Nadolski A (2004) Distributed largest-first algorithm for graph coloring. In: EURO-PAR 2004, Pisa, Italy

Chapter 34
Paying the Way of FPV-ROV with a Function of GSM-GPS Reverse Positioning

Mengnan Sun and Bing Zheng

Abstract ROV (Remote Operated Vehicle) is typically operated by a joystick and a LCD display screen on the water, and the information from the screen is just the underwater image and a few of bald date, voltage, and depth character information in a way of OSD (On-Screen Display). In this paper, a new type of ROV design is presented, the FPV (First Person View) idea in the field of aerial photo was applied to the traditional ROV display system. In one screen, more information like speed, depth, battery, head angle, pitch angle, roll angle, GPS (if on the water), etc., from the ROV were added in. Additionally, a GSM-GPS reverse positioning function was implemented. It is demonstrated that the operator can get an intuitionistic multidimensional information and visual experience under water. And through GSM-GPS, the GPS location of ROV can be sent to operator's phone at an emergency.

Keywords ROV · OSD · FPV · Positioning

34.1 Introduction

Marine scientists or ocean amateurs are fund to use underwater vehicle to explore ocean environment. The world's major maritime country has developed various ROV models, Japan and the United States take the leading position [1–3]. ROV is widely used because it possesses excellent performance stability and flexibility of operation. For most ROV drivers, the major interactive interface with underwater vehicle is the LCD screen, which displays the real-time image and some simple and bald date, voltage, depth character information in a way of OSD (On-Screen Display), like Fig. 34.1. What we have known is that the ROV is just regarded as a vehicle working under water, what we pay attention to is the image and data collection from the

M. Sun · B. Zheng (✉)
Department of Electric Engineering, Ocean University of China, Qingdao 266100, China
e-mail: bingzh@ouc.edu.cn

X. Wang et al. (eds.), *Advanced Technologies in Ad Hoc and Sensor Networks*, Lecture Notes in Electrical Engineering 295, DOI: 10.1007/978-3-642-54174-2_34,

Fig. 34.1 Universal ROV display scene with some simple and bald information (get from the web https://www.insecure.ws/2010/01/07/fpv-starting-blocks/)

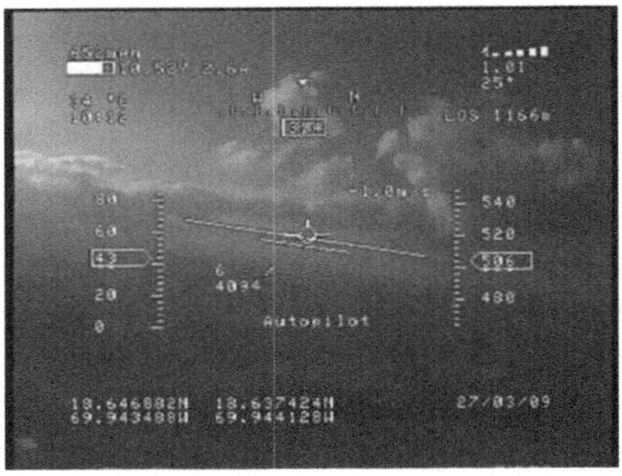

Fig. 34.2 Aerial photo FPV display with multidimensional information (get from the Web http://www.stinger.no/cases)

sensor that mounted on the ROV. There may be several sensors on ROV, and usually manufacturer also provides upper computer software to display sensor data, but every software window is independent. It is not enough for an ROV operator, because what he wants is the whole real-time image of underwater sensor information integration.

But in another field, Aerial Photo, through FPV technology, we can find some dramatic, useful application, and reference [4, 5]. As shown in Fig. 34.2, in an aerial photo image with the FPV idea, much more sensor data is overlaid on one screen. Contrasted with the traditional display screen in Fig. 34.1, aerial photo image more real-time information—speed, altitude, satellite number, battery power, voltage, current, head angle, pitch angle, roll angle, time, etc. In this case, operator do not need to pay much attention to the dashboard, just focus on the screen to obtain a more comfortable experience and can easily get any parameter from the only screen but not to be distracted to other place.

In this paper, a FPV-ROV with a function of GSM-GPS reverse positioning is implemented. We complete the basic design of ROV, and apply the FPV idea in the field of aerial photo to the image display of underwater vehicle. Additionally, with the use of GSM-GPS model, the controllability of the ROV achieves an excellent improvement at an emergency.

34.2 System Framework Description

34.2.1 ROV System Overview

Figure 34.3 shows the architecture of our ROV. The whole system has an ability of 420 kg displacement and 85 kg payload. It includes two parts in machine partition. The upper portion is the cavity, which is used to seal some electronic control unit. The lower portion is the load carrier. Two underwater cameras and four high brightness light sources is equipped downward to obtain the image of seafloor. In addition, two forward-looking cameras and two upward-looking cameras are fixed for the need of navigation. A 30 mW 532 nm line laser and a dual frequency digital chirp sonar is assembled for other imaging application. In terms of the dynamical system, three DC brushless thrusters of tecnadyne model 560 are selected as the control of vertical direction, and two model 1060 as the control of horizontal direction.

34.2.2 OSD Display System

Figure 34.4 shows a framework of the OSD system, which can provide the FPV display effect for the upper operator. It is divided into three levels. The bottom level consists of underwater sensor networks. In our ROV system, the unmature sensor, electronic compass, attitude sensor, GPS model, current sensor, altimeter are chosen, the information provided from all of these sensors will be presented on the same screen with the underwater camera image. The top level consists of a display screen, an underwater camera array, and an OSD chip responsible for the integration of the camera image and the sensor networks data. Both of these two levels are connected by a MCU (Microprogrammed Control Unit) chip, as a commander of the whole communication link.

34.2.3 GSM-GPS Reverse Positioning

When used in GPS mode, because of the intense attenuation of electromagnetic wave in the water, GPS signal can only be obtained above the water surface. So, in this case ROV is usually at a state of on-water, either because of the special application or at an emergency.

The GSM model is a communication model that can get accessed to the world GSM telephone communication networks. In this part, a GSM model is added in to help utilize the information of GPS data. As shown in Fig. 34.5, it demonstrates the principle of GSM-GPS reverse positioning. When needed in the situation stated above, the operator can use his phone to send some simple instruction messages to the GSM terminal on ROV through the GSM networks. After ROV receives parses the instruction message, it will send its location information from the GPS back to the operator, which is called “reverse positioning.”

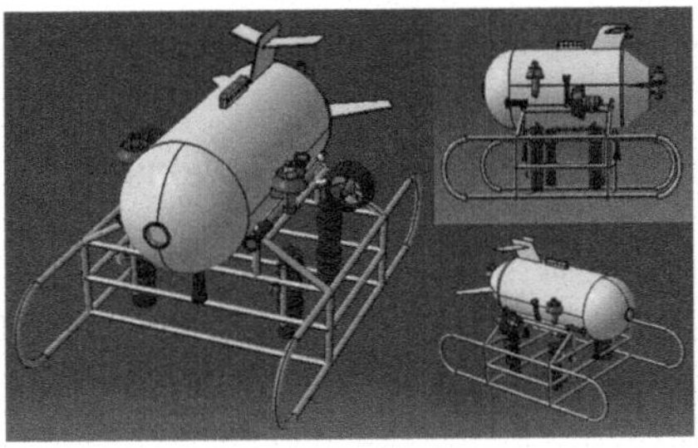

Fig. 34.3 The novel ROV system framework

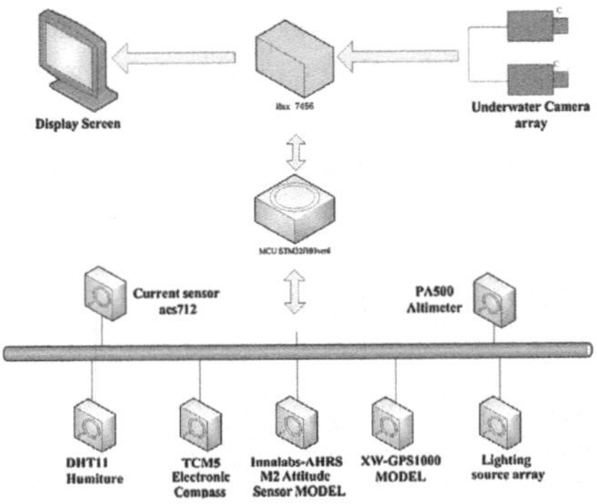

Fig. 34.4 The framework of the OSD system with multisensor networks

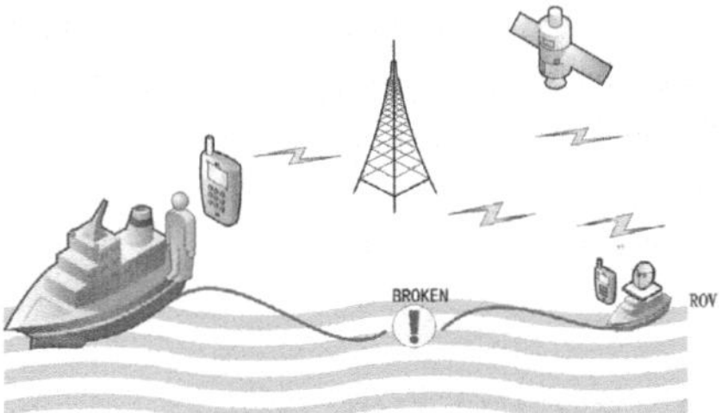

Fig. 34.5 The principle of GSM-GPS reverse positioning

34.3 Implementation

34.3.1 OSD Character Superimposition Design

Hardware design A MCU stm32f103vct6 of STMicroelectronics was used as the core controller. It has abundant communication interfaces, including 5 USARTs, 2 SPIs, 1 USB interfaces. Most of our chosen sensor can be connected with the USART and SPI interface protocol. Table 34.1 presents the lectotype of every sensor.

Display As the only interaction access to the underwater environment, the display screen plays an important role in terms of user experience. We are trying our best to

Table 34.1 Sensors Information Selected

	Instrument	Model	Interface
Top Level	LCD screen	Lenovo L2022wA	VGA
	Underwater camera	UWC-300	PAL
	OSD	Chip Max7456	SPI
Middle Level	MCU	STM32F103VCT6	/
Bottom Level	Humiture	DHT11	1-wire bus
	Electronic Compass	TCM5	RS232
	Attitude Sensor	Innalabs-AHRS M2	RS232
	GPS	XW-GPS1000	RS232
	Lighting source array	non-standard	digital
	Altimeter	PA200	RS232
	Current sensor	acs712	AD

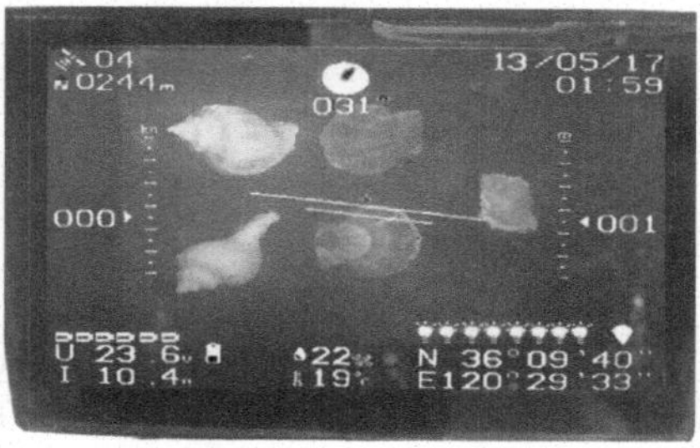

Fig. 34.6 The final FPV display scene in ROV system

transform the dull sensor data into a lively and visual graph, easy to understand and control.

The speed, depth, battery level, head angle, pitch angle, roll angle, GPS, voltage, current, temperature, humidity, camera state, light source state were displayed. They were separated into two parts: static characters and dynamic information. As shown in Fig. 34.6, the final display effect of this FPV system gets much improvement.

34.3.2 GSM-GPS Reverse Positioning Design

As shown in Fig. 34.7, a SIM300 model is used as the GSM communication agent. It works on three frequency bands—EGSM 900 MHz, DCS 1800 MHz, PCS 1900 MHz throughout the world. Up to 10 GPRS channels can be provided to communicate. In this system, its SMS message function were used. The "AT instructions" are sent or received by our master control MCU STM32 through the RS232 communication path. Additionally, we selected a GPS model XW-GPS1000 of Beijing Starneto Technology Co. Ltd. The GPS location information of ROV is obtained through parsing the data stream in NMEA form.

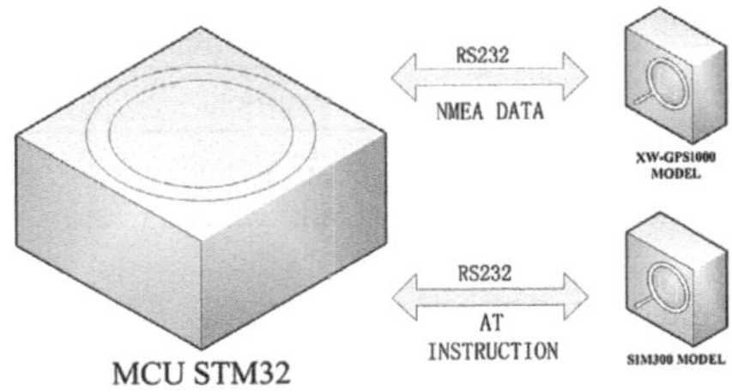

Fig. 34.7 The structure of GSM-GPS positioning system

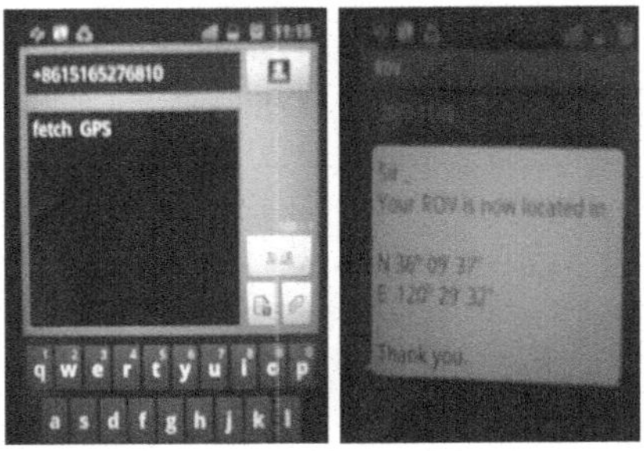

Fig. 34.8 The sent instruction and received GPS information from the operator's phone

Through the simulation scene of the rov's emergency, a "fetch GPS" SMS instruction was sent to the SIM model on ROV through the GSM networks, a SMS message containing the location information of the offline ROV was then received after several seconds. Fig. 34.8 presents the result of this testing.

34.4 Summary and Future Development

The FPV (First Person View) idea in the field of aerial photo was successfully applied to our new type of ROV design. Through testing, it showed that the OSD system and the GSM-GPS system worked well. The sensor networks information was converted to some vivid graph. In one screen, more information from the ROV like speed, depth, battery, head angle, pitch angle, roll angle, GPS, etc., were added. A GSM-GPS reverse positioning increased the capacity of control at an emergency. It is demonstrated that operator can get an intuitionistic multidimensional information and visual experience under water.

In future work, the light source array on this ROV will be calculated and designed to build an underwater inhomogeneous optic field [6], in which field our operator can get a clearer image under water.

References

1. Junku Y (2000) Design and control of autonomous underwater robots: a survey. Auton Robots 8(1):7–24
2. Bowen, Andrew D et al (2004) Exploring the deepest depths: preliminary design of a novel light-tethered hybrid ROV for global science in extreme environments. Mar Technol Soc J 38.2:92–101
3. Salgado-Jimenez T et al (2010) Deep water ROV design for the Mexican oil industry. OCEANS 2010 IEEE-Sydney. University of Sydney, Sydney
4. Mizosoe, Hiroki et al (2001) An advanced multimedia processing LSI suitable for HDTV applications. IEEE Trans Consum Electron 47.3:420–425
5. Senthil Kumar K, Ramesh G, Srinivasan KV (2011) First pilot view (FPV) flying UAV test bed for acoustic and image data generation
6. Hongmei XU (2009) Research of underwater imaging detection technology in inhomogeneous illumination Field [D]. Ocean University of China, Qingdao

Chapter 35
Underwater Image Feature Matching Based on Inhomogeneous Illumination

Congcong Xu and Bing Zheng

Abstract In this paper, the problem of underwater image matching was concerned. The accuracy of underwater image feature matching is significant for underwater vision research. Due to the attributes including absorption and scattering of water, three problems should be accomplished: image acquisition, image processing, and image matching. In response to the first problem, an inhomogeneous illumination method is developed to acquire clear underwater-image. Then contrast limited adaptive histogram equalization was used to process the original image. Finally, an effective algorithm based on SIFT features is used to achieve precise image matching.

Keywords Inhomogeneous illumination · CLAHE · Feature extraction · Image matching

35.1 Introduction

The precision of underwater image matching is subject to many factors such as the quality of water, the type of receivers when acquiring images, the effects of backscatter noise, and so on. Therefore, image acquisition, image processing, and image matching in turbid water become three research hotspots.

For the first question what's the crucial part is overcoming the backscattering. Laser scanning, structured light imaging, and polarized imaging, etc., solutions attract widespread attention. But these methods are ineffective to dispose the multiway scattering during the light transmission. Distinguish previous researches, an inhomogeneous illumination pattern which can decrease the strong backscattering

C. Xu · B. Zheng (✉)
Department of Electric Engineering, Ocean University of China, Qingdao 266100, China
e-mail: bingzh@ouc.edu.cn

X. Wang et al. (eds.), *Advanced Technologies in Ad Hoc and Sensor Networks*,
Lecture Notes in Electrical Engineering 295, DOI: 10.1007/978-3-642-54174-2_35,

for underwater imaging are proposed. Within this system, low backscatter noise and nonblind area can be realized when detecting target with wide FOV [1].

In the aspect of image processing, common techniques comprise image enhancement in the spatial and frequency domains. Simulating the PSF of water and using Wiener filter to restore the underwater images is also a prevailing approach. Taking the real-time and concision of the algorithm into consideration, Contrast Limited Adaptive Histogram Equalization (CLAHE) is selected to process image. From the results of experiments, this approach is effective and reliable with high resolution and definition.

The development of image matching by using a local interest points can be traced back to the work of Moravec on stereo matching using a corner detector, and the ground-breaking work in this field is invariant local feature matching arose by Schmid and Mohr (1997) [2]. This paper describes an approach of image matching using scale-invariant features which is suitable for matching different images proposed by Lowe (2004) [3]. This approach proposing a more effective measure to underwater image matching processes high robust and veracity.

35.2 Image Acquisition: Inhomogeneous Illumination Method

The attenuation of light arises from absorption which removing light energy and scattering which changes the light path's detection [4]. The key to acquire clear underwater image is eliminating backscatter effect as far as possible.

The inhomogeneous illumination system where the power density is inversely proportional to the light attenuation in water is put forward to resolve the problem. Monte Carlo method is used to solve the radiative transfer equation. In inhomogeneous illumination field, the power of light concentrates on the distributive light of central power, so the backscattering which will be attenuated during the transmission at long distance become too week to detect, meanwhile the scattering is feeble at short distance because of the low density.

The implementation of inhomogeneous illumination system is shown in Fig. 35.1.

According to the theory of power, density is inversely proportional to the light attenuation, the received power we obtain by the receiver under the inhomogeneous illumination is:

$$E(X, Y, Z) = C \cdot \frac{\gamma(X, Y, Z)^2}{D} \cdot e^{\alpha 2 \cdot \gamma(x,y,z)}$$

Where C is the constant parameter, $\gamma(X, Y, Z)$ denote the distance between the target and receiver, D is the diameter of the receiver, α is the attenuation parameter.

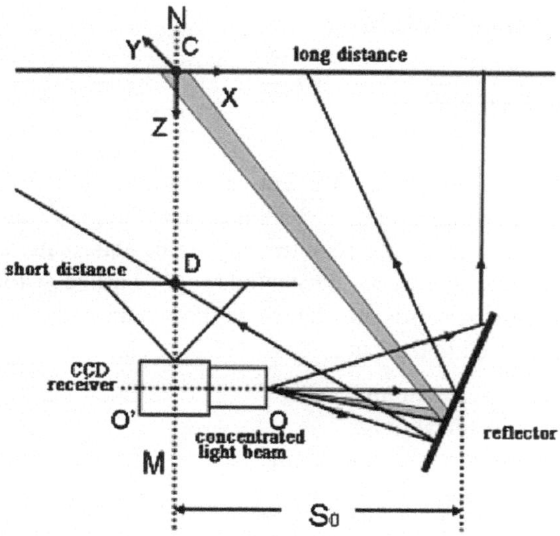

Fig. 35.1 The concentrated beam illuminate the space and be reflected by the reflector in underwater. The images are collected by the CCD locating the optical axis of the illumination source with a S_0 distance to the reflector

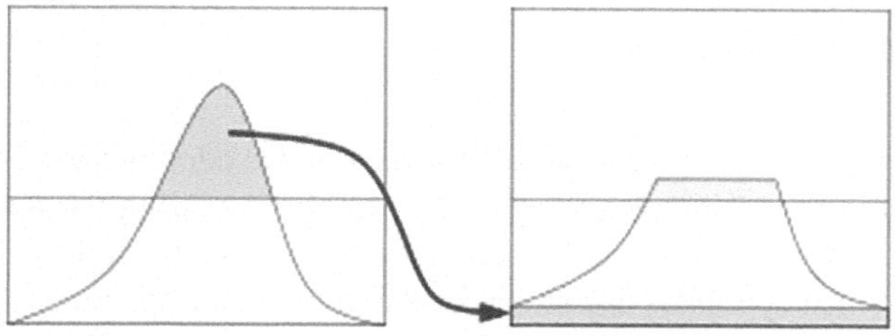

Fig. 35.2 The *blue* part is the clipped pitch which will be distributed to other histogram

35.3 Image Processing: Contrast Limited Adaptive Histogram Equalization

Contrast limited adaptive histogram equalization (CLAHE) is superior to the adapt histogram equalization (AHE) and histogram equalization in comparison. The distinction between CLAHE and (AHE) is the limitation for the contrast.

In CLAHE, every pitch is processed by the histogram equalization which the contrast is limited. After the equalizing, bilinear interpolation algorithm is adopted to combine adjacent domains in order to remove the boundary, which aroused because of the clipping of histogram equalization [5]. Generally, the clipped parts are distributed uniformly to other histogram equalization. The thought is shown in Fig. 35.2.

On the other hand, the other advantage of CLAHE is that its effective in overcoming the noise magnified in processing.

35.4 Image Feature Matching

35.4.1 Feature Extraction

As described in the introduction, an effective algorithm based on SIFT features is used to fulfill more reliable underwater image matching. According to David's research, the SIFT features are invariant to image scale and rotation, and are shown to provide robust matching across a substantial range of affine distortion, change in 3D viewpoint, addition of noise, and change on illumination [3].

Four steps are realized in this part:

- Detection of scale-space extrema
- Accurate keypoint localization
- Orientation assignment for each keypoint
- Generating the local image descriptor.

A 128-dimension is generated after accomplishing the above steps. According to the characteristics of the features, it is suitable for underwater image feature extraction.

35.4.2 Feature Matching

After generating the 128-dimension feature vectors, we adopt an measure to obtain the better candidates of matching keypoints, which comparing the distance of the closest neighbor to that of the second-closest neighbor. If this ratio is less than the set threshold r, the keypoint is discarded. In David's paper, r adopts 0.8. For the 128-dimensional vector, Best-Bin- First (BBF) algorithm is used to exhaustive search to identify the exact nearest neighbors of points in above step. Furthermore, Hough transform has much better performance on identifying correct subsets of matches with a consistent interpretation [3].

35.5 Experiment

35.5.1 Experimental Facilities and Image Capture

The experiment facilities are shown in Fig. 35.3.

The images we capture in same distance under normal illumination source and inhomogeneous illumination are shown in Fig. 35.4

According to the results of experiments, apparently, the image acquired under inhomogeneous illumination is more distinct than that of the normal illumination. It can overcome the backscattering effectively.

Fig. 35.3 The experimental target getting from clear water is in the picture a. In the *right* one, it is the inhomogeneous illumination underwater detection system

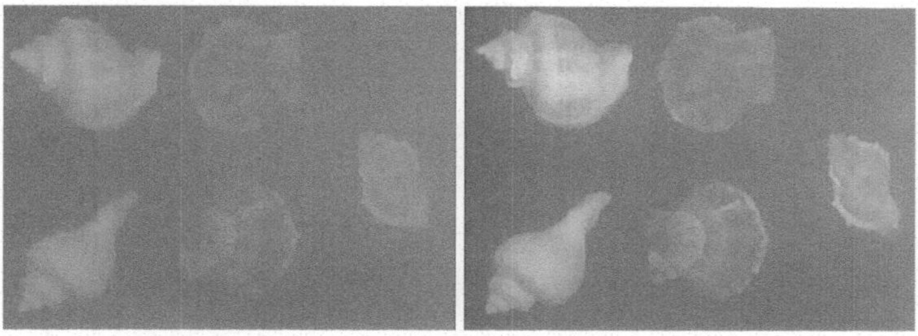

Fig. 35.4 The image captured under normal illumination on the *left* is ambiguous. The inhomogeneous illumination image on the *right* has a higher definition. The transparency is 70 cm and the target is at the distance of 0.6 time of visibility

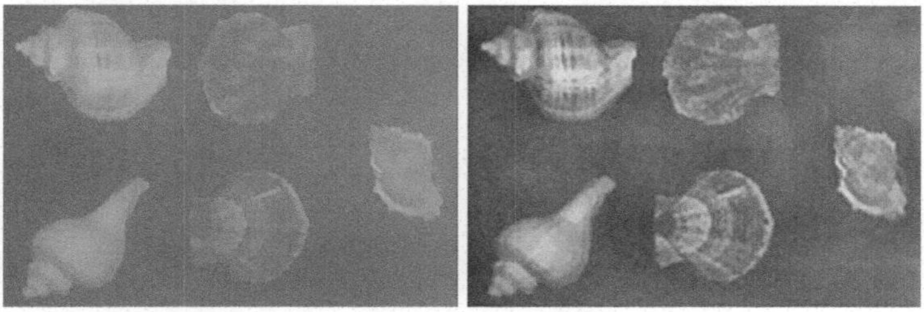

Fig. 35.5 The experimental environment is that transparency is 70 cm and the target distance is 0.6 time of visibility. The original image is shown on the *left* and the *right* is the processed image

35.5.2 Image Processing

CLAHE is used to process the image acquiring under the inhomogeneous illumination. The result of processing and histograms of the image are shown in Figs. 35.5 and 35.6.

In effect, the image processed by CLAHE get more improvement in definition and contrast, meanwhile more texture details are shown. So, we can draw a conclusion that this method is effective in underwater image processing.

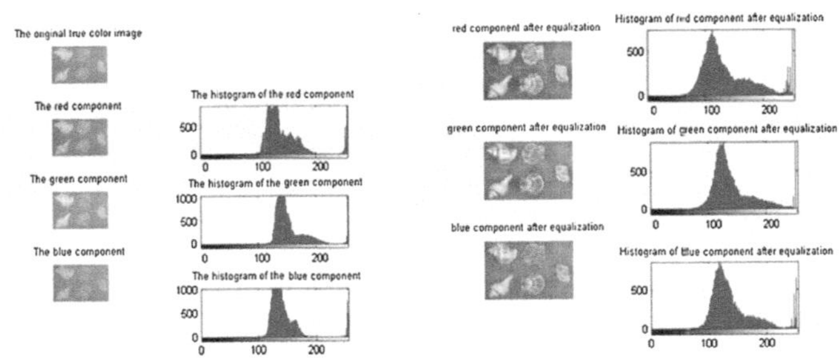

Fig. 35.6 The histogram of the original RGB image is shown on the *left* and the *right* is the processed image's histogram in three channels

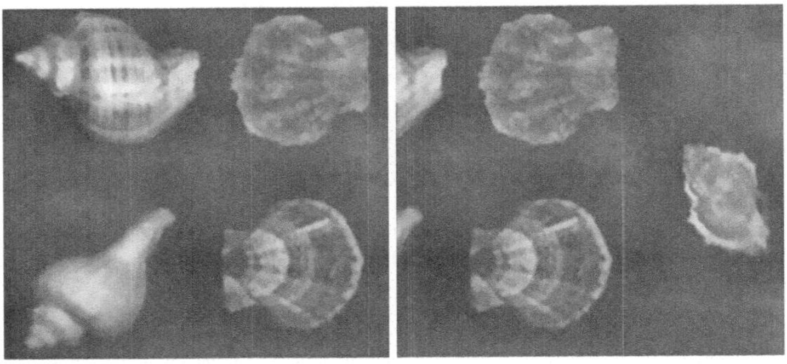

Fig. 35.7 The original images which are used to achieve extracting features and features matching. They are got after processing by CLAHE

Fig. 35.8 The results of extracting features. The number of keypoints on the *left* picture is 107, the other is 308

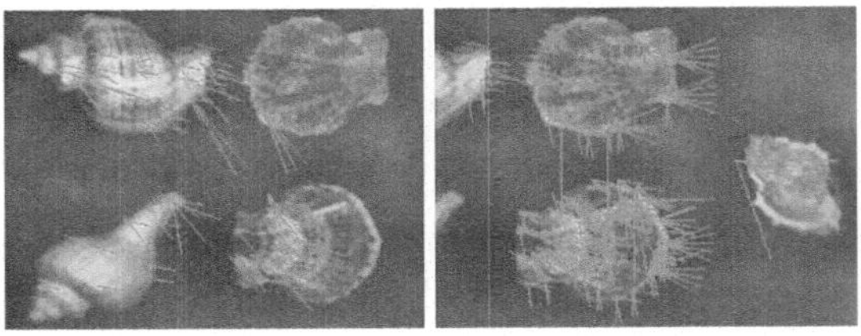

35.5.3 Underwater Image Feature Matching

The same components are contained in the images used to fulfill matching. The algorithm based on the SIFT is used to achieve feature extraction and matching between different pictures. The results are shown in Figs. 35.7, 35.8, and 35.9.

In the above experiments, the probability of matching exceeds 50 %, taking the image that extracted keypoint is less as standard. It can be able to stand the proof that the algorithm is brief, fast, and effective for underwater image feature matching, what's more, the matching accuracy is relatively high.

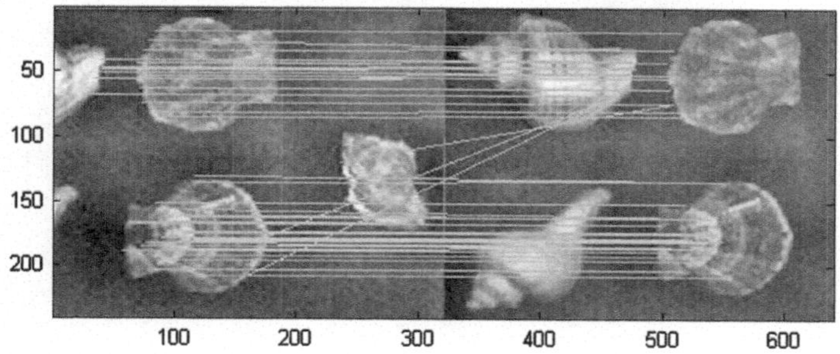

Fig. 35.9 The result of feature matching. The number of matched keypoints is 64. And the time-consuming to finishing matching is 0.110 s

According to the above matching results, this method of matching is suitable for the underwater image obtained by using inhomogeneous illumination system in the turbid water.

35.6 Conclusion and Future Works

In this paper, we realize the underwater image feature matching after processing the image by using contrast limited adaptive histogram equalization under inhomogeneous illumination.

The inhomogeneous illumination method is used to get the original underwater images, then the image was processed by CLAHE. The feature matching is achieved with the algorithm based on SIFT features.

In accordance with the results of experiments, the method Combining the inhomogeneous illumination theory which can remove the backscattering noise effectively and matching algorithm based on SIFT features presents high robust and reliable in this research.

The accuracy of matching is verified by experiments in this article. In future research, based upon the introduced feature-matching method, image stitching based on SIFT features will be tried to achieve precisely.

References

1. Hongmei Xu, Zhang Zhigang, Zheng Bing (2010) Distribution characteristics in the underwater inhomogeneous illumination field. Acta Photonica Sinica 39(9):16061610
2. Schmid C, Mohr R (1997) Local grayvalue invariants for image retrieval[J]. IEEE Transactions on Pattern Analysis and Machine Intelligence 19(5):530–534
3. Lowe DG (2004) Distinctive image features from scale-invariant keypoints. Int J Comput Vis 60(2):91–110
4. Mobley CD, Mobley CD (1994) Light and water: radiative transfer in natural waters. Academic press, San Diego
5. Zhang Q, Wang Z (2009) Proficient in MATLAB image processing. Publishing House of Electronics Industry, beijing, pp 140–141

Chapter 36
Application Software Model on Network Convergence Mode of Internet of Things

Jianguang Zhu, Shikui Zhai and Huina Xu

Abstract The Internet of Things realizes interconnection of physical objects in the real world by utilizing overall perception, reliable transportation, and smart processing. The data are delivered through multiple networks, which include Internet, Cable Television Network, and Mobile Communication Network. Through integrating three networks in the physical layer, the network resources can be shared, however, the applications built on different networks are hardly integrated due to nonstandard technologies. This paper provides an application Software model on triple-play based on web service. The simulation results shows that the model is efficient and effective.

Keywords Internet of things · Network · Triple-play · Software model · Module

36.1 Introduction

Internet of things (IOT) is a new direction of information industry, which gives things intelligence to communicate with human and other intelligent things. In order to achieve this goal, IOT should be capable of overall perception, reliable communication, and intelligent processing. Overall perception is fulfilled by leveraging on RFID, sensors, and two-dimensional code to acquire information of things, which includes user location, surroundings, individual preferences, and network status. Reliable communication takes advantages of network integration, business integration, terminal integration, and operations management integration to pass out information of things in real time and precisely.

J. Zhu (✉) · S. Zhai
Ocean University of China, Qingdao, China
e-mail: zhujianguang@qdhitech.gov.cn

H. Xu
Qingdao High-Tech Digital Zone Development Co. Ltd, Qingdao, China

X. Wang et al. (eds.), *Advanced Technologies in Ad Hoc and Sensor Networks*,
Lecture Notes in Electrical Engineering 295, DOI: 10.1007/978-3-642-54174-2_36,

The architecture of IOT can be divided into perception layer, network layer, and application layer. In the IOT, the network layer can pass out data gathered by perception layer in a high accessible, high reliability, and high security way. It solves the problem of perception layer gathered data transmission in certain range, especially in a long range. The network layer of IOT will assume a larger amount of data than existing networks and face a higher quality of service requirements. Therefore, the existing network cannot meet the needs of the IOT. This means that IOT need to expand and integrate the existing network by using new technologies to achieve more extensive and efficient connectivity features.

In order to achieve efficient, reliable, and fast transmission of IOT data, fusion of existing Internet, cable television network, and mobile communication network is needed, which fulfills network resources sharing, avoids repeat construction of basic pipelines, and builds multimedia basic platform with high adapt, easy maintenance, low-cost high-speed broadband [1–3]. After the triple-play, people can select communication operators and business as they need and access to communication network, Radio and TV network, and Internet through a network cable line or wireless assessment. The triple-plays will break the vicious competition in the telecommunications and broadcasting video transmission areas, prompting major operator fair competition for market.

Although the systems built on different kinds of networks are easy to use thanks to their friendly visual interfaces, they often suffer from several problems. First, the designed software only has designed functions and are not easy to extend their functions. Second, due to the use of nonstandard technologies, the system integration and interoperability of different kinds of networks are extremely complex. It is difficult to access the data from different systems in general interfaces with lots of users. Third, the designer has to spend lot of time on rewriting the codes and debugging them to adapt to the new triple-play environment, resulting in long development cycle, huge cost, and low efficiency.

In this paper, we propose an application software model on triple-play. We present a standardized development system model which could satisfy a variety of requirements. To the best of our knowledge, it is the first custom-built system model for triple-play environment. Our main contributions are as follows:

(1) We abstract the general models from different kinds of designed systems, which contribute to system integration and interoperability.

(2) We use the concept of component to fulfill the system expansion.

(3) Development of system is visual and rapid. The development procedure is turned into setting some parameters by given subsystems. The designers do not need the knowledge of programming language.

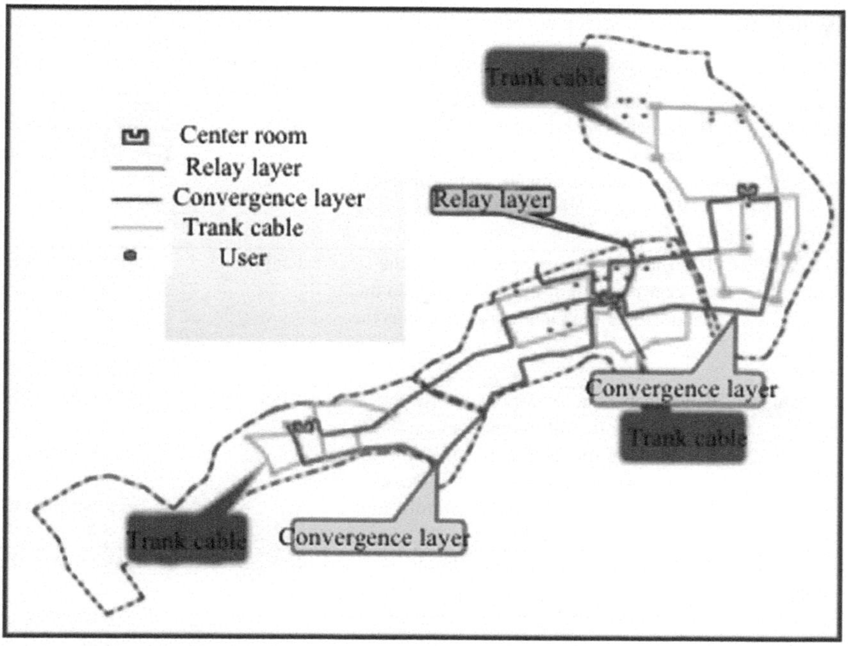

Fig. 36.1 Network structure for fiber optic cable

36.2 Background

36.2.1 Triple-Play

U.S first fulfills legislation of triple-play. In February, 1996, United States issued a new "1966 Communications Act," which lifts the ban on triple-play from policies and regulations [4]; Britain Telecom provides not only the Internet, telephone, and other communications services but also its own web TV channel; France's triple-play technology also has been rapid development [5, 6]. Market research firm Pyramid states in a recent report that as France operators accelerate the investment in fiber optic network, above 50 % families will choose triple-play service in 2014. As triple-play technology continues to mature, it will be gradually and widely applied in other countries [7–9].

36.2.2 Network Convergence Mode

Fiber optic cable network can be build stratified, including relay layer, convergence layer, and access layer. Access layer can be subdivided into the trunk cable, wiring cable, and introduction cable [10–13]. Figure 36.1 describes the whole fiber optic cable network (Qingdao High-tech Zone as an example).

Table 36.1 Build mode for different layer

Layer	Construction mode	Network structure	Agency
Relay layer	High-tech zone and users build, respectively, and connector box approach is adopted, and Optical Distribution Frame (ODF) is used to assign core cable to users.	Star	Telecomm, radio, and television
Convergence layer	High-tech Zone is responsible for the construction of units allocated to the needs of the ODF core cable into the side	Ring	Telecomm, radio and television, and government
Trunk cable	High-tech Zone is responsible for the construction of units allocated to the needs of the ODF core cable into the side. In the transfer of cable boxes, each divided into different units into demand-side module	Ring	Telecomm, radio and television, and government
Introduction cable	High-tech Zone is responsible for construction, laying a cable from the junction box to the user reserved room	Star	Telecomm, radio and television, and government

(1) Relay layer

Telecom operators, broadcasting, and High-tech Zone build trunk cable ring interconnected to meet the high-speed broadband transmission network carrying information services and various telecom operator business needs.

(2) Convergence layer

The main characteristics of convergence layer is large transmission capacity, high importance of nodes, and the number of nodes should not be excessive according to the needs of business and network development. Convergence layer structure should be clear, efficient, flexible, safe, reliable, easy maintenance, and management to meet the needs of business developments, meet the needs of the developing situation.

(3) Trunk cable

Trunk cable must have good scalability to adapt to the changes in the future network structure, network capacity, high reliability requirements for business security. The realization of the access layer technology is more, and cost and function is the main two aspects to determine whether technology is applicable.

(4) Introduction cable

In operation stage, introduction cable can be build according to user needs.

The construction mode and function of every cable level as shown in Table 36.1.

Compared to traditional modes of the construction of the operators alone, this new access model not only avoids the single operator resource monopoly, build the

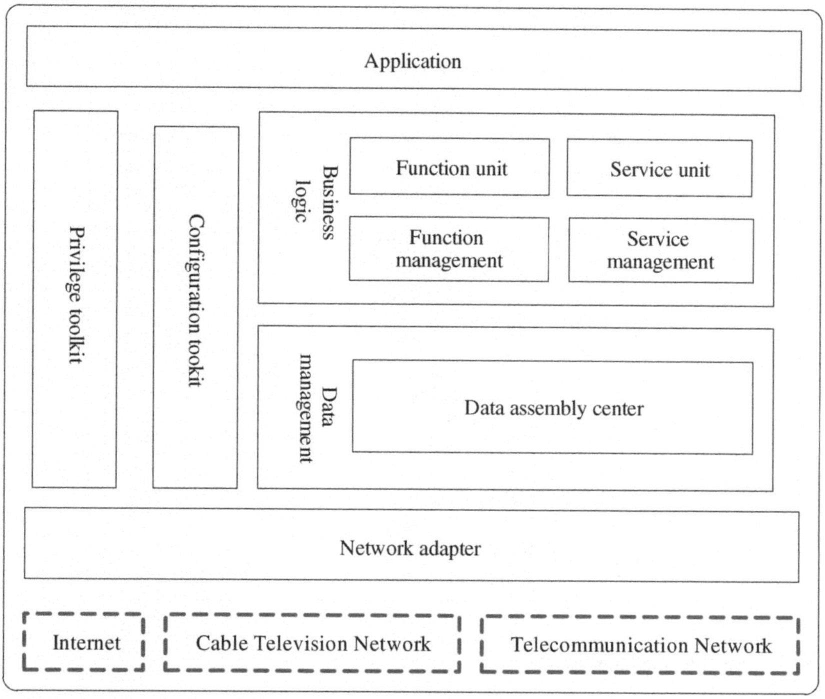

Fig. 36.2 Software model atchitecture

communication environment of fair competition, but also at least save more than 60 % of the basic network and to invest in pipelines.

36.3 Application Software Model

The designed software model consists of network adapter, data management, business logic, configuration toolkit, and security toolkit. As separate assemblys, each module is integrated by way of reference calling. Figure 36.2 describes the whole software model architecture.

1) Network adapter

As a core component, network adapter switches between different types of networks(Internet, cable television network, and telecommunication network) to provide uniform data flow for upper layer, and also it has good flexibility and scalability. Figure 36.3 describes the basic structure of network adapter.

Network type and network drive are constructed customly. As network identify, each network type is associated with a network drive, and network drive parses data flow coming from different network. Network type and network drive are associated

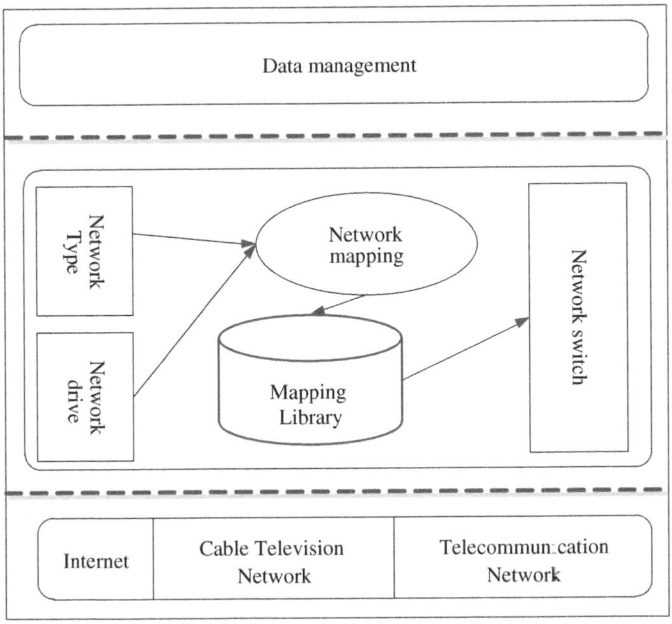

Fig. 36.3 Network adapter structure

by network mapping to form mapping library, which stores all related information between network types and drives and is responsible for selecting network drive to parse different data flow. Network switch is the main controller to call network drive to parse network data flow.

2) Data management

This module includes several Data assembly centers (DACs) preserving the collected data from the different network. These data are often distributed due to different regional. An efficient aggregation mechanism may be necessary between network adapter and DACs, in which proprietary protocols should be avoided. With the aggregation, the data structure can be unified by transforming data to DAC-compatible formats. In the aggregation process, it is very important to get comprehensive metadata from network adapter, including what contents are contained, when, where, and how the data are collected, who produced the data, which is helpful for data discovery in huge number of datasets. Each DAC may disseminate the data in different ways. For uniform and open access, an agreed consensus on the standard protocols should be achieved, which is also the responsibility of the business logic.

3) Business

We have designed reusable module library that consists of function units, function management, service units, and service management. Each of them is designed, implemented, and tested as a logical unit. The implementation details of each module are transparent for other modules. They communicate with each other by means of the predefined interfaces. Each of interfaces has its own parameters that can be

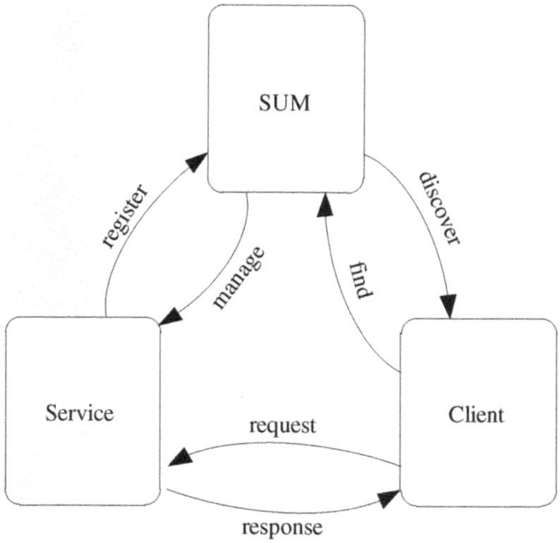

Fig. 36.4 Service and management

configured by the configuration toolkit. Applying these modules to compose the applications dramatically reduces the development cycle greatly instead of building it from scratch.

Figure 36.4 describes service units and service management. All service units register in SUM(Service Unit Management) with service name, parameters, address, provider, and so on. Through publishing all service units by SUM, each client can find and discover what it want. SUM is responsible for calling service unit when client sends requests to service unit, and also transports the response information to client.

4) Configuration toolkit

In order to assist the development of surveillance software, we have designed the configuration toolkit to support visual developments. The toolkit enables software customizability by adjusting configuration schemes. It consists of parameters setting tool and setting examination tool. The former accelerates the system development process greatly since the complex programming and debugging are turned into easy parameters configuration. The latter is used to examine whether there are errors in the configuration scheme. If the examination fails, the developers have to go back to find the faults and correct them. After this step, the developers can compose the system following the development procedure [14].

5) Permission toolkit

Software model presented in this paper contains a separate permission toolkit including three types of permission configuration: user permission, roles permission, and permission definition. All permissions can be associated with users and roles dynamically with good flexibility, which is developed on the basis of standard security service interface with a strong versatility. Data transports safely through data encryption and signature service interfaces.

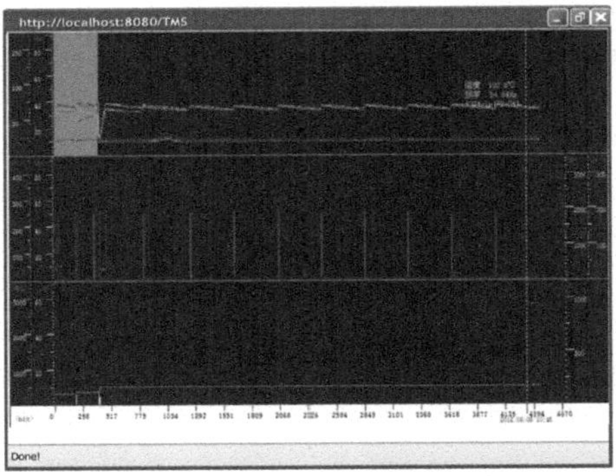

Fig. 36.5 UI of surveillance software

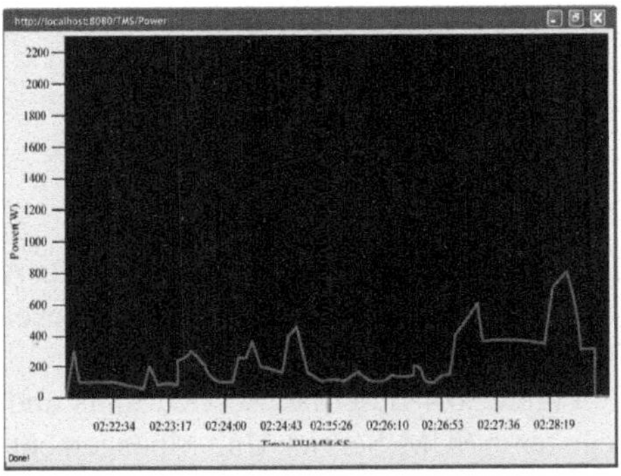

Fig. 36.6 Power surveillance curve

36.4 Applicaiton Example

The new system model has been validate by applying it to developing surveillance software for network equipments. Compared to traditional software model, software development process is shortened. The complexity of multioperators network supporting is overcome through dedicatedly considering the triple-play and being compatible with existing network. As shown in Fig. 36.5, the network equipments surveillance software presents the data gathered from Internet of Things, which are compsed of smart sensors. In Fig. 36.6, a power surveillance curve is demonstrated with data measured from smart power meter. The case has proved that the new system model is flexible enough and remarkably improves the software application's network compatibility.

36.5 Conclusion and Future Work

The implementation of the depth triple-play in the physical layer can optimize the effectiveness and performance of public services. This paper provides an application software model on triple-play which can integrate systems on different kinds of networks. The simulation result shows that the model is efficient and effective. The tentative observation of this architecture promises that it could adapt to other situations.

Nevertheless, many issues still remain to be explored. Our ongoing works are: (1) Confirm this architecture in other systems; (2) Develop more reusable component which could indicate other kinds of data result.

References

1. Tian L (2012) Applications and prospects of Internet of things based on mobile communication network. In: International conference on computer science and network technology. pp 1167–1170
2. Yokotani T (2012) Application and technical issues on Internet of things. In: International conference on optical Internet, pp 67–68
3. Zheng G, Song L, Shen Y (2013) The research on connection-oriented convergence of triple networks. Appl Mech Mater 318:235–238
4. Yunxia H, Yaoxing J (2011) Research and analysis of triple play at home and abroad. Inf Commun Technol 3:51–55
5. Wu XB, Chen GH, Das SK (2008) Avoiding energy holes in wireless sensor networks with nonuniform node distribution. IEEE Trans Parallel Distrib Syst 19(5):710–720
6. Dongzhi B (2010) Current Situation and Influence of the Three Network Convergence. China New Telecommun 4:5–10
7. Maiese Kenneth (2008) Triple play: Promoting neurovascular longevity with nicotinamide, WNT, and erythropoietin in diabetes mellitus. Biomed Pharmacother 62(4):218–232
8. Liu Changbin, Shi Lei, Liu Bin (2007) Utility-based Bandwidth allocation for triple-play services. IEEE Comput Soc 40(2):2699–2703
9. Lindsey S, Raghavendra CS (2002) PEGASIS: power-efficient GAthering in sensor information system. In: 2002 IEEE aerospace conference proceedings, pp 1125–1130
10. Harks T, Poschwatta T (2005) Priority pricing in utility fair networks. In: Proceedings of the 13th IEEE international conference on network protocols (ICNP'05), pp 311–320
11. Cave M, Correa L, Crocioni P (2006) Regulating for non-price discrimination—the case of UK fixed telecommunications? Compet Regul Netw Ind 7(3):391–415
12. George J (2004) Designing passive optical networks for cost effective triple play support. In: Proceedings of FTTH conference, pp 4–6
13. Shi L, Liu C, Liu B (2008) Network utility maximization for triple-play services. Comput Commun 31(10):2257–2269
14. Lau KK, Wang Z (2007) Software component models. IEEE Trans Softw Eng 33(10):709–724

Author Index

X. Wang et al. (eds.), *Advanced Technologies in Ad Hoc and Sensor Networks*,
Lecture Notes in Electrical Engineering 295, DOI: 10.1007/978-3-642-54174-2,

The manufacturer's authorised representative in the EU is Springer
Nature Customer Service Centre GmbH, Europaplatz 3, 69115 Heidelberg,
Germany. If you have any concerns regarding our products, please
contact ProductSafety@springernature.com

Printed and bound by CPI Group (UK) Ltd, Croydon, CR0 4YY
25/09/2026
02222839-0007